Total Quality Management: Key Concepts and Case Studies

Total Quality Management: Key Concepts and Case Studies

D.R. Kiran

BSP BS Publications
A Unit of BSP Books Pvt. Ltd., India.

AMSTERDAM • BOSTON • HEIDELBERG • LONDON
NEW YORK • OXFORD • PARIS • SAN DIEGO
SAN FRANCISCO • SINGAPORE • SYDNEY • TOKYO

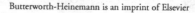

ELSEVIER   Butterworth-Heinemann is an imprint of Elsevier

Library of Congress Cataloging-in-Publication Data
A catalog record for this book is available from the Library of Congress

British Library Cataloguing-in-Publication Data
A catalogue record for this book is available from the British Library

ISBN 978-0-12-811035-5

For information on all Butterworth Heinemann publications
visit our website at https://www.elsevier.com/

 Working together
to grow libraries in
developing countries

www.elsevier.com • www.bookaid.org

Publisher: Joe Hayton
Acquisition Editor: Brian Guerin
Editorial Project Manager: Edward Payne
Production Project Manager: Vijayaraj Purushothaman
Cover Designer: Matthew Limbert

Typeset by SPi Global, India

Contents

About the Author

Prof. D.R. Kiran, B.Sc., B.E., M.Sc. (Eng.), (Ph.D.), FIE(I), FIIProdE, FIIPlantE, FITTE, FISNT, has a rich practical experience of forty years, both in industry and academia. Starting his career in 1968 with Larsen & Toubro, he held the top positions Planning Manager of Rallifan (CF division), World Bank Adviser/Instructor for Transport Managers in Tanzania, and the Principal of a Chennai based Engineering College.

In recognition of his services in the field of engineering education, he was presented with the coveted Bharat Excellence Award and Gold Medal for Excellence in Education in New Delhi in 2006. He is listed as an International Expert in Industrial Engineering and Management in the International Directory of Experts and Expertise. He is nominated for the post of Honorary Deputy Director General in India for International Biographical Center.

Earlier during the 1980s, he was introduced to Dr. Julius Nyerere, the then President of Tanzania as a Pioneer of Work Study in that country. He was one among few non-political foreigners to be interviewed by the government newspaper of Tanzania.

He started his academic career in 1979, and taught subjects such as Total Quality Management, Professional Ethics, Maintenance Engineering Management and Production Planning & Control at B.E. level. He has been the national Council member of National Institute for Quality and Reliability, which has been nominated by the Quality Council of India to be the nodal point for the ZED (Zero Effect Zero Defect) Cell program initiated by the Government of India. His active participation in NIQR activities helped him in organizing and participating in quality conventions and quality-related seminars. This provided the inspiration for him in planning for an integrated book on quality management, covering all quality-related topics and this holistic approach in this book *Total Quality Management: Key Concepts and Case Studies* is expected to be of immense help to the students, as well as the practicing engineers. His rich experience in this field enabled him to provide several case studies on the key topics.

Earlier, his experience as a World Bank Expert in maintenance field and his teaching of the subject at B.E. level helped him in contributing several case studies in his text book *Maintenance Engineering and Management: Precepts and Practices*. His teaching experience in Professional Ethics for Engineers and the lack of Indian case studies in other books motivated him to author a book on *Professional Ethics and Human Values* published by McGraw Hill (India).

Having started his career as an Industrial Engineer and as Planning Manager with Rallifan and National Bicycle Co., as well teaching this subject at the B.E. level, helped him in authoring a fourth book, *Elements of Production Planning and Control.*

He has published 23 papers in professional journals and seminar proceedings and was the Chief Guest in several Technical Meets. He was the Organizing Secretary for the successful 29th Production Convention of the Institution of Engineers held in August 2014. He is the National Vice President of Indian Institution of Plant Engineers, as well as Vice Chairman, Past Chairman, Secretary, Council Member, etc. of professional associations such as IIProductionE, Chartered Inst. of Transport, NIQR, and ISNT. He was responsible for the establishment of several student chapters for IIProdE and NIQR.

He is widely traveled, having visited over 30 countries and is a philanthropist.

Foreword

In today's world, factors influencing corporate actions globally are changing more and more rapidly. We see in today's scenario business acquisitions, mergers, and disappearance. This is because every organization wants to delight the customer in every aspect of quality, cost, delivery, and service, and make their products obsolete by substituting new products. We continually see these organizations, which do not satisfy their requirements and are unable to sell their products and services leading to selling their businesses.

The changing environment in which the companies are finding themselves can be characterized as follows:

- Change in values and structures in society
- Increasing technological progress
- Increasing demands on products and services
- Changing market structure from domestic to global
- Improving communications and transport technologies

Today, the world is becoming a borderless state and we see products designed in one continent, manufactured in another, and sold in yet another continent. So, the challenge for organizations is how to survive in this ever-changing scenario. Organizations have to move from product quality to process quality and then move to the next step on the ladder to innovative quality—finally building an innovative organization. This means every organization strives to bring zero defect, supported by excellent process control and innovative product and manufacturing scenarios. We see in this decade, many institutions/industries have started nurturing a Total Quality Management (TQM) culture in their organizations to stay ahead in business.

As it is seen in developed countries such as Japan and Europe, quantitative techniques play a very major role in helping organizations to move from simply recording data to data analysis and algorithms for scientific decision-making. This helps in cutting down cycle time—both planning and execution cost and statistics play an important role.

The author, Professor D.R. Kiran, with a professional career spanning 47 years, has come out with an excellent exposition on TQM. In fact, in the book, he has given readers a wide canvas on TQM starting from overview, evolution, and the contributions by Gurus. This phase is followed by leadership and TQM, principles of scientific management systems approach, supported by strategic planning.

In the third phase, he discusses the intricacies of quality cost, the way to organize TQM, and the importance of total employee involvement, supplier partnership, and total productive maintenance leading to customer satisfaction. He also touches upon the importance of quality circles and its role in Total Employee Involvement (TEI). In the same phase, on problem-solving areas, he relates in detail the basics of statistics, process capability, 7 QC tools, and 7 new tools of TQM, the importance of Kaizen, and the application of six sigma in problem-solving.

He also clearly spells out the various awards available in the world to challenge, as motivation for companies which want to practice TQM and raise their quality standards.

The book also deals with the design facet of quality—design for quality, value engineering, Quality Function Deployment (QFD), Failure Mode and Effect Analysis (FMEA), and other relevant models and tools.

I would like to mention that this book is one of the best expositions on TQM and will serve as a handbook for students as well as industries who want to understand the basics of TQM and start initiating TQM culture in their respective institutions.

I should place on record the excellent work done by Professor D.R. Kiran in sharing his knowledge and experience with the professional community, and I would like to congratulate him for this superb contribution.

Dr N Ravichandran

Preface

Almost during the entire period from the Industrial Revolution to around the Second World War, Germany had remained at the pinnacle of quality production of industrial goods, especially in the automobile sector, joined during the later stages by the United States and other European countries. Post-1950, Japan realized that quality production was the only means to increase their economy through export of industrial goods. Through the help of quality gurus like Shewhart and Deming, it overtook all other nations in quality production and achieved the goal of becoming the epitome of quality precepts and practices.

During the later years of the 20th century, the economic liberalization in India led to freer imports of industrial goods, and Indian industry has been forced to concentrate to a larger extent on quality output to face the international competition. This need, as well as the success story of Japan in overcoming the stiff competition from the United States and Europe, gave a momentum to the quality movement in India and other Asian nations. By 2010, India could overtake all other Asian nations next to Japan in garnering several Deming awards.

With this background, this book deals with the management principles and practices that govern the quality function. Though several books on quality management are available, there are very few that deal with all the aspects of quality control and management both in practice, as well as prescribed in university syllabi around the world. In view of this, I desired to bring all quality management-related topics specified in several university syllabi, and which I have been teaching, into a single book, providing case studies based on my experience, and this is the result of my effort. This book is hence aptly given the title of *Total Quality Management—Key Concepts and Case Studies*.

A special feature of this book is the importance given to the fundamentals of statistics, so that the reader is fully equipped to better understand the vital topic of statistical quality control. Apart from dealing with the Quality systems and standards of the ISO 9000 series, features of the environmental management systems per ISO 14000 series are also dealt with, to highlight the significant impact of the environment on quality production. A chapter on Lean Management is added to highlight that the scarce resources must be expended only for creating the required value to a product and not otherwise.

Being an industrial engineer myself having practiced methods study and value engineering, I have attempted to explain these traditional Industrial Engineering techniques, relating them to the modern concepts of Kaizen, DFSS, etc. At the end of each chapter, a humorous and anecdotic reference to the theme of the chapter is given as "On the lighter side."

Syllabi from around 20 Universities and Institutions of India, as well as International Universities were collected to draw the outline for this book.

Prof. D.R. Kiran

Acknowledgments

The author wishes to acknowledge his indebtedness to all the persons who were associated with him during his 45-year long industrial and academic career, and who helped him in bringing this book forward. Special acknowledgement is due to Dr. N. Ravichandran, Executive Director of Lucas TVS, Chennai, for his foreword to this book.

About the Book

This textbook, *Total Quality Management*, Key Concepts and Case Studies, brings together all the management principles and practices that govern the quality function of quality management, as well as related topics specified in several university syllabi, and which have been taught to the TQM students all over the world, into a single, comprehensive book. This book illustrates all the fundamental principles involved in the quality function and gives detailed explanation about the latest developments in quality concepts. This book is expected to be of immense help to practicing engineers to give them a comprehensive grasp of the subject and to give a clear-cut understanding of quality principles and practices. They can appreciate the need to maintain high levels of quality in their company's activities to sustain the progress.

Salient features:

- Syllabi from around 20 universities and institutions of India in addition to those from Russia and United States are collected for this book.
- This book emphasizes how traditional inspection methods have metamorphosed into Total Quality Management with all the tools of quality control methods professed by the Quality Gurus. Special attempt is made to explain in detail and with illustrations, the traditional and the modern tools of quality control.
- The chapters on the Systems Approach to Management, Decision Theory, and others, emphasize the fundamental management principles the quality professional should know, while the chapters on Quality Function Deployment, Taguchi's Loss Function, and others, emphasize the modern concepts of Total Quality Management.
- A special feature of this book is the importance given to the fundamentals of statistics, so that the reader is fully equipped to understand better the vital topic of statistical quality control.

Another feature of this book is the explanation given to traditional industrial engineering techniques relating them to the modern concepts of Kaizen, DFSS, and others. Several case studies and illustrations are provided based on the author's experience.

Chapter 1

Total Quality Management: An Overview

Chapter Outline

ABBREVIATIONS

DMAIC	define, measure, analyze, improve, and control
DMADV	define, measure, analyze, design, verify
IDOV	identify, design, optimize, and validate
DCCDI	define, customer, concept, design, and implement
DMEDI	define, measure, explore, develop, and implement
KISS Principle	keep it simple statistically
DFSS	design for six sigma
SDLC	software development life cycle
cGMPs	current good manufacturing practices

1.1 WHAT IS QUALITY?

Quality indicates the capability of all components of an entity to satisfy the stated and implied needs, that a quality item will perform satisfactorily in service, and is suitable for its intended purpose. Quality is referred to as "fitness for use," "fitness for purpose," "customer satisfaction," "conformance to the requirements," or has a pragmatic interpretation as the non-inferiority or

Total Quality Management: Key Concepts and Case Studies. http://dx.doi.org/10.1016/B978-0-12-811035-5.00001-5

superiority of something. In any case, to achieve satisfactory quality, we must be concerned with all three stages of the product or service cycle which include:

- The definition of needs
- The product design and conformance
- The product support throughout its lifetime

Total quality management (TQM) consists of organization-wide efforts and an integrated system of principles, methods, and best practices to install and make a permanent climate in which an organization continuously improves its ability to deliver high-quality products and services to customers.

1.2   QUALITY DEFINITIONS

There have been several definitions of quality as given by various quality organizations, as well as quality gurus. Nevertheless, all these definitions focus on the efforts put in by organizations to fulfill customer requirements as stated above.

1. TQM is a management philosophy and company practices that aim to harness the human and material resources of an organization in the most effective way to achieve the objectives of the organization.
   –British Standards Institution Standard BS 7850-1:1992
2. TQM is a philosophy for managing an organization in a way which enables it to meet stakeholder needs and expectations efficiently and effectively, without compromising ethical values.
   –The Chartered Quality Institute
3. TQM is a term first used to describe a management approach to quality improvement. Since then, TQM has taken on many meanings. Simply put, it is a management approach to long-term success through customer satisfaction. TQM is based on all members of an organization participating in improving processes, products, services, and the culture in which they work.
   –The American Society for Quality
4. TQM refers to management methods used to enhance quality and productivity in organizations, particularly businesses. TQM is a comprehensive system approach that works horizontally across an organization, involving all departments and employees and extending backward and forward to include both suppliers and clients/customers. It provides a framework for implementing effective quality and productivity initiatives that can increase the profitability and competitiveness of organizations.
   www.inc.com/encyclopedia
5. Quality is the fitness for the purpose or use. (Fitness as defined by the customer.)
   –Dr. J.M. Juran in 1988
6. Quality means getting everyone to do what he has agreed to do and do it right the first time. Quality is conformance to specifications.
   –Philip Crosby

7. Quality is the degree to which a set of inherent characteristics fulfills requirements.
   –ISO 9000
8. Quality is an effective system of integrating quality improvement efforts of various groups of the organization so as to provide products/services at a level which allow customer satisfaction.
   –A.V. Feigenbaum
9. TQM is a way of managing to improve the effectiveness, flexibility, and competitiveness of a business.
   –Prof. John Oakland
10. TQM is a management approach for an organization, centered on quality, based on the participation of all its members and aiming at long-term success through customer satisfaction, and benefits to all members of the organization and to society. One major aim is to reduce variation from every process so that greater consistency of effort is obtained.
    –Royse, D., Thyer, B., Padgett D., and Logan T., 2006
11. TQM is the total approach in every aspect of management.
    –Bill Creech
12. TQM is an evolving system of practices, tools, and training methods for managing companies to provide customer satisfaction in a rapidly changing world.
    –Alan Graham and David Walden
13. Quality is synonym to characteristic, trait, goodness, or excellence.
    –Roget's Thesaurus
14. Quality in a product or service is not what the supplier puts in. It is what the customer gets out and is willing to pay for.
    –Peter Drucker
15. Quality means products and services meet or exceed customers' expectations.
16. Quality is conformance to requirements or specifications or standards.
    –Crosby
17. Quality of a product includes a peculiar and essential character or feature. It may be noted here that the character applies to a peculiar and distinguishing quality of a thing or class, whereas the property implies a characteristic that belongs to a thing, essential nature and may be used to describe a type, carry a certain degree of excellence.
    –Webster's Collegiate Dictionary
18. Quality is the totality of features and characteristics of an entity, or a product or service that bear in its ability to satisfy stated or implied needs. The implied needs are usually very vague, indicated, or not indicated at all. Nonetheless they are expected by the customer sometimes more than the standard.
    –ISO 8402-1994
19. Quality is the management philosophy and company practices that aim to harness the human and material resources of an organization in the most

effective way to achieve its objectives. The objective of an organization may be customer satisfaction, growth, profit, and market position.
    –*BS 5750*

20. TQM is the application of quantitative methods of human resources to improve the materials and services, supplies to an organization; all the processes within an organization and the degrees to which the needs of the customer are at present and in future.
    –*U.S. Department of Defense*

21. TQM is both a philosophy and a set of guiding principles that represent the foundation of a continuously improving organization.

22. TQM is an approach for effective management of an enterprise through focus on its people and performance, safety, proper packaging, timely delivery, efficient technical service, and incorporating effective customer feedback.

23. TQM is a business philosophy founded on customer satisfaction.

24. Quality means productivity, competitive cost, timely delivery, and total customer satisfaction.

25. Product quality is decided by the customer needs, conforming to specifications, assured performance, safety, proper packaging, timely delivery, efficient technical service, and incorporating effective customer feedback.

26. TQM is an integrated organizational approach in delighting both external and internal customers by meeting their expectations on a continuous basis through everyone involved with the organization working on continuous improvement in all products, services, and procedures, along with proper problem-solving methodology.

1.3   QUOTES ON QUALITY

Apart from the above definitions, we can also cite some quotes from Mahatma Gandhi et al.

1. It is the quality of our work that pleases the God and not the quantity.
    –*Mahatma Gandhi*

2. I have offended God and mankind because my work did not reach the quality it should have.
    –*Leonardo Da Vinci*

3. 20th century was a century of manufacturing and 21st century is the century of Quality.
    –*J.M. Juran*

4. Quality is everybody's responsibility.
    –*Edward Deming*

5. Quality is never an accident. It is always the result of intelligent effort.
    –*John Ruskin*

6. When it is obvious that the goal cannot be reached, do not adjust the goal, adjust the action steps.
    –*Confucius*

7. No matter how many goals you have achieved, you must set your eyes on a higher one.
   —Jessica Savich
8. Problems are not stop signs, they are guidelines.
   —Robert Schuller
9. Good management consists in showing average people how to do the work of superior people.
   —John D Rockfeller
10. The new one is just an old thing that was forgotten.
    —A Russian proverb

1.4 THE SCALE OF QUALITY

Few companies are able to spell out their quality levels to precise detail. The general terms of reference to the quality manager are to achieve the customer satisfaction economically, rather than costly over-perfection. Fig. 1.1 below illustrates the different scales of quality by attributes.

Quality Level	Customer opinion on Quality Level	Attribute	Cost Factor
18	Meticulous	7 Sigma level	
17	Fantastic		
16	Amazing		Costly
15	Outstanding	6 Sigma level	
14	Exceptional		
13	Excellent	5 Sigma level	
12	Very good		Optimal
11	Good	4 Sigma level	
10	Acceptable		
9	Fair		
8	Troublesome	3 Sigma level	Cheap
7	Poor		
6	Extremely poor	2 Sigma level	
5	Nonexistent		
4	Shocking	1 Sigma level	
3	Hopeless		Bankruptcy
2	Disastrous		
1	Catastrophic	0 Sigma level	

FIG. 1.1 Scale of quality.

1.5   THE PARADIGM OF TQM

In their definition of TQM, the International Organization for Standardization (ISO) has specified the three paradigms of TQM to be composed of:

Total: Organization wide
Quality: With its usual definitions and all its complexities
Management: The system of managing with steps like Plan, Organize, Control, Lead, Staff, provisioning, etc.

We can elaborate ISO's above definition indicating that the following are involved in TQM:

Total

1. *All functions*
   - Design
   - Production
   - Marketing
   - Purchase
   - Maintenance
   - Quality Control
   - HR
2. *All levels*
   - Chairman and Managing Director
   - General Manager
   - Supervisor
   - Operator
3. *All persons having a stake*
   - Factory personnel
   - Corporate office
   - Shareholders
   - Suppliers

Quality

1. Customer satisfaction
2. Customer driven
3. Functional requirement of the product
4. Product specifications
5. Process parameters

Management

1. Effective direction, monitoring, and control
2. Continuous improvement
3. Effective utilization of resources
4. Executive commitment
5. Well-planned and effective decision-making
6. Employee empowerment

The TQM Company must think Quality, act Quality, and speak Quality to achieve Quality.

1.6 HOW CAN EFFECTIVE TQM CHANGE THE SITUATION?

TQM means organized change management effort.

From	To
Result-oriented	Process-oriented
Personalized	Process led
Product-oriented	Customer-centered
Owner-driven	Value-driven
Internally-focused	Stakeholder-focused
Making money	Reputation
Sustaining imitation/importation	Innovation/improvement
Tactic-oriented	Strategy-oriented
Expansion	Creation
Seniority-oriented	Competency-oriented
Motivation through fear and loyalty	Motivation through shared vision
"It's their problem" attitude	Accountability to all problems
"We always did it this way"	"Let us improve to suit the customer"
Decisions-based on assumptions and subjective judgment	Decision-based on data and facts
All begin and end with management	All begin and end with the customer
Routine management	Breakthrough management
Crisis management and recovery	Doing it right the first time

1.7 QUALITY OF DESIGN VERSUS QUALITY OF CONFORMANCE

There are two major aspects of quality: Quality of design and quality of conformance. Quality of design as defined by the Business Dictionary, is "the level of effectiveness of the design process in determining a product's operational requirements and their incorporation into design requirements." This involves the variations of a product or service in grades or levels of quality, and includes the types of materials used in construction, tolerance in manufacturing, reliability, etc. Quality of conformance is also defined by the Business Dictionary as "the level of the effectiveness of the design and production functions in conforming to the product manufacturing requirements and process specifications while meeting process control limits, product tolerances and production targets."

Quality of conformance signifies how well the product conforms to the specifications and tolerances required by the design and is influenced by the choices of manufacturing processes, training and supervision of the workforce, the type of quality-assurance system used, as well as the motivation of the workforce to achieve quality.

1.8   CHANGING CRITERIA OF QUALITY

CHANGING CRITERIA OF QUALITY

Criterion	Concept on the Past	Today's Concept
Definition of quality	Product quality	Customer requirement
Quality focus	Focus on product and services	Focus on all business tasks
Quality responsibility	With inspection/quality control department	With all employees
Organization	Quality is a function	Quality is a strategy
Result, or how you achieve the result?	Result is important	Process is important
Measurement	Quality is measured by product attributes	Quality is measured by the cost of quality
Quality champions	Quality champions were quality employers	Quality champions are top management teams
Human involvement	System is the main focus	Both systems and the human component are the main focus
Concept	Quality is considered a tool	Quality is process philosophy

1.9   THE FIVE APPROACHES TO QUALITY

Harvard professor David Garvin, in his book, *Managing Quality,* summarized five principal approaches to define quality.

1. Transcendent
   - Those who hold the transcendental view would say "I can't define it, but I know it when I see it."
   - Advertisers are fond of promoting products in these terms.
   - "Where shopping is a pleasure" (supermarket). "We love to fly and it shows" (airline). Television and print media are awash with such indefinable claims and therein lies the problem.
   - Quality is difficult to define or to operationalize. It thus becomes elusive when using the approach as the basis for competitive advantage. Moreover, the functions of design, production, and service may find it difficult to use the definition as a basis for quality management.
2. Product-based
   - Quality is viewed as a quantifiable or measurable characteristic or attribute. For example, durability or reliability can be measured and the engineer can design to that benchmark.
   - Quality is determined objectively.
   - Although this approach has many benefits, it has limitations as well. Where quality is based on individual taste or preference, the benchmark for measurement may be misleading.

3. User-based
   - It is based on idea that quality is an individual matter and products that best satisfy their preferences are those with the highest quality. This is a rational approach, but leads to two problems.
   - Consumer preference varies widely and it is difficult to aggregate these preferences into products with wide appeal. This leads to the choice between a niche strategy and a market aggregation approach, which tries to identify those product attributes that meet the needs of the largest number of consumers.
   - Another problem concerns the answer to the question "Are quality and customer satisfaction the same?" The answer is probably not. One may admit that a Lincoln Continental has many quality attributes, but satisfaction may be better achieved with a Ford Escort.
4. Manufacturing-based
   - Manufacturing-based definitions are concerned primarily with engineering and manufacturing practices and use the universal definition of "conformance to requirements." Requirements or specifications are established by design and any deviation implies a reduction in quality. The concept applies to services as well as products. Excellence in quality is not necessarily in the eye of the beholder, but rather in the standards set by the organization.
   - This approach has a serious weakness. The consumers' perception of quality is equated with conformance and hence, is internally focused.
5. Value-based
   - It is defined in term of costs and prices, as well as number of other attributes. Thus, the consumers' purchase decision is based on quality at an acceptable price. This approach is reflected in the popular *Consumer Reports* magazine, which ranks products and services based on two criteria: quality and value.
   - The highest quality is not usually the best value. That designation is assigned to the "best-buy" product or service.

1.10 PDCA CYCLE

PDCA (plan-do-check-act) is a four-step management method developed by Edwards Deming and is widely used in business for the control and continuous improvement of processes and products. It is also called the PDCA circle, PDCA cycle, or PDCA wheel. Having been developed and emphasized by Deming, it is also named the "Deming Wheel." Deming, however, referred to it as Shewart cycle, as a nod to his teacher, Walter Shewart. Subsequently, Deming replaced the word "Check" with "Study." This PDCA cycle has come to be known as Deming Cycle. However, in most books on quality including this one, PDCA cycle is shown in Fig. 1.2, and continued to be referred to as the Deming Cycle.

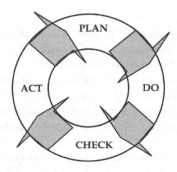

FIG. 1.2   The PDCA cycle.

The four components of this cycle are:

1. *Plan*: This step involves identifying and analyzing the problem. The objectives and processes are first established, keeping in mind the output expectations and quality requirements.
2. *Do*: This step involves implementing the plan, and executing the process to make the product. All data is collected and charted to be useful in the next step.
3. *Check*: This step involves measuring how effective the test solution was, and analyzing whether it could be improved in any way. Study the data results collected in the above step and compare against the expected results. The root causes for these differences are analyzed to ascertain if it is possible to improve the process to get better results. In Chapter 22 on Kaizen, this step is explained more in detail.
4. *Act*: In this step, the improved solutions are implemented by modifying the process or taking corrective actions on significant differences between actual and planned results by analyzing the differences to determine their root causes. Determine where to apply changes that will include improvement of the process or product. When a pass through these four steps does not result in the need to improve, the scope to which PDCA is applied may be refined to plan and improve with more detail in the next iteration of the cycle, or attention needs to be focused in a different stage of the process. Some authors use the term "adjust" instead of "Act," signifying improvement.
5. *Diversify*: Jablonski, in his five-phase guideline for implementing TQM, adds a fifth step—Diversification. In this stage, managers utilize their TQM experiences and successes to bring groups outside the organization (suppliers, distributors, and other companies that have impacted the business's overall health) into the quality process. Diversification activities include training, rewarding, supporting, and partnering with groups that are embraced by the organization's TQM initiatives.

1.11 WHEN TO USE THE PDCA CYCLE

The PDCA Cycle is an effective problem-solving tool. It can be applied:

- For repeated PDCA cycles when Kaizen, the Continuous Improvement tool can to be applied for new areas for improvement.
- For exploring a range of possible new solutions to problems, trying them out, and improving them in a controlled way before selecting one for full implementation.
- For planning data collection and analysis, so as to verify and prioritize problems or root causes.
- For avoiding the large scale waste of resources.
- Daily routine management—for the individual and/or the team.
- Other fields where PDCA can be used are:
  - Project management
  - Continuous development
  - Vendor development
  - Human resources development
  - New product development
  - Process trials

1.12 VARIATIONS OF PDCA TERMINOLOGY

1. *OPDCA*—Another version of this PDCA cycle is OPDCA. The added "O" stands for observation, or as some versions say, "Grasp the current condition."
2. *DMAIC*—as explained further in Chapter 24.
3. The *Baconian method*—Francis Bacon of the 16th century, in which his ideas for a universal reform of knowledge into scientific methodology and the improvement of mankind's state using the Scientific method are emphasized.
4. *Empiricism*—John Locke of the 17th century which states that knowledge comes primarily from sensory experience.
5. As explained in Chapter 22 on Kaizen, several industrial engineering methods use approaches similar to PDCA. Examples are, methods improvement studies, whose seven basic steps are SREDDIM, value engineering, CREW, etc.
6. *Manufacturing system*—The manufacturing system, as explained in Chapter 6, emphasizes all the aspects PDCA, in all stages of manufacturing, as illustrated by Fig. 6.7.

1.13 DEMING'S FOURTEEN POINTS TO IMPROVE QUALITY

After returning to the United States from Japan, Deming propounded the Deming Philosophy, with 14 principles which signified Total Quality Control. They are:

1. Create consistency to purpose towards improvement of product and service.
2. Adopt a new philosophy.

3. Cease dependence on mass inspection to achieve quality.
4. End the practice of awarding business on the basis of price tag.
5. Improve constantly and forever the system of production and service.
6. Institute modern methods of training and education on the job, including management.
7. Institute leadership.
8. Drive out fear.
9. Break down barriers between departments.
10. Eliminate slogans, extortions, and targets for the work force asking for Zero Defect and new levels of productivity.
11. Eliminate work standards that prescribe numerical quotas for both the work force and managers. In their place, use useful aids and supported supervision.
12. Remove barriers that impede hourly paid workers and managers from enjoying pride of workmanship.
13. Institute a vigorous program of education and retraining.
14. Structure top management to empower them to achieve the above 13 points.

1.14   DEMING SYSTEM OF PROFOUND KNOWLEDGE

Deming professed a system of profound knowledge that encompasses four interrelated dimensions.

1. *Appreciation of a system*: Understanding the overall processes involving all stake holders such as suppliers, producers, and customers (or recipients) of goods and services. For example, optimization of one subsystem or a part of the system is easier to achieve than optimization of the whole system. The former may give an impression of improvement, but in reality, it builds barriers that obstruct overall progress.
2. *Knowledge of statistical theories* that not only indicate the process variations in quality, but also the range and causes of variation. Deming believed that managers who lacked this understanding of the variation generally get confused between the random and significant variations, and he estimated that almost 98% of the quality losses are due to this confusion. Deming also illustrated the two kinds of mistakes that are normally committed by the above category of managers.
   2.1 *Mistake No.1*: To react to any fault, complaint, mistake, breakdown accident, shortage etc., as if they come from significant causes, when in fact they come from random or general causes. For example, an engineer may treat an occasional dimensional variation seriously and get the tool holder adjusted.
   2.2 *Mistake No.2*: To attribute to random causes if actually they come from significant causes. For example, the variations due to a loosely clamped tool holder may sometimes get dismissed as random causes.

3. *Theory of knowledge*: The concepts explaining knowledge and the limits of what can be known should be fully appreciated. A manager shall not only have profound knowledge on things but also should appreciate the theory behind such knowledge concerned with the nature and scope of knowledge, its presuppositions. We must know precisely what particular procedure to use in order to measure or judge something, and we need an unambiguous decision rule to tell us how to interpret and act on the result.

4. *Knowledge of psychology*: Concepts of human nature to understand people and the interactions between the leaders and the employees. Deming emphasized that people can be motivated more intrinsically than extrinsically.

1.15   JURAN QUALITY TRILOGY

Joseph Juran put forth his concept of the cross functional management for the quality function more popularly known as Juran's Quality Trilogy and is as follows (adapted from the website *forumqulaitygurus.com*):

1. *Quality Planning*: This is the activity of developing the products and processes required to meet the customers' needs. It involves the following steps:
   - Establish quality goals.
   - Identify the customers, those that would be impacted by the quality goals.
   - Determine the customers' need.
   - Develop product features with reference to the customers' needs.
   - Develop a process.
   - Prove process capability.

2. *Quality Control*: This activity sets up the control parameters as detailed below:
   - Choose control subjects.
   - Choose units of measurements.
   - Establish measurement.
   - Establish controls and transfer the plans to the operating forces.
   - Establish standards of performance.
   - Measure actual performance.
   - Interpret the difference.
   - Take action on the difference.

3. *Quality Improvement*: This activity emphasizes the raising of the performance to unprecedented levels and consists of the following steps:
   - Prove the need for improvement.
   - Establish the infrastructure needed to achieve annual quality improvements, including the training.
   - Identify specific projects for improvement.
   - Organize to guide projects.
   - Organize for diagnosis—for discovery of causes.
   - Diagnose to find the cause.

- Provide remedies.
- Prove that the remedies are effective under operating conditions.
- Provide for control to hold gains.

1.16 CONCLUSION

While the concept and practice of TQM evolved in the manufacturing industry, there is a marked shift of its emphasis into service and other industry categories, too, including health and social services as evidenced by the NIQR Awards, given even to the Dabbawalas of Mumbai. The real challenge today is to ensure that managers do not lose sight of the basic principles detailed in this book on which quality management and performance excellence are based.

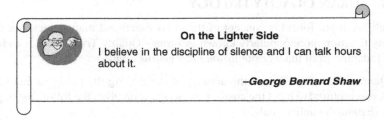

On the Lighter Side

I believe in the discipline of silence and I can talk hours about it.

–George Bernard Shaw

FURTHER READING

[1] Garvin D. Managing quality: the strategic and competitive edge. New York: Free Press; 1988.

[2] www.deming.org/theman/theories/profoundknowledge.

[3] W. Edwards Deming. Wikipedia, the Free Encyclopedia.

[4] wikipedia.org/wiki/Baconian_method.

[5] www.businessdictionary.com/definition/quality.html.

[6] asq.org/glossary/q.html.

Chapter 2

Evolution of Total Quality Management

Chapter Outline

2.1 INTRODUCTION

Although Total Quality Management (TQM) techniques were adopted prior to World War II by a number of organizations, the creation of the TQM philosophy is generally attributed to Dr. W. Edwards Deming, as illustrated in Chapters 1 and 3.

Before the Industrial Revolution, every operator generally produced the whole product himself and hence, was inspecting his own work after completing all the needed operations. However, the Industrial Revolution introduced the concept of specialization of labor, by which a worker made only a portion, or a single operation, and not the entire product. Since most products of this early period were not complicated, quality was not greatly affected. However, this necessitated inspection after each and every operation before the component moved on to the next operator. This initiated further controls as detailed below.

2.2 THE HISTORICAL DEVELOPMENT OF TQM

We can trace the development of TQM into five phases as propounded by A.V. Feigenbaum:

- Operative Quality Control
- Foreman Quality Control
- Inspection Quality Control
- Statistical Quality Control
- Total Quality Control

Total Quality Management: Key Concepts and Case Studies. http://dx.doi.org/10.1016/B978-0-12-811035-5.00002-7

Though Feigenbaum treated Sampling control as part of the Statistical quality control (SQC) and did not list it separately, this book treats it as a significant step between in the Inspection Quality Control and Statistical Quality Control, and it is described in Section 2.2.4.

2.2.1   Operative Quality Control

- Prevalent during the medieval era, but became prominent after the Industrial Revolution.
- Number of products was less and individually made.
- Quality is judged by the workmanship and aesthetic appearance.
- Dimensional accuracy not given importance as there are hardly any mating or interchangeable components and
- Workers were trained in craftsmanship to achieve higher quality viz. Workmanship.

2.2.2   Foreman Quality Control

- Until the 17th century as business increased, the foreman supervised the quality function more as a person controlling the work of several operatives. Quality conformance became one of his responsibilities.
- The concept of quality remained more or less the same as before.
- Here again, the dimensional accuracy is less critical than the functional quality, the workmanship, or the aesthetic appearance.

2.2.3   Inspection Quality Control

- During the Industrial Revolution, which saw the emergence of mass production, manufacturing operations were being broken down to produce small components and then assembling them.
- Hence, a need arose to maintain dimensional accuracy leading to product inspection procedures, basically for dimensional accuracy.
- Initially this inspection was done by the operator himself or the supervisor on line inspection.
- This led to deployment of highly skilled, trained operators as inspectors for better quality standards, initially as a decentralized function of production.
- Around this period, F.W. Taylor, who is called the *Father of Scientific Management*, put forward the principles of functional specialization for effective performance.
- This gave birth to the centralized inspection departments, who generally were performing 100% inspections at least for high value items.

2.2.3.1   Patrol Inspection

Initially the produced products were brought to the inspectors' table and sometimes operators had to wait until the products passed the inspection. To avoid this delay, a system of patrol inspection was started, whereby the inspectors kept walking together with the necessary tools along the operating machines and picking up the machined components to inspect them then and there, informing the operator or the foreman in case of rejections. This is called patrol inspection.

2.2.3.2 Process Control

The concept of controlling the process in addition to the machine operation started around 1945, when it was felt that it was not only the operator, but the process itself that gave rise to defective production.

2.2.3.3 Source Inspection

A recent trend in inward inspection is the source inspection conducted at the supplier's premises, after which the accepted goods go directly to the buyer's shop floor, thereby avoiding the intermediate storage. This is described in more detail in Chapter 19.

2.2.3.4 Supplier Partnership

An extension of the source inspection is the suppler partnership, wherein the goods as certified by the supplier go directly to the buyer's shop floor. This is based in mutual trust and is described in more detail in Chapter 12.

2.2.4 Statistical Quality Control

In view of the small size of the component and the high volume of production by each operator, this inspection for each and every component and its operation became very expensive, more than the cost of manufacture. These facts led to developing the concept of SQC.

- The inspectors were, in any case, required to maintain records and data to justify the rejections they made.
- A study of these records revealed the statistical conformance of these values.
- During World War II, due to very high demand of higher outputs in shorter duration, the managers started applying statistical principles in developing random inspection procedures based on the theory of probability to save inspection time and to enhance the productivity.
- This also helped in determining the trend of the rejections leading to the development of control charts, etc. Walter A. Shewart of Bell Telephones is credited for this in 1924, when in his book *Economic Control of Manufactured Product* showed that productivity improves when variation is reduced. These charts enabled the correction of the process before the rejections occurred.

2.2.4.1 Sampling Inspection

- The above developments led to further expansion of random sampling. Instead of inspecting all pieces of the items received at the receiving inspection, only a small number of samples chosen at random from the lot were inspected by the inspectors.

- All these developments revolutionized the inspection procedures making the inspection departments highly effective and quicker with results.
- This enabled the inspection departments to be more effective with quicker results. Higher quality could be achieved at lower costs.
- Hence, the departments are aptly renamed as Quality Control Departments and this process as SQC.

2.2.5 Total Quality Control

- After the US bombing of Japan, the latter wanted to teach the Americans a lesson, not by driving them out, but by capturing American markets as explained in the next paragraph.
- The quality consciousness was developed as a national spirit among the Japanese workforce to achieve the above goal.
- Deming, an American quality expert was invited to Japan to study their production systems.
- He developed several principles, mostly adapted from traditional management principles and explained them in simple language, and terms that were easily understood and remembered by the Japanese workforce, helping to convert their quality consciousness to quality commitment as a national fervor.

2.3 QUALITY MANAGEMENT IN THE JAPANESE SCENARIO

- Before World War II, Japan was not a highly industrialized nation. Most of the electrical and electronic goods were imported from the United States and Europe. So Japan was playing second fiddle to the United States in commerce and trade.
- Japan's decision to side with Hitler alienated them against the United States and Japan's raid on Pearl Harbor infuriated the United States, resulting in the dropping of atom bombs on Hiroshima and Nagasaki, one of the most inhumane acts ever committed in the history of mankind.
- Consequently, the Japanese wanted to pay back the Americans in their own coin.
- They knew it couldn't be in a war and hence, decided to beat the Americans in world trade, by producing more quality goods and capturing the international market currently rules by Americans. The Japanese being highly patriotic by nature, this desire percolated into the minds of every national, specifically into all categories of personnel in Japanese industry.
- Higher productivity became the initial buzz word. Because it was realized higher productivity without quality products would take the nation nowhere, the subsequent buzz word was high quality production. Quality in all aspects of manufacture was given high priority.

- As a first step, Japan sent a delegation of productivity personnel on a study tour of the United States and then invited Deming as a consultant in quality improvement and training in Japan.
- While Deming's chief task was to lecture on quality control methods, he talked more on the theory of systems and cooperation and professed team spirit and more operative-level participation in shop floor decisions. This, in fact, being the very spirit of the Japanese, they paid more attention to his teachings and it gave them a different approach to problem-solving than what they were following.
- Deming also professed that quality is more a management responsibility than the inspector's responsibility, and developed the systematic approach, the PDCA (Plan–Do–Check–Act).
- The whole quality revolution in Japan during the post-war era can be accredited to the above principle, as well as other practices, such as quality circles.

2.4 POST-DEMING/JURAN QUALITY SCENARIO

- In 1952, Kaoru Ishikawa propounded the Cause and Effect diagram that aided analysis-oriented quality control. In 1960, he formalized "quality circles"—the use of small groups to eliminate variation and improve processes.
- The concept of Quality Assurance emerged.
- In 1983, A.V. Feigenbaum coined the words "Companywide Quality Control," that later became "Total Quality Management."
- Genichi Taguchi, who received his PhD at the age of 18, conceived the Taguchi Loss Function Mode.
- In 1990, Taichi Ohno, the initiator of Toyota Production Systems, together with Shigeo Shingo, developed the concept of Single Minute Exchange of Dies (SMED).
- In 1993, Michael Hammer and simultaneously, Shigeo Shingo, developed the concept of Business Process Reengineering.
- Bill Crosby propounded the Zero Loss principle, saying "Do it Right the First Time."
- Masaki Imai is the exponent of Kaizen, which he explained very clearly in his book on Kaizen.
- Noriaki Kano gave us the Customer Satisfaction Model.

In 2015, the Indian Government initiated the slogan "Zero Effect, Zero Defect (ZED)," signifying that together with achieving zero defect level, the Indian industry should ensure that its processes have zero effect on environment.

Most of these are explained in more detail in later chapters.

2.5 CONCLUSION

The roots of TQM can be traced back to the early 1920s when statistical theory was first applied to product quality control. Quality management has evolved from the operative quality control of the pre-1920 days to the TQM of today. This metamorphosis provides an interesting story, tracing through the quality journey and illustrating the Japanese style management approach to quality improvement.

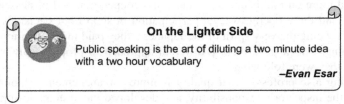

On the Lighter Side

Public speaking is the art of diluting a two minute idea with a two hour vocabulary

–Evan Esar

FURTHER READING

[1] http://asq.org/.
[2] en.wikipedia.org/wiki/Total_quality_management.
[3] http://www.bpir.com/total-quality-management.
[4] www.businessballs.com/dtiresources/quality_management_history.pdf.

Chapter 3

Quality Gurus

Chapter Outline

The significance accorded to quality has been profound in recent decades more than in the past. Like the Industrial Revolution of the 18th century, we can call this era the quality revolution era. Because quality is more a philosophy and work culture, several theories have been propounded by experts in the quality movement. These theories are considered today as TQM philosophies and the experts who propounded these are known as Quality Gurus. The Oxford Dictionary defines a Guru as "a respected and influential teacher or authority." In a similar manner, we can extend this definition of quality guru as a charismatic individual whose concepts and approaches to quality within business and life have made a major and lasting impact. This chapter summarizes the brief particulars and theories of some of the gurus who made a mark in the TQM movement.

3.1 WILFREDO PARETO

One of the earliest economists (1848–1923), Wilfredo Pareto has indirectly contributed to the vital TQM concept of selective control which is one of the best utilized concepts in inventory control. During his studies on the distribution of wealth, he found that very few people possessed a major portion of the

Total Quality Management: Key Concepts and Case Studies. http://dx.doi.org/10.1016/B978-0-12-811035-5.00003-9

country's wealth, while the combined wealth of a majority of people was a small fraction. He called this vital few and trivial many, and illustrated this study by arranging the individuals in order of their wealth, and drawing a graph called the Pareto Diagram, indicating the individual wealth against the individual which is shown in Fig. 3.1.

This has become the principle behind the selective control in Materials anagement called ABC analysis, where the A items form 10% of the items by number, but 70% by value are given a high level of planning and control with close monitoring in parameters such as lead time, purchase lot quantities, inviting sealed tenders, and price fixation, etc., whereas, items of C category where procurement costs are higher compared to their annual value. Hence, they are procured in larger quantities with little consumption control.

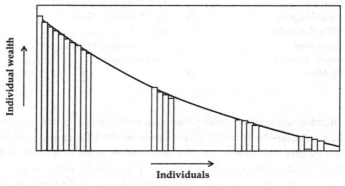

FIG. 3.1   An illustration of a Pareto diagram.

In quality control, this ABC system of control can also be applied. It helps in investigating the vital few processes which result in major defects, both in number and extent. These processes require closer control, whereas the many other processes resulting in trivial defect occurrence do not require such close control. This helps us in selecting processes that require closer control from those that require general control.

3.2   WALTER A. SHEWHART

Walter Shewhart is perhaps the oldest known quality guru after Wilfredo Pareto. Born around 1890, he received his doctorate in physics in 1917 and taught at the Universities of Illinois and California. He later worked with Western Electric (1918–1924) and Bell Laboratories of AT&T (1925–1956).

He is the originator of the statistical process control and control chart theory.

He also propounded the Plan-Do-Study-Act (PDSA) cycle for improvements in the process. This was later modified as PDCA cycle by Deming.

In 1931, he published a book *Economic control of Quality of Manufactured Product*, which is still considered a complete and thorough exposition of the basic principles of quality control.

3.3 EDWARDS DEMING

W. Edwards Deming (1900–1993) is perhaps the most acknowledged quality guru of the world. The whole quality revolution in Japan during the post-war era is accredited to him, and he is generally called The Father of Japanese Quality. Born in 1900, he got his PhD at the age of 28. He was an ardent student of Walter Shewhart. After the devastation of Japan during World War II, JUSE (Japanese Union of Scientists and Engineers), under the leadership of K. Konayagi, invited Deming in helping with the reconstruction of Japanese industry. While his chief task was to lecture

on quality control methods, he talked more on the important concepts which he referred to as the theory of systems and cooperation. He even said "the control chart is no substitute to the brain." The Japanese realized that his teachings made sense and gave a different approach to problem-solving than what they were following.

He professed that quality is more a management responsibility than the inspector's responsibility and developed the systematic approach, the PDCA (Plan-Do-Check-Act) cycle in his consultancy works, as well as in TQM lectures. He named this PDCA cycle after his teacher as the Shewhart cycle. (though some authors attribute the development of PDCA cycle to Walter Shewhart, Samuel Kho states that this was, in fact, developed by Deming and and named the Shewhart cycle, as a tribute his favorite teacher.)

Subsequently, he replaced the word *Check* with *Study*. This PDSA cycle had come to be known as the Deming Cycle. In 1960, he was honored by the Japanese Emperor with the Second Order of the Sacred Treasure.

Evans and Lindsay make an interesting observation that despite Deming's popularity in Japan, he remained virtually unknown, even in his home town of Washington until 1980. Only when NBC telecast a program called "If Japan

Can, Why Can't We?", that highlighted Deming's contributions to Japan and his later works with Nashua Corporation, his name was often on the lips of US executives. Corporations like Ford, GM, and Proctor & Gamble invited him to work with them as a consultant and improve their quality.

After returning to the United States, he studied the management attitude of managing quality vis-a-vis the Japanese attitude, and propounded the Deming Philosophy, with 14 principles that really made all quality experts around the world orient their thinking from Total Quality Control point of view, rather than a Quality Control point of view. The Deming Philosophy focuses on continual improvement in product and service quality by reducing uncertainty and variability in design, manufacturing, and servicing processes, driven by the leadership of the top management. His thinking can be best expressed as *Management by Positive Cooperation*, or in other words, the quality control is not the function of quality engineers or the production department alone, but by all levels of management with their appreciation and cooperation.

In recognition of his services in developing Japan as a major industrial and economic force during the post-war period, the Japanese government, under the auspices of JUSE, instituted the Deming Awards, given to top industrial organizations of Asia that are successful in the quality movement. It may be interesting to note that the Indian Industrial Corporations have excelled among the Asian corporations coveting a majority of these awards year after year. The criteria, the parameters, and other details of the Deming Awards are discussed in detail in Chapter 14.

3.4 JOSEPH JURAN

Joseph Juran, born in 1904 in Rumania, is well known for the *Juran Trilogy*, which is an approach to cross functional management that is composed of three managerial processes-planning, control, and continuous improvement. He focused on quality control as an integral part of the management control in his lectures to the Japanese in the early 1950s. He believed that quality does not happen by accident, but has to be planned and there can be no shortcuts to quality. The Quality Control handbook edited and published by him is still considered a significant reference book. He, too, went to Japan in the mid-1950s to lecture and conduct seminars to top and middle-level executives. Specializing in managing for quality, he authored hundreds of papers and 12 books. He was honored by the Japanese Emperor with the Second Order of the Sacred Treasure.

Apart from his famous Quality Trilogy as detailed in Chapter 1, Juran also proposed the following 10 steps for quality improvement.

- Build awareness of the need and opportunity to improve.
- Set goals for that improvement.
- Create plans to reach the goals.
- Provide training.
- Conduct projects to solve problems.
- Report on progress.
- Give recognition for success.
- Communicate results.
- Keep score.
- Maintain momentum.

3.5   ARMAND FEIGENBAUM

Armand Feigenbaum is the originator of the Total Quality concept. Involving not only the quality control personnel and the machine operators, but also all the employees of the organization at all levels and of all functions. He propounded this concept in his book *Total Quality Control*, which today is still a good reference book. In fact, his concept of Total Quality Control has been re-christened without much change in the principles later as Total Quality Management, which is today's industrial buzz word, internationally.

As per Feigenbaum's definition: Total Quality Control as an effective system for integrating the quality development, quality maintenance and quality improvement efforts of the various groups in an organization so as to enable the production and service at the most economical levels which allow full customer satisfaction.

Feigenbaum's three steps to quality philosophy:

1. *Modern Quality Technology*: A majority of the quality problems can be solved, not by the traditional quality controlling methods, but by effective integration of office staff, production/plant engineers, as well as the shop floor operatives in the process of continually evaluating and implementing new techniques to satisfy the customers in the future. This is signified by the term "Total" in *Total quality Control*.
2. *Organizational Commitment*: The continuous training and motivation of the entire workforce, as well as the integration of quality in strategic planning and provision of the means, infrastructure, and other facilities in including it in all aspects of management. This is signified by the term "Quality" in *Total Quality Control*.

3. *Quality Leadership*: The management emphasis should be based on sound planning rather than reaction to failures. Management must maintain a constant focus on leading the quality effort. This is signified by the term "Control" in *Total Quality Control*.

The Japanese, especially Ishikawa, grasped this concept of total quality control as the basis for the Companywide Quality Control (CWQC).

Feigenbaum is also known for his concept of "hidden plant," which means that in every factory a certain portion of its capacity, which may be as high as 40%, is wasted through not getting it right the first time. Many of his ideas remain embedded in the contemporary thinking and have become some important elements of the Malcolm Baldridge National Quality Award (MBNQA) criteria.

3.6   PRASANTA CHANDRA MAHALANOBIS

Born during the last decade of the 19th century, P.C. Mahalanobis was one of the most internationally renowned statisticians of his time. In 1924, he made some important discoveries pertaining to the probable error of results in agricultural experiments. He contributed significantly in the development of experimental statistics in India. He founded the Indian Statistical Institute in 1931, which was declared by the Indian government in 1959 as an Institution of National Importance. In 1957, he became the Hony. President of the International Statistical Institute and in 1961, he was elected a fellow of the American Statistical Association. Together with Genichi Taguchi, he developed the Mahalanobis-Tagichi Strategy for multidimensional systems. Apart from several awards and honors, he received Padma Vibhushan from the President of India in 1968. He has written several books, including as a coauthor of *Mahalanobis-Taguchi Strategy* and *Mahalanobis-Taguchi System*.

3.7   SHIGEO SHINGO

Born in 1909, Shigeo Shingo first worked with the Taipei Railway Company in Yokohama, where he raised the productivity by 100% by his industrial engineering techniques, and then as a consultant with Toyota and Matsushita, and saw their development as integrated systems. His successful performance along with Taichi Ohno at the Toyota Motor Corporation has become an established Quality-oriented system as Toyota Production System. As a part of this system, he developed the Single Minute Exchange of Dies (SMED). The widening gap between Toyota and other Japanese companies opened the eyes of others in Japan and elsewhere in the world to this phenomenon of Toyota Production System and it began spreading rapidly globally.

He is also credited with the development of Poka-yoke, described more in subsequent chapters.

Along with Michael Hammer, Shingo is credited with the development of Business Process Reengineering (BPR) that has globally created a revolution in method improvement and industrial engineering techniques, concentrating more on the review of the overall organizational improve-

ment, rather than individual operational improvements. In recognition of this, the National Convention of Industrial Engineering of 1994 in Chennai named one of its halls as Shingo Hall. The Japanese government, as a tribute to him, has instituted the Shingo award for excellence in manufacturing. He died in 1990.

Shigeo Shingo has written 14 books and over 100 papers on several fields of manufacturing.

3.8 TAICHI OHNO

Taichi Ohno (1912–1990) is credited as the initiator of Toyota Production Systems, together with Shigeo Shingo especially for SMED. He is also credited with the concept of the three M's—Muda, Muri, and Mura.

3.9 KAORU ISHIKAWA

Ishikawa (1916–1989) is well known for his Cause and Effect Diagram that helps in determining the root cause of quality problems. This diagram is known as the Ishikawa diagram after him, and also as the Fishbone Diagram after its appearance. This is explained in more detail in Chapter 20. He also professed the application of the seven statistical tools namely:

- Pareto analysis
- Cause and effect diagram
- Control charts
- Stratification
- Checklist

- Histograms
- Scatter diagrams

His other achievements are:

1. Developing in 1949 the first basic quality control course for JUSE.
2. Being credited with initiating the Company-wide Quality Movement in 1962, (described in more detail in Section 3.5), by building upon Feigenbaum's concept of Total Quality Control.
3. Contributing to the Quality Circle Movement in Japan.
4. Being awarded the honorary membership of American Society of Quality in 1986. Evans and Lindsay say that by 1986, only four persons were awarded this honorary membership by ASQ, the other three being Armand Feigenbaum, Edwards Deming, and Joseph Juran.
5. Receiving the Walter Shewart-MBNQA Medal in 1988 for his outstanding contribution to the development of quality control theory, principles, techniques, and standardization activities for both Japanese and world industry, which enhanced quality and productivity.
6. Having a National Medal named after him by ASQ in 1983, followed by a similar medal named Harrington - Ishikawa Medal instituted by the Asian Pacific Quality Organization.
7. Being honored by the Japanese Emperor with the Second Order of the Sacred Treasure.
8. Publishing more than 600 papers and 31 books including:
   - *Introduction to Quality Control*
   - *What is Total Quality Control, the Japanese Way*
   - *Guide in Quality Control*

3.10   GENICHI TAGUCHI

Born in 1924, Genichi Taguchi got his Ph.D. at the age of 18. He is known for his Quality Loss function model which explains the economic value of reducing variation as explained further in Chapter 31.

He emphasized the need for incorporating quality and reliability at the design stage, prior to production. He developed a methodology which is fundamentally a prototyping technique that enables the design and development engineer to produce a robust design that can survive repetitive manufacturing in order to deliver the functionality needed by the customer. He advocated the techniques of experimental design to identify the most important design variables in order to minimize the effects of uncontrollable factors on product variations. His approach attacked quality problems early in the design stages, rather than reacting to problems that might arise later in production. This methodology is highly adapted by industrial organizations in the United States under the name of Taguchi Methods.

These are Taguchi's three stages of product development which are further explained in Chapter 31.

System design stage, which is the non-statistical stage for engineering, marketing, and gathering customer knowledge.

Parameter stage, which is the analysis of how the product should perform against defined parameters and a robust solution of cost-effective manufacturing, irrespective of the operating parameters. Here is where Taguchi argued that the performance requirements of the system are generally underspecified and that are given too loose tolerances to allow for the process variations, resulting in the quality loss function as described earlier.

Tolerance design stage, which involves the effect that the various parameters have on performance, resources can be focused on reducing and controlling variation in the critical few dimensions.

3.11   PHILLIP B. CROSBY

Born in 1926, Phillip B. Crosby (1926–2001), started his career as a reliability engineer. He participated in the Martin Missile experiment that initiated his famous zero defect concept. He is also well known for his concept of *Do it Right the First Time.* He gave the four absolutes of quality as:

1. The definition of quality is conformance to requirements, not only conformance to specifications.
2. The system force causing quality is preventive, not appraisal.
3. The performance standard must be zero defect, not that's close enough.
4. The measurement of quality is the price of non-conformance not indices.

Like Deming, he, too gave 14 principles as steps to quality improvement.

1. *Make it clear that the management is committed to quality.* This can be accomplished by:
   - A corporate policy on quality needs to be issued.
   - Quality should be made the first item on the agenda of regular meetings.
   - The CEO and other officials need to compose clear quality speeches in their minds.

2. *Form quality improvement teams* with senior representatives from each department.
   The purpose of the team is to guide the process. This team requires clear direction and leadership from top management. Institute leadership among employees.

3. *Measure the processes* to determine where current and potential quality problems lie.
   Measurement is the habit of seeing how we are going along, by using simple patterns.

4. *Evaluate the costs of quality* and explain its use as a management tool.
   Cost of quality is defined as the cost of non-conformance, or the cost of doing things wrong.

5. *Raise the quality awareness and personal concern* among all employees.
   The expression "quality is a must" needs to be spread around. Awareness must be adapted to the culture of the company.

6. *Take actions to correct problems* identified through previous steps and eliminate them forever.

7. *Establish progress monitoring* for the improvement process.

8. Train supervisors to actively carry out their part of the quality improvement programs.
   The management has to develop a quality education system that would provide a standard message, and could be taught by anyone who is trained to use it. The time and money spent in training is worth it since it will result in quantum leaps in improvement.

9. *Hold a zero defect day* to reaffirm management commitment.
   Zero Defect Day makes the management stand and make the commitment in front of all employees. It shows everyone that the management is serious.

10. *Encourage individuals to establish improvement goals* for themselves and for their group. Breakdown the inter-departmental barriers.

11. *Encourage employees to tell management about obstacles to improving quality*:
    Drive out the fear among employees and create an environment of trust so that everyone can communicate with the seniors more freely.

12. *Recognize and appreciate* those who participate in this quality movement. Establish awards programs.

13. *Establish quality councils*, meet and communicate on a regular basis.
    The purpose of these quality councils is to bring the quality professionals together and let them share their experiences and learn from one another.

14. *Do it all over again* to emphasize that the quality improvement program is continuous and never ends.
    A typical quality improvement program takes anywhere from 12 to 18 months. By that time, the changing situation and scenario might need further planning of the quality programmed with newly-defined policies.

Some of Phillip B. Crosby's books are:

1. Quality without Tears
2. Quality is Free
3. Quality is Still Free
4. Quality and Me
5. The Art of Getting Your Own Sweet Way, More things
6. The Eternally Successful Organization and Leading
7. The Art of Becoming an Executive

3.12 YOSHIO KONDO

An admirer of Kaoru Ishikawa, Yoshio Kondo has done considerable work in popularizing the Concept of CWQC. He identified that quality is more compatible with human nature than cost and productivity. He developed a four-point approach to motivate which makes it possible for work to be reborn as a creative activity.

He received the Nikkei Prize in 1967, the Deming Prize in 1971, the American Institute of Mining Award in 1971, and the Tanigwa-Harris Prize of Japan Institute of Metals in 1981. He was elected Honorary Member of ASQ in 2004. He authored several books including:

- Companywide Quality Control
- Quality in the 21st Century

3.13 SHIGERU MIZUNO

Shigeru Mizuno of the Tokyo Institute of Technology is credited with the development of Quality Function Deployment (QFD), together with Yoji Akao. He applied QFD at the Heavy shipbuilding industry of Mitsubishi Heavy Industries in 1972. After 4 years of further refinement and development, it was again applied in Toyota 1977, with astounding results, with a 60% cumulative reduction in the startup COSTS. QFD, explained more in Chapter 30, is a disciplined approach to product design, engineering, and production with an objective assessment of customer requirements and company facilities, and translation of the customer requirements into product development and production process. He is also credited with the development of three TQM tools viz. relationship diagram, affinity diagram, and systematic tree diagram.

3.14   YOJI AKAO

Yoji Akao is credited with the development of QFD together with Shigeru Mijuno. During the 1960s, he was exploring ways to apply powerful problem-solving algorithms to designing products. Initially using the Ishikawa diagram, his more complex analysis led to matrices to identify the design elements which would impact customer satisfaction. With the help of Shigeru Mizuno, this work led to the development of QFD in the 1970s. He is currently the chairman of the International Council for QFD.

Yoji Akao is also credited with the initiation of *Hoshin Kanri,* which is the annual planning and policy deployment, and literally means control or management of the way of setting direction. Hoshin Kanri was used at the Komatsu Company in 1965.

3.15   NORIAKI KANO

Noriaki Kano is renowned for his Customer Satisfaction Model and the concept of the customer's expectations of the quality of the product or service which is described more in Chapter 10.

Professor Kano is currently the Chairman of the Deming Awards Committee. As explained earlier, the Deming Awards are awarded to the top few Asian organizations successful in the quality movement. It may be interesting to note that a majority of the organizations are from India. In this connection, Professor Kano has visited India a good number of times and interacted well with quality professionals of India. As a council member of the National Institution for Quality and Reliability, which has hosted several of his lecture meetings, this author had the opportunity of interacting with him personally, especially in connection with his Customer Satisfaction Model.

He has delivered several lectures, both in Japan and abroad and has authored 2 books:

- *Theory of Attractive Quality and Packaging*
- *TQM in Service Industries*

3.16 MASAAKI IMAI

Masaaki Imai (born in 1930), is an exponent of Kaizen. He established the Kaizen Institute to help Western Corporations to introduce the Kaizen concept. He authored the best-selling book, *Kaizen: the Key to Japan's Competitive Success*. His other book is *Gemba Kaizen*.

3.17 CLAUS MÖLLER

Claus Möller developed the concept of personal quality, a central element of TQM. His 12 golden rules for quality improvement are:

1. Set personal quality goals.
2. Establish your own personal quality account.
3. Check how others are satisfied with your efforts.
4. Regard the next link in the customer chain as a valued customer.
5. Avoid errors.
6. Perform tasks more effectively.
7. Utilize resources optimally.
8. Be committed.
9. Learn to finish what you start.
10. Control your stresses.
11. Be ethical.
12. Demand quality.

3.18 BLANTON GODFREY

Blanton Godfrey, a fellow of the American Statistical Association, has been the Chairman and Chief Executive officer of the Juran Institute. He is the co-editor of Juran Quality Control Handbook. He was involved in the creation of the MBNQAs and the US delegate for the ISO Technical committees. In 1992, he was awarded the Edwards Medal for outstanding contribution to the science and proactive of quality management. In 2001, he became the founding editor of Six Sigma Forum magazine. He also authored a book, *Curing Health Care* in 2002.

3.19 CLARENCE IRWING LEWIS

Born in late 19th century, he is considered the father of modern modal logic. His writings on the relationship between information, experience, theory, and knowledge, were greatly referred to by several quality professionals, including Dr. Deming.

3.20 DAVID GARVIN

He is credited with identifying the eight dimensions of quality, as detailed in Chapter 7. He professed that the customers have a different perception of quality than that of the managers.

3.21 DORIAN SHAININ

Born in 1914, Dorian Shainin developed a discipline called *Statistical Engineering*, also known as the *Dainin Technique*. It is successfully used in quality improvement, product development, product reliability, analytical problem solving, etc. He became a fellow of American Society for Quality in 1949 and received the Shewhart Medal in 1989.

In recognition of his services to ASQ, The Dorain Shainin medal was instituted in 2005 for the development and application of creative or unique statistical approaches. His books are:

1. *Managing Manpower in the Industrial Atmosphere*
2. *Tool Engineer's Handbook*
3. *Quality Control Handbook*
4. *New decision Making Tools for Managers*
5. *Quality Control for Plastic Engineers*
6. *Manufacturing, Planning and Estimating Handbook*
7. *Statistics in Action*

3.22   EDWARD DE BONO

Edward de Bono, born in Maltais, is a leading authority in the field of creative thinking. He is known for his two techniques, Six Thinking Hats (STH) and Direction Attention Thinking tools (DATT) frameworks. He is the founder of the World Centre for New Thinking, which acts as a platform to make visible new thinking from any source, also called Hypothesis Development. He has written over 70 books and has been invited to lecture in about 58 countries.

3.23   ELIYAHU M. GOLDRATT

Eliyahu Goldratt, born in Israel in 1948, is the originator of the Theory of Constrains (TOC), which provides a framework for managing enterprises with a holistic and focused approach to solve the conflicts between the local operating level decisions. A physical constraint may be like the machine capacity, where as a non-physical constraint may be like the demand of a product, a corporate procedure, or an individual's approach to a problem. This includes a five step methodology.

1. Identify the system's constraints.
2. Decide how to manage the constraints within the system.
3. Identify a majority of the resources that are not constraints. Arrange them in tandem with those groups identified in the first step.
4. Elevate each constraint by analyzing the logical need for the constraint. Most constraints in the industrial corporations are policy constraints, rather than physical constraints. Thus a review of the policy rules and procedures that have developed over a time would give a clue for breaking these constraints and elevating them into non-constraints.
5. After breaking the constraints, return to step 1, till all constraints are either eliminated or reduced to a minimum.

3.24   EUGENE L. GRANT

Eugene Grant (1897–1996), professor at Stanford University did extensive work on statistical quality control. Apart from receiving the Shewhart medal in 1952, Founders' Award from AIIE in 1965, the Wellingon Award in 1979, etc., ASQC and AEEE honored him by instituting the annual Eugene L. Grant Awards for distinguished contribution in quality control education. As a tribute to him, Juran is quoted as saying "Eugene's contribution to statistical methodology is more instrumental in advancing quality, much greater than that of Deming." His books are:

1. *Handbook of Industrial Engineering and Management* (Co-editor with Ireson)
2. *Principles of Engineering Economy*
3. *Statistical Quality Control*
4. *Depreciation*
5. *Basic Accounting and Cost Accounting*

3.25   BILL CONWAY

He professed that quality management is very much linked with other functions like purchasing, manufacturing, and distribution.

He listed another set of quality improvement tools, including:

1. Human relation skills
2. Statistical surveys
3. Simple statistical techniques of seven basic tools
4. Statistical process control
5. Engineering
6. Industrial Engineering

3.26   YASUTOSHI WASHIO

Past President of the Japanese Society for Quality Control, Yasutoshi Washio, is a globally renowned expert in TQM. He was a Deming Prize winner in 1993 and won the Indian Dronacharya award in 2004, for his expertise in world-class manufacturing through TQM practices, SQC, and SPC. He did extensive consultancy in Indian industries, especially in Sundaram Clayton, whose MD acknowledged him as the mentor for his company to strengthen the TQM processes within the company, and winning the Deming Award in 1998.

Yasutoshi Washio with NIQR members in Chennai. The author is in the RIGHT.

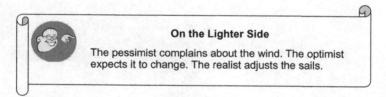

On the Lighter Side

The pessimist complains about the wind. The optimist expects it to change. The realist adjusts the sails.

FURTHER READING

The website of *forumqulaitygurus.com* gives details of other professionals who have excelled in the field of quality.

Chapter 4

Leadership and TQM

Chapter Outline

4.1 WHAT IS LEADERSHIP?

The success of any organization depends on the performance of the workforce. Their performance depends upon their morale level and their relationship with the bosses, apart from their own skills, experience, etc. This employee morale level is created by good leaders. All managers starting from the CEO must show themselves as good leaders, create clear values and high expectations for performance excellence, and then build these into the company's processes. They should strive to inspire and motivate the workforce. Leaders must create an order in the organization that is concurrent, predictable, and controllable. Leadership is the ability to positively influence people and systems under one's authority, have a meaningful impact, and achieve important skills.

Total Quality Management: Key Concepts and Case Studies. http://dx.doi.org/10.1016/B978-0-12-811035-5.00004-0

4.2   DEFINITIONS FOR LEADERSHIP

The following quotations from some authors would explain this concept better.

Senior management shall serve as role models to inspire and motivate the workforce and encourage involvement, learning, innovation, and creativity. Leadership is the ability to positively influence people and systems under one's authority with a meaningful impact to achieve important skills.

John Evans and William Lindsay

Leadership is a continuous management emphasis on sound planning, rather than reaction to failures. Management must maintain a constant focus and lead the quality effort.

Feigenbaum

A leader is one who instills purpose, not the one who controls by brutal force. Leaders and followers raise one another to higher levels of motivation and morality.

James McGregor Burns

A leader is someone who can take a group of people to places where they do not think they can go.

Bob Eason, CEO of Daimler Chrysler

Leadership is "we," and not "me" mission, not my show; vision, not division; and community, not domicile.

Rick Edgeman

Inventories can be managed, but people must be led.

Ross Perot, Texas billionaire

Perhaps the best of the explanatory definition is given by the Malcolm Baldridge National Quality Award, as stated below:

An organization's senior leaders should set directions and create a customer focus, clear and visible values, and high expectations. The directions, values, and expectations should balance the need of all the stakeholders. The leader should ensure the creation of strategies, systems, and methods for achieving excellence, stimulating innovations, and building knowledge and capabilities. The values and strategies should help guide all activities and decisions of the organization. Senior leaders should inspire and motivate the entire workforce and should encourage all employees to contribute, develop and learn, to be innovative, and to be creative.

Senior leaders should serve as role models through their ethical behavior and their personal involvement in planning, communication, coaching development of future leaders, review of organizational performance, and employee recognition. As role models, they can reinforce values and expectations while building leadership, commitment, and initiative throughout the organization.

From these definitions, we can represent the relationship between the leader and his subordinates as shown in Fig. 4.1.

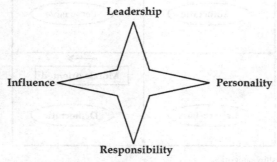

FIG. 4.1   The four elements of quality leadership.

4.3   THEORIES OF LEADERSHIP

Because the concept of scientific management was developed during the postindustrial revolution period, the major and fundamental components of management theories have been the leadership theories. While the early theories can be traced to the early 20th century, the leadership theories related to quality management can be traced to the past four decades, especially after 1970. Whereas dozens of such theories have been developed, we will consider the following classification of leaders for the purpose of understanding the differences between various leadership styles and concepts.

The critical role played by senior managers in the quality movement is emphasized by the Malcolm Baldridge Award, where strategic management leadership is given priority. So is the case with other assessment frameworks like Deming Award, European Foundation for Quality Management (EFQM), Canada Award for Business Excellence (CABE), ISO 9000, etc.

4.4   LEADERSHIP CATEGORIES

The early management books have highlighted the leadership types and traits as:

- *Autocratic*, where the leader forces his subordinates to achieve the targets as he perceives them without giving any scope for thinking and judgment by them.
- *Persuasive*, where the leader does discuss with his subordinates and persuades them to achieve the targets either by incentives or by snoopervision, a word coined to imply snooping supervision.
- *Democratic*, where the leader gives high value to the feelings of the employee and gives prime importance to employee morale.
- *Laissez-faire*, where the leader is too passive and gives full freedom to the subordinates without having any control over them. This would sometimes cause miscommunication and disputes among the employees, but the leader makes no serious attempt to solve these disputes.

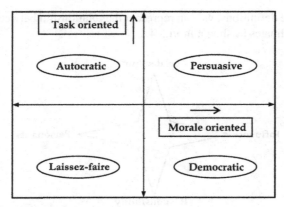

FIG. 4.2   Leadership categories.

These categories are represented in the following Fig. 4.2.

To summarize, a perfect quality-conscious leader should be more of the persuasive type and democratic to a lesser extent and certainly neither autocratic nor a laissez-faire type.

4.5   LEADERSHIP AND GOAL SETTING

The essentials of SMART goals can be explained by the first letters of the words comprising the acronym SMART.

SMART means:

- *S*pecific
- *M*easurable
- *A*mbitious
- *R*ealistic
- *T*ime-bound

We can also say that the 5Ds of goal setting are:

- *D*irection
- *D*edication
- *D*etermination
- *D*iscipline
- *D*eadline

This can be represented by the Fig. 4.3.

There can be three categories of managers with regards to the concept of goal setting:

1. Those who plan and write their goals and follow them systematically are successful in life with money and fame.
2. Those who have goals, but neither write them, nor follow them are just better off.
3. Those who have no goals and do not even know what is meant by setting goals.

SMART means	5D's of Goal setting
Specific	Direction
Measurable	Dedication
Ambitious	Determination
Realistic	Discipline and
Time bound	Deadline

FIG. 4.3 Principles of SMART and goal setting.

There can also be three categories of managers with regards to their work commitment:

1. The first type does not work at all, but procrastinate their duties.
2. The second type work according to the rules, but no more.
3. The third type work, doing not only what is expected of them, but more.

Similarly, the success and failure of a manager can be attributed to:

Success

- *Sense of direction*
- *Undertaking*
- *Courage*
- *Clarity*
- *Esteem*
- *Self Confidence*
- *Self-Acceptance*

Failure

- *Frustration, Futility*
- *Aggressiveness*
- *Insecurity*
- *Loneliness*
- *Uncertainty*
- *Resentment*
- *Emptiness*

The above can also be represented by the Fig. 4.4.

SUCCESS means	FAILURE means
• Sense of direction	• Frustration, Futility
• Undertaking	• Aggressiveness
• Courage	• Insecurity
• Clarity	• Loneliness
• Esteem	• Uncertainty
• Self Confidence	• Resentment
• Self-Acceptance	• Emptiness

FIG. 4.4 The components of success and failure.

4.6   CHARACTERISTICS OF QUALITY LEADERS

Evans and Lindsay summarize the characteristics of true quality leaders as follows:

1. They focus on creating and balancing value for customers and all other stakeholders that serve as a basis for setting business directions and performance expectations at all levels of the organization.
2. They create and sustain a leadership system and environment for empowerment, innovation, agility, and organizational learning.
3. They set high expectations and demonstrate substantial personal commitment and involvement in quality, often with missionary-like enthusiasm.
4. They integrate quality values into daily leadership and management and communicate extensively through the leadership structure and to all employees.
5. They review organizational performance—including their own performance as leaders, to assess original success and progress and translate review findings into priorities for improvement and opportunities for innovation for the organization a whole, as well as their own leadership effectiveness.
6. They create an environment that fosters legal and ethical behavior and a governance system that addresses management and fiscal accountability and protection of stockholder and stakeholder interests.
7. They integrate public responsibilities and community support into their business practices.

   Besterfield et al., add the following to these characteristics,

8. They emphasize improvement rather than maintenance. Leaders use the phrase, "If it is not perfect, improve it" rather than "If it is not broke, don't fix it." There is always room for improvement, even if the improvement is small.
9. They emphasize on prevention. "An ounce of prevention is better than a pound of cure," they say.
10. They encourage collaboration rather than competition. When functional areas, departments, or work groups are in competition, they may find subtle ways of working against each other or withholding information. Instead, there must be collaboration among units.
11. They train and coach, rather than direct and supervise. They appreciate the importance of human resource development and help their subordinates learn to do a better job.
12. They learn from problems. When a problem exists, it is treated as an opportunity for improvement, rather than something to be minimized or covered up.
13. They continually demonstrate their commitment to quality.
14. They recognize and encourage team effort. Rewarding of teams and individuals is important because people like to know that their contributions are appreciated.

The following acronyms can also explain other basic characteristics of a *leader*:

- *L*istening (learn by listening)
- *E*mpathy
- *A*rticulation
- *D*ecision-making
- *E*mpowerment
- *R*ole model (be a role model for others)

4.7 WARREN BENNIS PRINCIPLES OF GREAT TEAMS

Warren Bennis studied teams that worked in organizations like Apple Computers, Palo Alto Research Center, Lockheed Skunk Works, and Walt Disney Animation Studios. He listed his findings in the form of Warren Bennis Principles of Great Teams. They are:

- *Shared dreams*: The great teams generally shared the dream to transform the world for better living. They are obsessed with what they are doing and do not consider their work as just a salary earning job. Their sharing this dream and belief gave them the cohesiveness and the energy needed to work.
- *Mission is bigger than ego*: Great teams placed mission way above individual egos as a building stone for success.
- *Protection of the team members* from external comments and criticisms would keep them satisfied and help in achieving the mission.
- *Fostering enmity*: Here enmity implies a healthy competition environment in an implicit mission, such as destroying the enemy in terms of competition is more motivating than an explicit mission.
- *Dare to be different*: As Bennis says, their sense of operating on the fringes feeds their obsession to succeed.
- *Pain and suffering*: The nature of the work would be such that the team members generally go through intense pain and suffering, but an ideal team leader keeps up their morale by his sympathetic involvement.

4.8 THE SEVEN HABITS OF HIGHLY EFFECTIVE LEADERS

Stephen Covey in his book, *The Seven Habits of Highly Effective People*, highlighted the following habits that should be instilled by people who want to be highly successful leaders:

Habit 1	Be proactive	Principles of personal vision
Habit 2	Begin with the end in mind	Principles of personal leadership
Habit 3	Put first things first	Principles of personal management
Habit 4	Think win/win	Principle of interpersonal leadership
Habit 5	Seek first to understand, then to be understood	Principle of communication
Habit 6	Synergize	Principles of creative communication
Habit 7	Sharpen the saw	Principles of balanced self-renewal

Covey, in his later book, *The Leader in Me*, added the 8th habit that states "Find your voice and inspire others to find theirs."

4.9 THE TEN COMMANDMENTS OF cGMPs (CURRENT GOOD MANUFACTURING PRACTICES)

1. Have good SOP (standard operating practices)
2. Compliance to procedures
3. Document the work
4. Validate the manufacturing and testing procedures
5. Design and build proper facilities and equipment
6. Upkeep facilities and maintain equipment
7. Improve competency of the people
8. Cleanliness as a daily habit
9. Process control
10. Continuous monitoring by audits

4.10 FIFTY INSIGHTS FOR CEOs

Gita Piramal and Jennifer Netarwala, in their book, *Smart Leadership Insights for CEO's*, indicate the following fifty insights for a successful leader.

1. Begin with vision
   A vision is a moving target and a target you set yourself.
2. Get buy-in
   A vision without alignment is just a theory. Set in the organizational structure so that there is a personal goal for each employee aligned to the company's vision.
3. Find leaders at all levels
   A leader creates leaders, not followers.
4. Plan future leaders
   The challenge is to identify leaders from the new crops coming in.
5. Spread leadership culture
   It should be a part of the leader's agenda.
6. Set an example
   It is the CEO's example that sets the tone.
7. Be in the know
   People who are successful and become leaders are those who have their ears glued to what's happening.
8. Create entrepreneurs
   There is very little difference between an entrepreneur and being an entrepreneur in an organization.
9. Encourage innovation
   Create as many connections inside the organization as possible to encourage innovation.

10. Reward innovation
    Innovation is not about doing different things, but about doing things differently and better. When people do differently, they shall have to be recognized, rewarded, and the entire process supported.
11. Think regional but act local
    It is always carried on with the local talent.
12. Get people to think strategy
    It is imperative that people across the organization be encouraged to think out of their own mind that they can develop broader appreciation of the complexity of the problem and trade-offs involved.
13. Hire extraordinary people
    Talented people are the key determinant of business success.
14. Discuss and debate
    Every employee should be encouraged to speak out.
15. Encourage nonconformists
    Organizations need their share of mavericks because they invariably account for a disproportionately large number of breakthroughs and creative ideas.
16. Park people
    Learn how to deal with stagnating people.
17. Adhere to standards
    Meritocracy is important.
18. Be fair consistently
    Once you introduce meritocracy, people will start leaving gracefully.
19. Discuss changes
    As a CEO, you should sit with subordinates and discuss what you are doing and why you are doing it.
20. Understand motivation
    In any takeover of an organization, price is not the biggest concern.
21. Let go of ego
    A lot of acquisitions go wrong because of the ego of the person at the top.
22. Let people adopt to their roles
    While integrating people after acquisitions, it does not make sense to the management to ask people to leave immediately.
23. Address emotions
    After an acquisition, first address emotions.
24. Be upfront
    When a company is sold, normally there is a good reason for that, which should be well known to all concerned in the company.
25. Take time
    While making decisions under uncertainty, do not hurry up. Take your time to consider all the facts before you. Adapt the Indian rule of "*Nidanam Pradhanam.*"
26. Don't penny pinch
    Training people and sharing expertise requires a lot of traveling and this can hurt the bottom line.

27. Share more
    Free flow of information allows you to exchange technology, access strengths, and transfer expertise to the subsidiaries.
28. Have a good mix
    You can't sell a low-cost product with bad quality.
29. Set time limits
    Set a time limit for every decision by having reverse sign-offs.
30. Don't procrastinate
    If you have an opportunity, don't hesitate.
31. Get quicker
    To be world-class there has to be continuous and quantifiable improvement in performance.
32. Anticipate
    Anticipate everything. Success depends on how quickly you move into those places and that you can reach the target in anticipation of the needs.
33. Don't wait for perfection
    If you wait to get everything absolutely right, you ate not going to take off.
34. Give direction
    When a company is in a turnaround phase, everyone from the lowest level to the highest level employees must work in the same direction.
35. Create ambition
    For any organization, what is very important is that you first create an ambience, and then you create an ambition among people.
36. Initiate change
    For change to happen there is a need to change the mindsets.
37. Impose mild penalties
    Funding and budget is one way to reward and punish.
38. Manage conflicts
    Maybe you have a point, maybe we can sleep on it, maybe we can talk about it tomorrow, and let us do some homework on it.
39. Handle controversy
    In a controversy, the key thing is to understand the essence of the controversy and where does the organization figure in it.
40. Encourage teams
    Remember "Together Everyone Achieves More."
41. Stretch continuously
    You have to see how you can stretch the boundaries. If you achieve the target, the target must be raised again.
42. Establish trust
    Trust and fairness are the two most important values in an organization.
43. Take risks
    Most of the time, people are not going to applaud when you take risks, but still you have to do it with gut instinct.

44. Incentives work
    How do you help people be the best that can be?
45. Experiment
    Create an atmosphere where one can have tough discussions and walk out of the meetings and come back the next day without sulking.
46. Manage performance
    Have clear individual appraisals, where employees set out what they want to do in the next one or two years, and there shall be a continuous assessment of how they are performing vis-à-vis the goals.
47. Manage disappointments
    When the benevolent and ever truthful king Harishchandra was forced to lose his kingdom, due to his verbal commitment given long ago to the cunning sage Viswamitra, his minister advised him.
    Don't worry about the failures, but know the causes.
48. Listen closely
    If a person comes to you with a problem, don't say you do not have the time. The problem does not disappear but the person disappears.
49. Establish a wavelength
    Getting everyone on board is a democratic process.
50. Benchmark with the best
    World-class benchmarks are impotent. Creating benchmarks also helps people to know what they will be judged against.

4.11   FIFTEEN THOUGHTS OF CHANAKYA

Our own Chanukya in his Chanuka Sastra, had expressed these thoughts that should provide an insight into our mind to be adapted for becoming a successful leader.

1. Learn from the mistakes of others, you can't live long enough to make them all yourself.
2. A person should not be too honest. Straight trees are cut first and likely so, honest people are screwed first. A classic example for this thought is illustrated below from Hindu mythology:

   Prana vitta mana bhangambu landu bonka vachunu aghamu pondadu Adhipa (in Telugu), meaning, When it comes to the loss of life or money or honor, you can lie and you will not go to hell, O King.

3. Even if a snake is not poisonous, it should pretend to be venomous.
4. There is some self-interest behind every friendship. There is no friendship without self-interests. This is a bitter truth.
5. Before you start some work, always ask yourself three questions—"Why am I doing it?," "What the results might be?" and "Will I be successful?" Only when you think deeply and find satisfactory answers to these questions, go ahead.
6. As soon as the fear approaches near, attack and destroy it.

7. The world's biggest power is the youth and beauty of a woman.
8. Once you start working on something, don't be afraid of failure and don't abandon it. People who work sincerely are the happiest.
9. The fragrance of flowers spreads only in the direction of the wind. But the goodness of a person spreads in all direction.
10. God is not present in idols. Your feelings are your god. The soul is your temple.
11. A man is great by deeds, not by birth.
12. Never make friends with people who are above or below you in status. Such friendships will never give you any happiness.
13. Treat your kid like a darling for the first five years. For the next five years, scold them. By the time they turn sixteen, treat them like a friend. Your grownup children are your best friends.
14. Books are as useful to a stupid person as a mirror is useful to a blind person.
15. Education is the best friend. An educated person is respected everywhere. Education beats beauty and the youth.

4.12   WILKIE'S LEADERSHIP QUALITIES

David J. Wilkie, in his contribution to management, *Handbook for Plant Engineers*, lists the general qualities required for a leader as:

1. Drive
2. High intelligence
3. Technical knowledge given by education and practical experience
4. Acceptability
5. Maturity
6. Management skills
7. Loyalty
8. Human relations
9. Leadership and responsibility

Of these, the most important is the leadership and responsibility. The following points further identify their characteristics.

- People take on the leadership role by giving satisfaction to others.
- The leader is obligated to achieve positive results.

4.13   LEADERSHIP RESPONSIBILITIES

The website http://whj.hubpages.comsums up the responsibilities of a leader as:

1. Disseminating the idea of total quality.
2. Setting standards like zero failure.
3. Monitoring quality performance (quality costs).
4. Introducing a quality system based on "prevention" rather than detection.
5. Introducing process control methods like Statistical Process Control.

4.14 MORAL LEADERSHIP

Today, professional ethics and human values are given high importance in leadership and the awareness of moral leadership is hence, quite significant to any leader. This leadership may take several forms and in general, it indicates the success story in moving a group towards the goal. This success is based on character ethics and it is the leadership attitude that distinguishes a good leader from a bad leader, as indicated by the following illustration.

Hitler was a leader. He could successfully control a large German army and move them towards his goal of conquering Europe. But his goal was not just and fair. Similar was with the case of Joseph Stalin of Russia and Idi Amin of Uganda. Because of their evil goals and dictatorial attitude, people obeyed them out of fear, but not out of regard. After their fall, all, including their own followers were happy to be relieved of their tyranny.

On the other, hand Mahatma Gandhi, too, could lead a large humanity in achieving the goal of independent India. He never used power and his charismatic personality, his conviction, honesty, integrity, and the ability to guide people towards the goal made him respected all over the world even today. This is moral leadership.

4.15 CONTRIBUTORS FOR MORAL LEADERSHIP

The main contributors to moral leadership are:

(a) *Respect*: Respecting other people irrespective of their position is the very basis of moral leadership.

(b) *Behavioral pattern*: As indicated in Chapter 7, as the leader ascends up the ladder of the hierarchy of needs, he is driven by this unconscious desire to win, to be loved, to be appreciated, to be perfect, and to be successful. If this desire in the leader to win takes precedence over the need to be fair and reasonable, which would adversely affect the moral leadership?

(c) *Style*: Every leader has his own style of functioning and dealing with his subordinates or peers, which will have direct bearing on the responsiveness of others, and consequently, his success as a moral leader.

(d) *Habit*: Habits are behaviors a man picks up over the years. He may not be conscious of these habits, but they certainly affect the way he interacts with others. They may be positive or negative habits. Examples are the use of certain expressions, jokes, or gestures.

(e) *Intention*: Here the leader consciously and intentionally acts in ways that harm the others, either physically or mentally. This can take two forms:
- Behavior in which harming others is accepted by the leader as necessary to achieve something valued by him.
- Behavior in which the leader has no bad intentions, but does not see or realize the harm his actions do to others.

4.16 ROLE OF TOP MANAGEMENT IN QUALITY MANAGEMENT

Having discussed the qualities of leaders in general, let us now see what characters and attitudes a leader should possess to sustain the quality culture of an organization.

1. The CEO should have the commitment and take the lead in the process of change to infuse a quality culture in the organization.
2. The CEO has to establish continuous improvement in quality as the organization's primary objective at all levels, starting with himself. Personal change is a prerequisite for organizational quality improvement.
3. Top management should learn and teach others on principles and managing for continuous improvement.
4. Provide guidance and steadfast determination to continuous improvement through teamwork.
5. Provide long-term stability to the leadership of the organization through building up the cadres down the line.
6. Inspire strong team spirit and orientation among all the employees, ie, from Top Management to grass-roots level employees.
7. Ensure change from the old system of "product and function driven" to "customer and process driven." Production and Quality targets are to be oriented to Quality and Customer satisfaction.
8. Ensure free flow of communication and open management by sharing information on targets/goals so that there is one common language in the whole organization and everyone is motivated to work in teams/groups for common objectives.
9. Build positive working relationships with the workforce and trade union. Remove barriers.
10. Provide sufficient budget support for ongoing training activities and for implementation of new methods.
11. Redesign the organization's motivation and reward systems to generate employees' continued commitment to long-term goals of continuous improvement.

4.17 LEADERSHIP AND KNOWLEDGE OF PSYCHOLOGY

In July 2014, a blogger in ASQ's quality management discussion in LinkedIn groups, the W. Edwards Deming Institute had blogged the following concepts on why the knowledge of human psychology is essential for quality leaders.

One of the four cornerstones of Dr. Deming's management system is an understanding of psychology. Managers should be learning psychology to improve the management of the human systems in their organization and in working with customers, suppliers, and other stakeholders. We frequently have issues in our organizations that grow out of faulty theories about how people think (often based on beliefs of much more coldly rational thought than the research shows is really found in people).

4.18   CASE STUDIES ON LEADERSHIP QUALITIES

Two case studies on great leaders are cited below:

1. Go out of the way to help people, meaning do your duty plus something more.

   A youngster was once pushing the wheelchair of an old lady around 1870. Though rubber was known at that time, the rubber tires were unknown. Since the wheelchair had metal wheels, the boy realized how uncomfortable the lady felt due to the jolts. He took a rubber strip and stuck it around the rim. This increased the comfort level considerably and this youngster, John Boyd Dunlop, went on with his invention to become the world's first and topmost rubber tire manufacturer.

2. Continuous learning and knowledge gathering is a never-ending process for the true leader. It is a system of profound knowledge as professed by Deming (see paragraph 1.14 of Chapter 1)

   During the early forties, a young boy who used to go round door to door on a bicycle selling cloths, once sat with a cloth merchant making comparative notes about the varieties and values of cloths he was dealing with, in a new note pad. After a fortnight when the boy visited him again, the merchant noticed the note pad to be full with lots of scribbled notes of information not only about the cloth, but also about the processes, the raw-materials and the byproducts, including the related petroleum byproducts and their manufacture in a very logical order. The merchant was very impressed with this young boy and blessed him to be a great leader in textile manufacture. This boy not only became India's leading textile manufacturer, but also used the knowledge gained by his quest on other related materials like petroleum, only to widen his activities to the petroleum refining. Yes, this boy was Dhirubhai Ambani, who built India's largest industrial empire next only to the Tatas.

4.19   SOME QUOTATIONS ON LEADERSHIP

1. He who learns how to obey, will know how to command.
   –Solon
2. Leadership is the ability to turn vision into reality.
   –Warren Bennis
3. Leaders don't create followers. They create more leaders.
   –Tom Peters
4. One of the true tests of a leader is that he recognizes the problem before it becomes an emergency.
   –Arnold H Glasgow
5. That some can achieve great success, is proof to all that others can achieve as well.
   –Abraham Lincoln

6. The greatest danger for most of us lies not in setting a too high standard and falling short, but in setting our aim too low and achieving our mark.
   –Michael Angelo

7. Action without study is fatal, and study without action is futile.
   –Mary Beard

8. Attitude is a little thing that makes a big difference.
   –Winston Churchill

9. If ethics are poor at the top, that behavior is copied down through the organization.
   –Robert Noyce

10. The time is always right to do what is right.
    –Martin Luther King Jr.

11. The difference between impossible and possible lies in a person's determination.
    –Tommy Lasorda

12. A failure establishes only this—that our determination to succeed is not strong enough.
    –Christian Nestell Bovee

4.20   CONCLUSION

Leadership is a function which is important at all levels of management. At the top level, it is important for getting cooperation in the formulation of plans and policies. In the middle and lower levels, it is required for interpretation and execution of plans and programs framed by the top management. It is not an overemphasis to say that the role played by the leader of a group is the forerunner for the success of the group, as illustrated in this chapter.

On the Lighter Side

1. If you think your boss is stupid, remember that you wouldn't have got the job if he was any smarter.

2. **Put your foot down not to put your foot down**
   While riding a two wheeler in water logged roads in inching traffic, you have to stop frequently but that would spoil your shoes. So you would prefer to do a lot of balancing act to avoid stepping on the road. This attitude of putting your foot down (determination) not to put your foot down is the very quality of a great leader.

FURTHER READING

[1] Besterfield DH, et al. Total quality management. Upper Saddle River, NJ: Prentice Hall; 2003.

[2] Feigenbaum AV. Total quality control. 3rd ed. New York: McGraw-Hill; 1983.

[3] Deming WE. Out of the crisis. Cambridge, MA: Massachusetts Institute of Technology; 1982.

[4] Evans JR, Lindsay WM. The management and control of quality. Mason, OH: Thomson South Western Publ; 2005.

[5] Juran JM. Leadership for quality, an executive handbook. New York: Free Press; 1989.

Chapter 5

Scientific Management

Chapter Outline

5.1 TQM AND SCIENTIFIC MANAGEMENT

The urge to do things in a better way to please a companion or companions has always been present in man ever since he started living as a social animal. This urge to do things better made the man learn how to manage his jobs and other things better. However, these attempts were very much unsystematic and chance-oriented, and hence, could not be attributed as steps in the development of quality management.

Nevertheless, the history of quality management is closely related to the history and development of scientific management, whose origin can be traced back to the industrial revolution that took place in the 18th and 19th centuries.

5.2 THE INDUSTRIAL REVOLUTION

The start of industrial revolution is generally attributed to the period around 1750, corresponding to the reign of King George II in England, when the development of domestic woolen units had accelerated the economic changes. In fact, the more marked period of industrial revolution can really be attributed to the period between 1815 and 1875, when the mechanical steam powered devices began to replace the spinning wheel and handlooms, and this had a remarkable impact on the British cotton and textile industry when the rate of industrial growth was higher than ever before.

In the early 19th century, a remarkable impetus for industrial growth was provided by the railways, which began operating in England by about 1830. While until around 1830, the industrial revolution centered around England, the rest of Europe and America soon caught up with this momentum and in some spheres, even overtook England.

5.3   EVOLUTION OF MANAGEMENT THINKING

A significant contribution of the industrial revolution was the large-scale industrialization and the outcome of factories. While this undoubtedly paved the way for overall development and better standards of living, it created new management problems, because the number of persons supervised by a single person and the levels of supervision had considerably increased. This initiated management thinking.

5.4   PHASES OF GROWTH OF MANAGEMENT THINKING

The growth of management thinking can be split into three phases.

(a) The first phase covering late 17th to mid-19th century mostly centered in Europe.
(b) The second phase from mid-19th century to mid-20th century. The United States contributed considerably to this thought, Frederick Winslow Taylor's Scientific Management thought being the significant outcome of this phase. In fact, it was during this period that work study really developed.
(c) During the third phase after the mid-20th century, specifically after 1960, considerable changes in management thinking have taken place. Taylor's scientific management thinking of job specialization has given way to modern thinking of total involvement of all employees in all the related function. Thus the concepts of total quality management, total productive maintenance, decentralization of routine functions, etc., have emerged. The customer and his views are today given the maximum weight and quality is given the topmost priority.

5.5   EARLY PIONEERS IN MANAGEMENT THINKING– PRE-19TH CENTURY

In chronological order, the following can be acknowledged as some of the early pioneers in management thinking, though there might be a host of others not cited here.

- *Richard Arkwright* (1732–1792) realized the value of job training to the operators in operating a new spinning jenny. He also developed a code for factory discipline.

- *J.R. Perronet* of France made some systematic overall timing of the manufacture of pins and arrived at standard production rate as early as 1760.
- *Mathew Boulton* started a factory in 1762 that was full of mechanical inventions and had a body of highly skilled craftsmen. His son Mathew Robertson Boulton, in conjunction with James Watt (Jr.), preplanned and expanded this factory into a closely integrated modern engineering plant over twocenturies ago. As early as 1770, he arranged weekly meetings of his partners and managers to examine business orders, pricing, or any other issues related to production.
- *Robert Owen* (1771–1858) worked for improved labor relations and employment conditions and was largely responsible for the introduction of the factory act of 1818.
- *Charles Babbage* (1792–1871) recognized the necessity for detailed costs and invented the early calculating machine. In his book *The Economy and Machinery and Manufacture* in 1832, he reflected the need for re-establishing general principles on management business undertakings. He was aware of some of the dangers of time study and printed his own blank formats for collection of data in main investigations.
- *Henri Fayol* (1841–1925) was responsible for the concept of the principles of organization, as well as the principles and functions of management.
- *Frederick Winslow Taylor* (1856–1915) is undoubtedly a significant figure in the history and development of management thought and is generally acknowledged as the founder of the scientific management, a term coined by Taylor himself. He opposed the traditional line type of organization and stressed that every supervisor must specialize in a specific function rather than being "a jack of all trades, but master of none." He stressed the maximizing of production by following the three basic principles, reflecting them in his book, *Shop Management* (1903).
- A definite task, determined by the definition of the job, leading to the best operation sequence.
- A definite time, established by stopwatch time study or estimated from standard data, and
- A definite method developed by detailed analysis and recorded in the instruction charts.

5.6   CONCEPTS OF SCIENTIFIC MANAGEMENT

Management is defined as the art and science of directing, coordinating, and controlling human effort, so that the established objectives of an enterprise can be achieved in accordance with the established policies and procedures.

Scientific methods means the systematic and logical procedure of observing and ascertaining the facts, followed by reasoning and critical analysis conducted in such a way that the same can be fully utilized in building up a system of co-operation founded on agreement, justice, and integration of ideas.

We can, hence, define scientific management as the application of systematic and logical procedures in the ascertainment of the facts, reasoning and critical analysis in finding rational solutions, in coordinating and controlling human efforts and resources so that the established objectives of an enterprise can be achieved in accordance with the established policies and procedures.

5.7 SPECIFIC AIMS OF SCIENTIFIC MANAGEMENT

1. To identify waste and its causes which may be in the form of inappropriate materials, manpower, machinery, manufacturing processes, or of capital. These may either be due to lack of coordinating efforts or unhealthy relations between employees and employers.
2. To eliminate waste after ascertaining the reason for it.
3. To unify the larger interests of labor, management, and ownership and also to lower the costs by systematizing each and every process of the production, distribution, and administration.
4. To unite higher wages with reduced labor cost and thus, to ensure benefit both to the employee and employer
5. To ensure a higher standard of living for the worker in proportion to his efficiency.
6. To increase the purchasing power of the consumer by lowering the selling price, thereby benefiting the community at large. It is thus an important social and economic movement.
7. To steady the level of employment of labor and capital by gauging the industrial and market fluctuations.

5.8 ADVANTAGES OF SCIENTIFIC MANAGEMENT

A. Advantages to the employer:
   (i) Scientific management lowers the cost of production.
   (ii) It minimizes labor troubles, strikes, and lockouts.
   (iii) It guarantees prompt deliveries of goods.
   (iv) It increases production with marginal increase in capital investment.
   (v) It improves the quality of products through better and more efficient supervision.
   (vi) It increases the demand for the product.
B. Advantages to the workforce:
   (i) The workforce gets higher reward for their higher performance.
   (ii) They work shorter times, but still get higher output.
   (iii) Working conditions are greatly improved and the workforce feels they are well looked after.
   (iv) The system ensures steady employment, because the business is steady.
   (v) The work fatigue is reduced.
   (vi) Due to increased wages, the standard of living increases.

C. Advantages to the community:
   (i) The employees, who constitute a greater part of the community, get direct benefits.
   (ii) The system makes the community wealthier, healthier, and more content.
   (iii) Absence of strikes and lockouts result in greater industrial stability.
   (iv) Export trade increases and the inflow of money from abroad enhances the general prosperity.
   (v) Periodical trade depressions are reduced to a minimum.

5.9   MISCONCEPTIONS OF SCIENTIFIC MANAGEMENT

Unfortunately the term "Scientific Management" soon developed into a matter of another dispute. This term, coined by Taylor, implies a precision which cannot always be present when dealing with humans. This allowed a variety of interpretations wherein the chief principle of Taylor was distorted by unscrupulous managers wanting to ride over the workers in the name of scientific management. All essential requirements, such as work load fixation, production targets, and pay earned were determined scientifically by time study and arithmetic calculations. They felt the workers had no cause for complaint under this system of control, and the latter were ultimately treated just as one among other factors of production. Such a situation gave rise to the following misconceptions of scientific management.

1. It was felt that scientific methods had been adopted to promote exploitation of workers as explained above. That is, they are applied only to techniques like work study and the problems of industrial relations were overlooked.
2. Scientific techniques were presumed to be consisting only of long mathematical formulae which can be applied to all situations. Nevertheless, the first step, that is, the scientific method of observation of factors and their reasoning, is one which greatly influences the logical solution that can be given for a given situation.
3. The initial concept of scientific management presumes that the detail with which the problem is analyzed is identical to the whole. In fact, the whole is not only the sum of the parts, but also interrelationships, plus the individuality of the whole as well as that of the part. For example, in time study, the worker's output does not only depend on the operator's physical movements, but also on his aptitude, his relations with the fellow employees, etc.
4. The misgiving that the scientific methods can be applied to only measurable characters resulted in the exaggerated and overstretched development of micro motion principles, ignoring altogether the psychological factor.
5. One immediate evil result of this concept was to chase the worker for higher productivity, with no or little improvement in factory layouts, working conditions, fatigue, etc.
6. Taylor's emphasis on extreme job specialization was subsequently felt unrealistic and created new psychological problems.

5.10   RESISTANCE TO SCIENTIFIC MANAGEMENT

Despite the above mentioned advantages, as explained in Section 5.8, scientific management soon developed into a matter of disrepute. Taylor's emphasis on precision in measuring work (time study) cannot always be present when dealing with humans. This allowed a variety of interpretations as distorted by unscrupulous managers wishing to ride over the workers in the name of scientific management. Naturally these efficiency experts, who had little knowledge of human relations met with strong opposition from labor. The disturbances spread to such an extent that the interstate Commerce Commission of the United States investigated and reported against time study, resulting in government banning time study in its undertakings for some period. This opposition to scientific management is similar to that against changes in operational methods. The same thing happened to other practices propounded by Taylor's principle of job specialization, such as patrol inspection, patrol maintenance, lubricating/machine cleaning gangs, etc. How this changed over to the operator himself taking care of basic inspection, basic machine maintenance, etc., are discussed more in detail elsewhere in this book.

5.11   CONCLUSION

Although scientific management as a distinct theory or school of thought was obsolete by the 1960s, the resistance to it illustrated in the previous paragraph, most of its themes including analysis, synthesis, logic, rationality, work efficiency, and elimination of waste, standardization of best practices, etc., are still important elements of industrial engineering and management today.

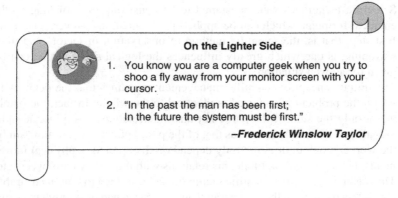

On the Lighter Side

1. You know you are a computer geek when you try to shoo a fly away from your monitor screen with your cursor.

2. "In the past the man has been first; In the future the system must be first."

–Frederick Winslow Taylor

FURTHER READING

[1] Drucker P. Management challenges for the 21st Century. New York: Harper Collins; 1999.

[2] Ho SK. TQM an integrated approach. London: Kogan Page Publ; 1995.

[3] en.wikipedia.org/wiki/Scientific_management.

[4] scientific-management-theories-principles-definition.html.

[5] www.businessdictionary.com/definition/scientific-management.html.

[6] Proceedings of the general management seminar for the transport managers, Dar Es salaam, Tanzania, 1982.

Chapter 6

System Approach to Management Theory

Chapter Outline

Total Quality Management: Key Concepts and Case Studies. http://dx.doi.org/10.1016/B978-0-12-811035-5.00006-4

6.1   DEVELOPMENT OF SYSTEM APPROACH

In the past, the departmental mangers tried to solve their problems by considering them as isolated situations, independent of other activities of the organization. For example, if a certain product manufacturer noticed a sales decline and traced it to the lack of aggressive sales effort, it was treated purely as a sales management problem and action was taken against that particular salesman or the department. The contribution of other departments to this effect, such as the quality control, design, credit policy, or the advertising, were never given much thought. Life study of any object must rely on the method, of analysis involving the simultaneous variations of mutually dependent variables. As early as the 1950s, a new integrated approach which considers the management in its totality has been developed and this is called the system approach.

6.2   WHAT IS A SYSTEM?

We are surrounded by and live in several systems. In fact, man is a system by himself. Some systems are natural like the planetary systems, animal systems, and environmental systems. Others are man-made, like the business systems, production systems, material handling systems, social systems, and total quality management systems. Whichever is the system, it is basically characterized by three components which form the input-process-output system (Fig. 6.1), with,

- a set of two more elements forming the *input*,
- seeking a common goal operating on certain data or matter or energy, forming the *process*, and
- yield matter or data in a time frame, forming the *output*.

6.3   DEFINITION OF A SYSTEM

Originating from the Greek word *sustema*, the word system has three definitions in the Oxford dictionary as below,

1. Group of things or parts working together, like the digestive system, railway system, or the control system or in computer field, a group of related hardware units, or programs, or both. Its adjective is *systemic*, meaning related to a system as a whole.
2. A set of ideas, theories or principles, or the social order like a political system or a government system.
3. Orderliness or methodical or an organized manner of working. Its adjective is *systematic* and the word *systematize* means arranging things according to a system.

The American Management Association in its AMA Management Handbook explains these three groups of definitions as follows:

1. The first group related systems to living creatures, including human beings, emphasizing the impact of personality and behavioral sciences.

2. The second group emphasized the general meaning of the system as a set of two or more elements like people, things or concepts, which are related or joined together to achieve a common goal, operating on data, matter and energy, to yield a result with a time reference. For example, a system of government, a system of measurement, or a system of classification.

3. The third group of definitions connects systems with specific science techniques like philosophy, mathematics, or biography.

Whatever may be the definition, a system approach is a purposeful and powerful means for accomplishing an objective. It provides the management with analytical framework with which it can identify, describe, and interrelate the process and the components that make up a particular system, as explained and illustrated in the following paragraphs. The study of any object or process must rely on the methods of analysis involving the simultaneous variations of mutually dependent variables.

6.4   TYPES OF SYSTEMS

This concept of system approach is universal and can be applied to any situation, whether manufacturing or service provision, or agriculture, or even education. As illustrated below by the following figures:

1. *Manufacturing system*: A group of *men* working on *machinery* and *materials* with the aid of *money* and a good *management*, (*5 Ms of input*), working per work-methods, schedules, and specifications (*process*) to yield the specified products and information by the date the customer wants them (*output*) (Fig. 6.1).

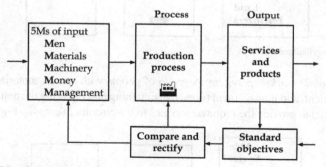

FIG. 6.1   Manufacturing system.

2. *Management information system*: A group of people with a set of manuals, data processing, and recording equipment (*input*), working on selecting, storing, processing, and retrieving data (*process*) to yield the information needed by the managers (*output*) (Fig. 6.2).

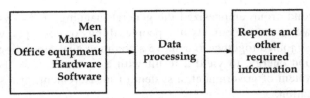

FIG. 6.2   Management information system.

3. *Business organization system*: A group of people with a vision and mission (*input*), performing production, design, finance, marketing, etc. functions (*process*) towards achieving the goal of optimal profit for the business (*output*) (Fig. 6.3).

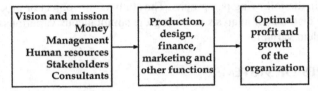

FIG. 6.3   Business organization system.

4. *Agricultural system*: A group of people with a set implements, fertilizers, and land (*input*), performing the farming activity (*process*) to yield the agricultural produce (*output*) (Fig. 6.4).

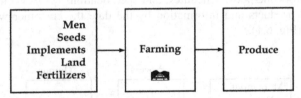

FIG. 6.4   Agriculture system.

5. *Automobile workshop system*: A group of people with spares, materials, tools, equipment, and management (*input*), maintaining and repairing the automobiles (*process*) to provide the required services to the customer (*output*) (Fig. 6.5).

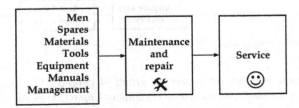

FIG. 6.5   Automobile garage system.

6. *Educational system*: Faculty aided by college infrastructure, textbooks, visual aids, etc. (*input*) teaching and training the students (*process*) to provide enlightenment (*output*) (Fig. 6.6).

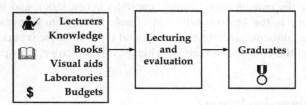

FIG.6.6 An educational system.

6.5 COMPONENTS OF A SYSTEM

Systems are most often represented in flow charts and block diagrams. The major components of a system are input, process, and output. While Fig. 6.1 indicates the process in its simple form, Fig. 6.7 below illustrates all the other controlling factors.

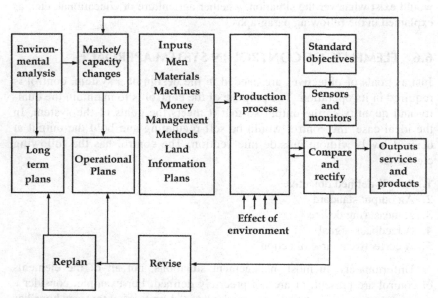

FIG. 6.7 The complete production system.

6.5.1   Input

Input to a system can be classified basically in three categories. The first is comprised of the physical inputs, including the materials that flow through the conversion process where work is performed on them. The second is the nonphysical materials, but that are required for the performance of the process, like the management, money, capital, energy, labor, and land. They third category is the environment that effects the system's operation. The design of the amount, placement, timing, and types of these inputs will have an impact on the conversion facility, whether it is a factory, a hospital, or an office.

6.5.2   Conversion Process

The second basic part if the system is the conversion process through which the inputs flow to produce the desired outputs. To be effective and efficient, systems must be designed so that the correct process acts on the inputs at the proper time.

6.5.3   Output

The third major component of the system after the conversion process is the output, which comprises the desired accomplishment of the system. In the automobile industry, the output is the number of completed cars of a desired quality produced within a specified time frame. This input—process—output elements would exist whatever the situation, whether agriculture or educational, etc., as explained in the following paragraphs.

6.6   ELEMENTS OF CONTROL IN SYSTEM APPROACH

Just as goals or objectives are needed in the design of a system, control is required in its operation. The purpose of the control is to maintain the quality and quantity of the output so that it meets the goals of the system. In the ideal case, the control would be self-regulating and hold the output at desired levels without outside intervention. The control has the following elements.

1. A well-defined objective
2. An output standard
3. A measuring device
4. A feedback signal
5. A corrective course of action

Unfortunately, in most management situations, not all of the elements of control are present, or are not precisely defined. For example, consider a production process in which the standard of the units of output per hour has been established for a product. However, when the output from the process is monitored, it is found that only 80% of the units are produced per hour. In this situation, there is an objective, an output standard, a measuring device, and a feedback signal, but no corrective action has been specified.

The control system can either be automatic or nonautomatic, as illustrated in the following figures.

6.7 EFFECT OF ENVIRONMENT ON THE SYSTEMS

A system's environment is composed of all activities outside the system which, if changed, will affect the system or which will be changed by any changes in the system itself. In this context, the general level of economic activity can be considered as environmental for most production activities, because any change in it will affect the production levels.

6.8 OPEN AND CLOSED SYSTEMS

Every system has a boundary which functions to maintain proper relationships between the system and the environment. A closed system has no environment around the system. No outside systems have any effect on this system and the inputs and outputs have definite mathematical relationships like a chemical process inside a hermitically sealed container (Fig. 6.8A).

The open system has an environment with which it relates exchanges and communicatoin, such as human systems or a production system (Fig. 6.8B).

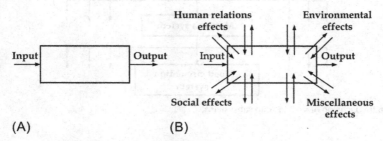

FIG. 6.8   (A) Closed system, (B) an open system.

6.9 SYSTEMS AND SUBSYSTEMS

Because of the complexity of most real systems, it is extremely difficult to work with or to understand the entire system. In order to overcome this problem, systems are broken into smaller subsystems and still smaller sub-subsystems to comprehend them better. For example, the whole universe is a system, while the solar system is its subsystem. We may still break it down to the earth system, world system, a group of nations, a country system, a social system, man as a system, blood circulation system, etc., as illustrated in Fig. 6.9 shown alongside.

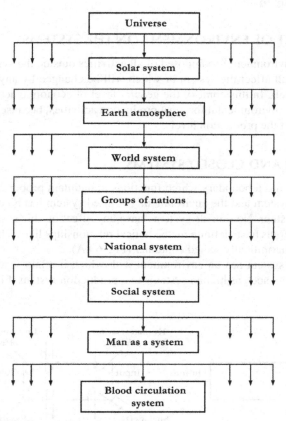

FIG. 6.9   Division of systems into subsystems.

6.10   RELATIONSHIP BETWEEN THE SYSTEMS AND SUBSYSTEMS

Each system and its subsystems are mutually related, some more and some less, some directly and some indirectly. This relationship is in the context of the whole and is complex. Any change in one part would affect the other to a varying, but a predictable degree. Fig. 6.10 illustrates the relationship between the systems and Fig. 6.11, the different generations of subsystems.

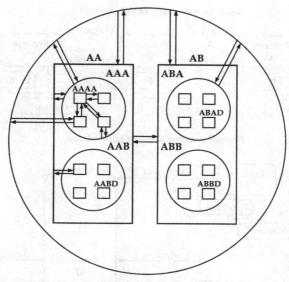

FIG. 6.10 Interrelationships between systems and subsystems.

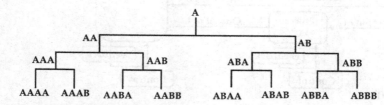

FIG. 6.11 Generations of subsystems.

6.11 COMBINATION OF SUBSYSTEMS

Two or more subsystems can be integrated in either series, or in parallel, or in combination, as illustrated in Fig. 6.12.

6.12 THE MANAGEMENT CUBE

The management activity is influenced by three aspects:

1. The management processes, including the planning, controlling, and execution.
2. The management functions such as design, planning, maintenance, marketing, purchasing, personnel, and finance.
3. The management level, such as operatives, supervisors, engineers, managers, directors, president, and Chairman.

Series combination of subsystems

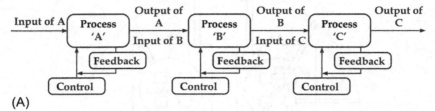

(A)

Parallel combinations of subsystems

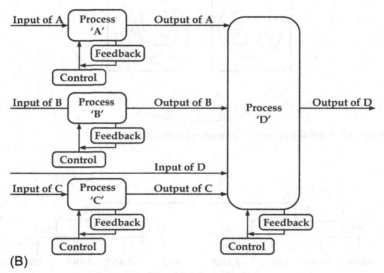

(B)

Intergrated combination of subsystems

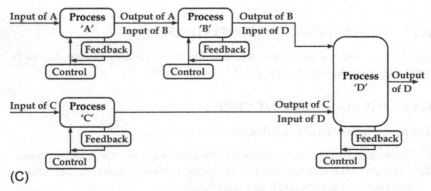

(C)

FIG. 6.12 (A) Series Combination of Subsystems, (B) Parallel Combination of Subsystems, (C) Integrated combination of Subsystems.

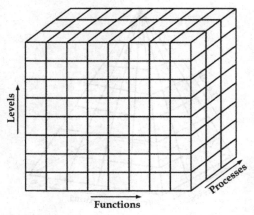

FIG. 6.13   Management Cube.

These three activities can be illustrated by Fig. 6.13, which is called the management cube. It shows how the managers of different functions perform the processes at different levels.

6.13 PLANNING PYRAMID

The management cube considers the same number of managers at each level. But in fact, the number of persons in each level becomes less at a higher level because of the span of control. For example, the planning of the vision and mission, or the policy-making is done at the Chairman level, while the day-to-day operational planning is done at the manager level. As one ascends higher in the hierarchy, the level of planning is escalated and simultaneously, the number of individuals responsible narrows down.

Hence, these activities are better represented by a pyramid signifying this aspect and showing what type of planning is done at each level in the organization, as illustrated in Fig. 6.14.

6.14 SUMMARY OF THE FEATURES OF MANAGEMENT AS A SYSTEM

All the points described in the previous paragraphs can be summarized below:

1. *Management is a system*: Management can be defined as a system, and unlike the biological and mechanical systems, it has the characteristics of a social system. It has subsystems which are integrated as a whole.

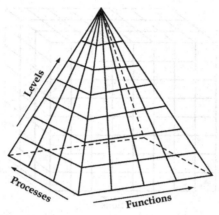

FIG. 6.14   Planning pyramid.

2. *Management is an open system*: It has an environment and with interaction, management takes its resources or inputs, allocates, and combines the resources to produce the desired outputs to the environment.

3. *Management is dynamic*: The equilibrium of the organization is ever-changing, moving toward the growth and expansion by preserving some of its energy. Thus, the organizational effectiveness is determined by the ratio of energy exchanges.

4. *Management is probabilistic*: A deterministic model always specifies the use of a situation in a condition with predetermined results. Management being probabilistic, points out only the probability and never the certainty of performance and the consequent results. Nevertheless, for aiding decision-making, the management assumes a certain amount of deterministic models to arrive at some conclusions and then do the analysis in combination with the probabilistic approach.

5. *Management is multidimensional*: System approach points out complex multilevel and multidirectional characters. At the macro level, management can be applied to the business systems as a whole. At the micro level, it can be applied to an organized unit or any if its elements.

6. *Management has multiple variables*: There is no simple cause-effect phenomenon like in physics. An event may be a result of so many variables which themselves are interrelated and interdependent.

7. *Management is adaptive*: The survival and growth of an organization in a dynamic environment demands an adaptive system that continuously adjusts to a changing environment through a feedback mechanism. The feedback mechanism provides information to take corrective action for achieving the desired results.

8. *Management is multidisciplinary*: Management being a system draws integrated knowledge from several disciplines and schools of thought. In fact, integration of the relevant aspects of various disciplines is the real contribution to the management as a system.

6.15 DECISION THEORY

While the previous paragraphs illustrate the systems approach to decision-making, the following paragraphs illustrate the aspect of decision-making as relevant to the manager.

The day-to-day administration of the affairs of any business enterprise requires an endless sequence of decisions. Decision-making is an integral part of a team, given the responsibility of meeting and discussing among themselves toward the achievement of a goal. They may either be routine or unusual, demanding the manager's closest attention and intelligence.

Actually the terms—to decide, to determine, to settle, and to conclude are generally used as synonyms, but we may distinguish between them as follows;

- *To decide* implies the bringing to an end any doubt, dispute, vacillation, or wavering between choices, by making up one's mind on what to do, the course of action, etc.
- *To determine*, in addition to the above, suggests or fixes precisely the form, character, functions, scope, etc.
- *To settle* stresses finality in decision, often the one arrived at by an arbitrational process and implies the termination of all doubts and controversy.
- *To conclude* means to decide after careful investigation reasoning.

In the past, when it was mostly a sellers' market, profit maximization was perhaps the sole motto of an enterprise, and accordingly, all decision-making was simple. But today, the manager is influenced by a hoard of factors.

1. Increased regulation of business activity by the local or national governments.
2. Growing separation of ownership and management in industry resulting in the management becoming more and more professional.
3. Steadily increasing importance given to high-level education for managers.
4. Development of new processes and equipment.
5. Expansion of market areas in view of improved communication and transport logistics.
6. Increased competition from other suppliers.
7. Growth of labor unions and more stringent labor laws.
8. Growing social pressure.

The following paragraphs detail the decision theory with regards to identifying its rationality, uncertainties, and other issues relevant in a given decision, as well as the resulting optimal decision.

6.16 PROBLEM ANALYSIS AND DECISION-MAKING

Traditionally, problem analysis must be done first, so that the information gathered in that process may be used towards decision-making. Wikipedia highlights the criteria for problem analysis and effective decision-making as follows.

6.16.1   Problem Analysis

- Analyze performance, what should the results be against what they actually are.
- Problems are merely deviations from performance standards.
- Problem must be precisely identified and described.
- Problems are caused by a change from a distinctive feature.
- Something can always be used to distinguish between what has and hasn't been affected by a cause.
- Causes of problems can be deducted from relevant changes found in analyzing the problem.
- Most likely cause of a problem is the one that exactly explains all the facts.

6.16.2   Decision-Making

- Objectives must first be established.
- Objectives must be classified and placed in order of importance.
- Alternative actions must be developed.
- The alternative must be evaluated against all the objectives.
- The alternative that is able to achieve all the objectives is the tentative decision.
- The tentative decision is evaluated for more possible consequences.
- The decisive actions are taken, and additional actions are taken to prevent any adverse consequences from becoming problems and starting both systems (problem analysis and decision-making) all over again.
- There are steps that are generally followed that result in a decision model that can be used to determine an optimal production plan.
- In a situation featuring conflict, role-playing is helpful for predicting decisions to be made by involved parties.

6.17   CHARACTERISTICS OF DECISION-MAKING

Decision-making involves the following characteristics

- Decision-making is a process of selection and the aim is to select the best alternative.
- A decision is aimed at achieving the objective of an organization if it is made in the organizational context.
- It involves evaluation of available alternatives, because only by this evaluating can one know the best alternative.
- Decision-making is a mental process and the final decision is made after thoughtful consideration.
- A decision involves rationality because by this, one can better one's happiness.
- Decision-making involves certain commitment. This commitment may be for a short run or long run, depending upon the type of decision.

6.18 SITUATIONS UNDER WHICH DECISIONS ARE TAKEN

6.18.1 Decision-Making Under Certainty

Decision-making under certainty, when we know with confidence what will occur. In this case, we have to consider only one possibility of occurrence for every alternative and decision-making is very simple.

6.18.2 Decision-Making Under Uncertainty

Decision-making under uncertainty, when there are a number of alternatives and no past data is available to compute or estimate the probability of occurrence. The estimation of market demand for new and untested products is an illustration.

6.18.3 Decision-Making Under Risk

Decision-making under risk, when there are a number of possible alternatives and the probability of occurrence of each alternative, based on past data and past experience, is known. This is controlled by the theory of probability and the result may either turn out to be correct or wrong, depending upon the accuracy of the data or the evaluation procedure.

6.18.4 Decision-Making Under Conflicts

Decision-making under conflicts, when the different individuals involved directly or indirectly in decision-making have contradictory opinions or data. The discussion among the team members for arriving at a decision is an illustration.

Illustration on the lighter side for the *Decision-making shown below*:

A sea captain on a rough sea may make a decision depending upon his attitude. An optimist expects the weather to change and takes no action. A pessimist complains about the wind, prays for the heavenly abode (moksham) for all, and keeps the crew on tenter hooks. A realist adjusts the sails, takes control of navigational aids, and takes emergency precautions.

6.19 CLASSIFICATIONS OF DECISIONS

There are several ways of classifying decisions in an organization.

6.19.1 Organizational and Personal Decisions

In an organization when an individual makes decisions as an executive for the organization they are called *organizational decisions*. The authority for making such decisions can be delegated from a superior to a subordinate. Such decisions directly affect the functioning of the organization. On the other hand, an executive can also make *personal* decisions that affect his and his family's personal life, though sometimes these decisions may affect the organization also. In this case, the decision-making power cannot be delegated.

6.19.2 Routine and Strategic Decisions

Routine decisions are made in the context of day-to-day operation of the organization. They are mostly repetitive in nature. They do not require much analysis and evaluation and can be made quickly. Authority of making these decisions is generally delegated to the lower level of workforce. *Strategic decisions* are those primarily for the future. They affect the organizational structure, objectives, facilities, finances, etc. These decisions are mostly nonrepetitive in nature and are made after careful analysis and evaluation of various alternatives and are generally made at higher levels of management.

6.19.3 Policy and Operative Decisions

This classification is generally similar to the above, except that they refer more to the factory-level production planning decisions. Decisions like plant location, plant layout, volume of production, machinery selection, sale and purchase decisions can be called *policy decisions*. These are sometimes published as a policy manual to become the basis for other *operative decisions*. Operative decisions are production-line decisions, such as the scheduling, routing, etc. and can be called operative decisions.

6.19.4 Programmed and Nonprogrammed Decisions

Programmed decisions are generally repetitive in nature and are made within the broad policy structure. They have short-run impact and are made by lower level managers. *Nonprogrammed decisions* are those made whenever specific problems arise unexpectedly due to certain circumstances. They are generally nonrepetitive in nature.

6.19.5 Individual and Group Decisions

This classification is based on the persons involved in the decision-making process. *Individual decisions* are made by a single person, generally the head of the institution, with or without consulting others affected. These are taken in the context of routine programmed decisions, where the analysis of various alternatives is simple and for which broad policy manuals are provided. Sometimes important nonprogrammed decisions also are made by individuals. *Group decisions* are made by groups or teams constituted for this purpose. These decisions are, in general, very important to an organization. They have certain positive values, such as greater participation of individuals and quality of the decision, but at the same time have certain negative values like delay in decision-making and difficulty in fixing responsibility for the decision-making process and the follow-up action.

6.20 DIFFERENT APPROACHES TO DECISION-MAKING

The different approaches adapted for managerial decision-making offer an insight into the theory of decision-making useful in teamwork.

6.20.1 Intuitive Decision-Making

In history there have been specialist decision-makers such as the tribal chiefs, medicine men, priests, and kings who depended mostly on their hunches or intuitions based on their training and experience. They actually involved un-structured gathering and classifying of historical data, followed by subjective evaluation.

6.20.2 Trial and Error Decision-Making

This is an adaptation of the intuitive decision-making, when the problem is sub-jectively isolated, defined, and a course of action is selected, and followed. If the results are favorable, the decision is allowed to stand. However, if the results tend to become failures, a second course of action is followed.

6.20.3 Follow-the-Leader Decision-Making

Here the decision is based on similar issues by other firms or an accepted trade practice, emphasizing that, "if it works fine with them, why not for us?" While this approach minimizes the risk, its main disadvantage lies in the fact that solu-tions depend very much upon the specific situations, which may vary from case to case and individual to individual. And besides, the variables in the same situ-ation may change from time to time.

6.20.4 Scientific Decision-Making

Here each problem is analyzed considering each of the alternative solutions. This process of analysis has been adapted in the physical services for centuries. The main four steps are:

- Define the problem,
- develop a hypothesis,
- test the two hypotheses, and
- prove or disprove the hypotheses.

This is better understood by Fig. 6.15.

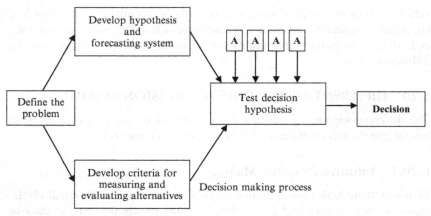

FIG. 6.15   Decision-making process.

6.20.5   Systematic Decision-Making

This is similar to the above, except that a holistic and systematic approach is given considering all the related factors as part of the whole system. The steps to be undertaken for this approach would be:

1. State the problem
2. List the options
3. Think about the possible benefits and consequences of each option
4. Consider your own value and beliefs
5. Weigh the option and then decide which one to take. If possible, share your list with a friend or adult
6. Act
7. Evaluate the results

6.21   BIAS IN DECISION-MAKING

Wikipedia lists the following instances of bias that creep into to the decision-making process, which should be kept in mind by the decision-makers.

1. *Selective search for evidence*—Our mind is set for certain conclusions. Subconsciously we tend to gather facts that support these conclusions, but disregard other facts that support different conclusions. This can also be called as Choice-supportive bias.
2. *Incomplete search for evidence*—We tend to accept the first alternative that looks like it might work.
3. *Inertia*—We have a built-in resistance to change from our thought patterns that we have used in the past.
4. *Selective perception*—Referring to para, our brain has two halves, the right brain which is creative and thinks of all alternatives, while the left brain

thinks logically and screens out information that appears illogical, even though the same apparent idea could later be developed into a relevant alternative for effective decision-making.

5. Over-optimism—many a time we tend to want to see things in a positive light and ignore the negative impacts.

6. *Recency*—We tend to place more attention on more recent information, and either ignore or forget more distant information. This is why exponential smoothing is done for forecasting as explained in Chapter 17.

7. *Repetition bias*—We tend to believe that what we are told more times or by more people is truer than what we are told occasionally, or by fewer people.

8. *Anchoring and adjustment*—Decisions are unduly influenced by initial information that shapes our view of subsequent information.

9. *Groupthink*—Even though this is opposite of what is stated in point 4 above and sometimes provides the needed moderation (two heads are better than one), this groupthink sometimes makes groups come into agreement at the cost of critical thinking, that is when most of the group does not want to spend time and energy in analyzing and thinking.

10. *Source credibility bias*—We reject something if we have a bias against the person and accept a statement by someone we like.

11. *Incremental decision-making* and escalating commitment—We look at a decision as a small step in a process and this tends to perpetuate a series of similar decisions. This is also called a *slippery slope*.

12. *Attribution asymmetry*—We tend to attribute our success to our abilities and talents. But we attribute our failures to bad luck. It is like saying "We won because of me. We lost because of him."

13. *Self-fulfilling decisions*—We conform to the decision-making expectations that others have of someone in our position.

14. *Underestimating uncertainty*—We wrongly presume to have control to minimize potential problems in our decisions.

6.22   DECISION TREE

Decision Tree is a pictorial representation of a decision situation, normally found in discussions of decision-making under uncertainty or risk. It shows decision alternatives, states of nature, probabilities attached to the state of nature, and conditional benefits and losses.

Investopedia Financial Dictionary defines decision tree as a management tool to clarify and find an answer to a complex problem. The structure allows users to take a problem with multiple possible solutions and display it in a simple, easy-to-understand format that shows the relationship between different events or decisions. The furthest branches on the tree represent possible end results. The tree approach is most useful in a sequential decision situation and can be illustrated by the logical testing procedure cited in Chapter 8 of the book *Maintenance Management—Precepts and Practices* (Fig. 6.16).

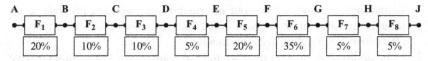

FIG. 6.16   A simple series circuit with known failure probabilities.

If a process consisting of a group of units connected in a series as per Fig. 6.16 above fails, then individual units have to be checked for failure since the exact unit which has failed is not known, the decision tree illustrated in Fig. 6.17 would help.

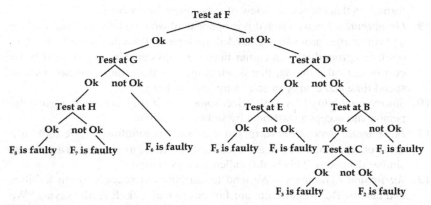

FIG. 6.17   Decision tree for the midway check method for the above situation.

6.23   SYSTEMATIC DECISION-MAKING

1. Outline your goal and outcome
2. Gather data
3. Develop alternatives (ie, brainstorming)
4. List pros and cons of each alternative
5. Make the decision
6. Immediately take action to implement it
7. Learn from and reflect on the decision

6.24   PROPER MANAGEMENT DECISION AND PROPER ENGINEERING DESIGN

The precepts as described in the previous paragraphs have given us new concepts and terminology for distinguishing the managerial and engineering decisions such as proper management decision (PMD) and proper engineering design (PED), which is emphasized more in books on Processional Ethics.

PMD is a decision that should be made by mangers or at least governed by management considerations, because:

- It involves factors relating to the well-being of the organization such as cost, scheduling, marketing, and employee morale and welfare.
- The decision does not force engineers or other professionals to make unacceptable compromises with their own technical practices or ethical standards.

PED is a decision made by the engineers or at least governed by professional engineering practices because it either,

- involves technical matters that require engineering expertise or
- falls within the ethical standards embodied in engineering codes, especially those requiring engineers to protect the health and safety of the public.

6.25 CONCLUSION

The chief purpose if the systems approach is to provide management with an analytical framework by which it can identify, describe, and interrelate the process and components that make up a particular system. In other words, the systems approach enables a manager to maintain a perspective of the whole process while he analyzes the parts. Again, in the day-to-day administration of the departmental affairs, any manager, especially the planning manager is required to make an endless sequence of decisions. His success depends on his effective decision-making. A clear understanding the precepts and practices of decision theory as detailed in this chapter would help the planning manager. This chapter discussed both of these interrelated precepts.

FURTHER READING

[1] Chenhall RH, Langfield-Smith K. Accounting, Organizations and Society. Elsevier; 1998;23(3): 243–264.
[2] Kiran DR. Maintenance engineering and management: precepts and practices. Hyderabad: BS Publishers; 2014.
[3] Kiran DR. Total quality management, an integrated approach. New Delhi: BS Publishers; 2016.
[4] Kiran DR. Professional ethics and human values. New Delhi: McGraw Hill Higher Education; 2012.
[5] Jenkins GM. A systems approach to management; 1968.
[6] www.yourarticlelibrary.com/management.
[7] en.wikipedia.org/wiki/Approaches_of_management.
[8] study.com/.../systems-approach-to-management-theory-lesson-quiz.html.

Chapter 7

Strategic Planning

Chapter Outline

7.1 INTRODUCTION

The need for organization to plan for the future is as old as business itself. Peter Drucker in his book, *The Practice of Management,* noted that managers must determine not only *what the business of the organization is* but also determine *what will be the business in the future.* This future planning when done systematically is called strategic planning. Strategic planning assumes significance in view of the changing environment in which there are many forces operating on the organization as illustrated in Fig. 7.1.

7.2 BUSINESS PLANS

Business planning is the formulation of a formal statement of a set of business goals, highlighting why they are considered attainable, and including the background information about the organization, or the team working towards the attainment of these goals.

Total Quality Management: Key Concepts and Case Studies. http://dx.doi.org/10.1016/B978-0-12-811035-5.00007-6

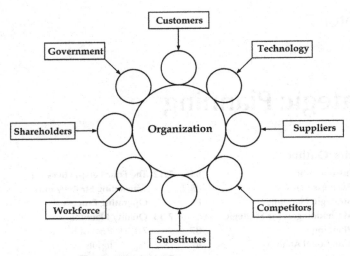

FIG. 7.1   Work environment of an organization.

The business plans can be in two categories, depending upon their importance for the day-to-day operations.

- Strategic planning and
- Operational planning

Strategic planning involves the determination and development of policies used in achieving the goals of the company, while the *operational planning* is the planning of the day-to-day operations and procedures in conforming to the policies set by strategic planning.

7.3 STRATEGIC PLANNING

Strategic planning is an organization's process of defining its strategy or direction, and making decisions on allocating its resources to pursue this strategy, including its capital and people. The following are some of the business analysis techniques used in strategic planning.

Strategic plans are decision-making tools. They may not have any fixed content for the strategic plan, but the format is determined by the goals and the audience. The goals must have a plan or method with resources for their achievement, and must be based on the statistical conformation and shall not be based purely on assumptions. Strategic planning is also called policy deployment by certain authors. The following are some of the tools for the performing analysis that would be helpful for strategic planning.

- *SWOT* analysis (Strengths, Weaknesses, Opportunities, and Threats)
- *PEST* analysis (Political, Economic, Social, and Technological analysis)

- *STEER* analysis (Socio-cultural, Technological, Economic, Ecological, and Regulatory factors)
- *EPISTEL* analysis (Environment, Political, Informatics, Social, Technological, Economic, and Legal).

7.4 METHODOLOGIES FOR STRATEGIC PLANNING

As explained above, SWOT analysis forms the best tool for the performance analysis. Its basic methodology is to apply the following steps.

- *Vision*—define vision and mission statements and set the hierarchy of goals and objectives.
- *SWOT*—conduct SWOT analysis of the above according to the desired goals.
- *Formulate*—design the processes and actions to be taken to achieve these goals.
- *Implement*—implement these processes and actions.
- *Control*—establish control systems and monitor them based on feedback.

Different authors have proposed other methodologies as below:

(a) *Three-step* process may be used:
- *Situation*—evaluate the current situation and how it came about.
- *Target*—define goals and/or objectives (sometimes called ideal state).
- *Path*—map a possible route to the goals/objectives.

(b) *Draw-see-think-plan* is another alternative approach:
- *Draw*—what is the ideal image or the desired end state?
- *See*—what is today's situation? What is the gap from ideal and why?
- *Think*—what specific actions must be taken to close the gap between today's situation and the ideal state?
- *Plan*—what resources are required to execute the activities?

(c) *See-think-draw* approach:
- *See*—what is today's situation?
- *Think*—define goals/objectives.
- *Draw*—map a route to achieving the goals/objectives.

7.5 SITUATIONAL ANALYSIS

When developing strategies, analysis of the organization and its environment as it is at the moment and how it may develop in the future, is important. The analysis has to be executed at an internal level as well as an external level to identify all opportunities and threats of the external environment, as well as the strengths and weaknesses of the organizations.

There are several factors to be assessed during the external situation analysis:

1. Markets (customers)
2. Competition
3. Technology

4. Supplier markets
5. Labor markets
6. The economy
7. The regulatory environment

Let us now consider the following case study as cited in *The World is Flat* by Thomas L Freidman.

> *ASIMCO of US first purchased Federal Mogul Corporation, a camshaft manufacturing company with a high-end customer profile. But when it went under bankruptcy, its associate company started manufacturing the semi-finished camshaft in China, but did the finishing operations in the United States, supplied them to US car companies without losing the goodwill name made by the former company.*

How does it rate in business ethics? Is it an unethical practice or a win-win tactic?

7.6   HOSHIN KANRI (方針 管理)

The Japanese term for strategic planning is Hoshni Kanri, which can be broken down into four parts, Ho-shin-Kan-ri.

Ho means direction, while *shin* means a shining needle, as used in compass. So the word *Hoshin*, means a compass needle equivalent to the word *diksuchi* in Sanskrit, representing the progress toward a goal.

Kan means control or channeling the progress (akin to the term kan in kanban) while *ri* translates into reason or logic. So *Kanri* refers to administration, management, control, charge of, or care for.

Thus taken altogether, Hoshin Kanri means management and control of the organization's direction, focus, or goal. It can be thought of as the application of Deming's Plan-Do-Check-Act cycle to the management process.

Or in other words, Hoshni Kanri represents the management planning and control toward the achievement of the goal. It is a method devised to capture and cement strategic goals as well as to provide insight about the future, and develop the means to bring these into reality. It is a systems approach to the management of change in critical business processes using step-by-step planning, implementation, and review process so as to improve the performance of business systems. As Dr. Yoji Akao puts it,

> *With Hoshni Kanri, the daily crush of events and bottom line pressures do not take precedence over strategic plans, rather, these short-term activities are determined and managed by the plans themselves.*

7.6.1   Nichijo Kanri

As a contrast, *Nichijo* means daily routine and *kanri* means management and control, similar to *kanri* of *hoshin kanri*. Thus, *nichijo kanri* covers all the day-to-day aspects of operations and planning and is complementary to *hoshin kanri*, which refers to the long-range or strategic planning.

7.7   DEFINITIONS OF STRATEGIC PLANNING

There are several definitions available in the web. Some of them are reproduced below to give an insight into the process.

Strategic planning is an organization's process of defining its strategy, or direction, and making decisions on allocating its resources to pursue its goals.

en.wikipedia.org/wiki/Strategic_planning

Strategic planning is a process to determine or reassess the vision, mission, and goals of an organization and then map out objective (measurable) ways to accomplish the identified goals.

planning.nmsu.edu/taskforce/glossary.html

Strategic planning in information technology is the first stage of the planning model. It aligns the system's strategic planning with overall organizational planning by assessing organizational objectives and strategies, setting its mission, assessing the environment, and setting its policies, objectives, and strategies.

www.wiley.co.uk/college/turban/glossary.html

Strategic planning is a top-down approach concerned with the long-term mission and objectives of an organization, the resources used in achieving those objectives, and the policies and guidelines that govern the acquisition, use, and disposition of those resources.

www.quantum3.co.za/CI%20Glossary.html

Strategic planning is the planning activity through which one confronts the major strategic decisions facing the organization. A decision is not rendered strategic merely by being important.

www.unisa.edu.au/poas/qap/planning/glossary.asp

Strategic planning is the process of answering the questions: "Where are we going? What should we be doing? And how will we do it?"

www.mgrush.com/content/view

Strategic planning identifies the medium to long-term goals integral to the institution's mission; general principles are fairly fixed, but the means for implementation are flexible.

www.scoea.bc.ca/glossary2001.html

Strategic planning is the process of developing long-range goals and plans for an organization.

www.thecomputerfolks.com/S.html

Strategic planning is a decision-making process in which decisions are made about establishing organizational purposes/mission, determining objectives, selecting strategies, and setting policies.

www.services.eliteral.com/glossary/decision-support-systems-glossary.php

Strategic planning is the process of thinking of and determining specific goals, objectives, and actions to move from one place to another.

www.interlinktc.com/public_html/definitions.html

Strategic planning is the determination of the steps required to reach an objective of achieving the optimum fit between the organization and the marketplace.

www.bayarearadioadvertising.com/terms/P400/

Strategic planning is developing short and long-term competitive strategies using tools such as SWOT Analysis to assess the current situation, develop missions and goals, and create an implementation plan.

www.gemba.com/resources.cfm

Strategic planning is the process of planning for a set of managerial decisions and actions that determine the long-term performance of an organization.

ecommerce.etsu.edu/Glossary.html

Strategic planning is the identification and ratification of business directions, visions, goals, directions, and objectives to ensure the business is appropriately positioned given its capabilities, and markets. The strategic plan provides a framework for development of tactical plans to achieve desired intentions.

www.icreb.com/compprof.html

7.8   STRATEGIC PLANNING ELEMENTS

The strategic plans involve the following, which are further explained in subsequent paragraphs.

- Business values
- Markets to be served
- Customer needs
- Customer positioning
- Predicting the future
- Product diversification
- Alignment of the plans with the vision, mission, and the concepts of the organization
- Gap analysis—to identify the gaps that exist between the present and future state of the organization.
- Investment in machinery and equipment
- Manpower

- Budgets
- Strategies for improving profits, etc.

7.9   BESTERFIELD'S SEVEN STEPS OF STRATEGIC PLANNING

Besterfield et al. illustrate the above in the following seven basic steps for strategic planning.

1. *Customer needs*: The first step is to discover the future needs of the customers. Who they are? What do they want? How should the organization meet and exceed their expectations? This step is further explained in a later chapter on quality function deployment.
2. *Customer positioning*: To determine which customer market the company should serve, and which types of products or services.
3. *Predicting the future*: Tools like demographic projections, economic forecasts, technical assessments, etc., shall be used to predict the future.
4. *Gap analysis*: The strategic planners shall identify and pinpoint the gaps between the current state and future state of the organization by analyzing the core values and concepts.
5. *Closing the gap*: By establishing more realistic goals and responsibilities, the plans should be developed to close the gaps.
6. *Alignment*: As the plans are developed, they must be aligned with the mission, vision, and core values of the organization for without these, the plans will have little success.
7. *Implementation*: Resources must be allocated to collecting the data, designing the changes, and monitoring the progress being made.

7.10   STRATEGY DEVELOPMENT AND STRATEGY DEPLOYMENT

The strategic planning should be aimed at achieving the following in two phases:

(a) Strategy development:
1. Set strategic directions and determine key planning requirements for long-term well as short-term goals.
2. Integrate customer and marketing expectations with the company's vision and mission.
(b) Strategy realization:
1. Allocate resources to ensure realization of the plans and goals.
2. Determine key performance measures and indices for tracking progress.
3. Adapt strategies for meeting the competition challenge.

7.11   EFFECTIVENESS OF THE STRATEGIC PLANNING

Strategic planning is an essential element of organizational management, but will be significant only if implemented effectively. The biggest failure of strategic planning is not in identifying what is to be done, but in wrongly executing the same. Michael Porter in his article in the Harvard Business Review supplemented this view by noting that strategy and effectiveness are both essential for superior performance.

> *As Larry Cassidy noted, "When companies fail to deliver their promises, the most frequent explanation given is that the CEO's strategy was wrong. But in fact, the strategy and its planning may not be the real cause. It is in the ineffective execution that the strategies most often fail and things that are supposed to happen do not happen."*

> *Edwin Bliss observed that "Success does not mean the absence of failures; it means the attainment of ultimate objectives. It means winning the war, not every battle." Shiv Khera, too, observed that "most people fail not because of lack of ability and intelligence, but because of lack of desire, direction, dedication, and discipline."*

7.12   THE FOUR PERSPECTIVES FOR TRANSLATING STRATEGY INTO OPERATING PROCESS

(a) Financial perspective
(b) Customer perspective
(c) Internal perspective
(d) Innovation and learning perspective

This is represented diagrammatically in Fig. 7.2.

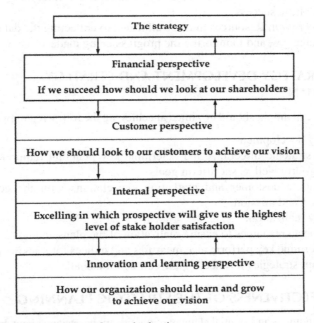

FIG. 7.2   The four perspectives of strategic planning.

7.13   QUALITY PLANNING

In the context of TQM, quality planning can be equated to operational planning as described above. For quality planning, customer satisfaction should be the goal. The important elements in quality planning are similar to that indicated under the strategic planning:

- Establish quality goals
- Identify the customers
- Discover customers' needs
- Develop product features
- Develop process features
- Establish process control and transforming operations

Road map of quality planning: The following worksheet (Fig. 7.3), listing the several initiating activities and the linking activities based on their outputs, would help in the preparation of a comprehensive checklist for the operational planning.

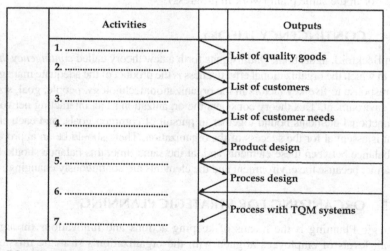

	Activities		Outputs
	1.		List of quality goods
	2.		List of customers
	3.		List of customer needs
	4.		Product design
	5.		Process design
	6.		Process with TQM systems
	7.		

FIG. 7.3   Worksheet for activity planning.

7.13.1   8 Ms of Resource Inputs

We know from books on productivity the following 8 Ms of resource inputs for any manufacturing activity.

1 Materials
2 Management
3 Men
4 Machinery and equipment
5 Money

6 Market
7 MIS systems
8 Monitoring and control

Armond Feigenbaum adds Motivation as the 9th M factor effecting quality.
Some of the quality planning and control activities during the production
cycle apart from PP&C:

A. Design control
- Engineering of quality products
- Planning of quality processes
- Establishing quality standards
- Conducting quality research, safety studies, etc.

B. Incoming material control
- Establishing purchase parameters and standards
- Conducting vendor surveys and vendor development
- Controlling receipt of materials and parts
- Controlling materials and parts processed by other plants (outsourcing) or in the same plant (work in progress).

7.14 CONTINGENCY THEORY

John Beckford, in his book *Quality,* puts forth a new theory called *contingency theory,* in which the organizational effectiveness is the product of the adequate managerial response to five key factors in the organization: technology, people, goal, size, and environment. This theory considers the organization as an interacting network of functional elements bound together in pursuit of common goals, and each element is essential for the success of the organization. There should be an appropriate balance between these elements and, at the same time, this balance should be dynamic, because the environment and the elements are continuously changing.

7.15 ORGANIZING FOR STRATEGIC PLANNING

Strategic Planning is the means of keeping actions and innovations throughout all levels of employees aligned with the organization's strategic mission. Fig. 7.4 illustrates the coordinated responsibilities of all levels of management for effective strategic planning.

7.16 LEAVITT'S DIAMOND

In 1965, Dr. Harold Leavitt of Stanford University proposed that every organizational system is made up of four main components: People, Task, Structure, and Technology, as illustrated in Fig. 7.5. Any change in any one of these elements will have a direct effect on all the other elements. The way in which each of these main components interacts with the others can help determine the success or downfall of an organization.

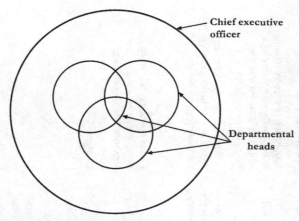

FIG. 7.4 Coordinated responsibilities in strategic planning.

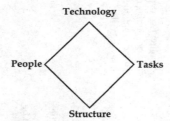

FIG. 7.5 Leavitt's diamond.

The inter-relationship between these four elements can be explained by Table 7.1 below:

7.17 MISSION AND VISION STATEMENTS

The vision statement defines how you want your business to be seen by the outside world, including the investors, clients, suppliers, the market, and even competitors. It would answer the question "What do we want become in the future?" and provide inspiration for setting future goals. It should be as short as possible, preferably in one sentence. It describes what the company believes are the ideal conditions for the community.

http://ctb.ku.edu/, the website community tool box emphasizes the following characteristics of vision statements:

- Understood and shared by members of the community
- Broad enough to include a diverse variety of local perspectives
- Inspiring and uplifting to everyone involved in your effort
- Easy to communicate

TABLE 7.1 The Inter-Relationship Between Leavitt's Elements

How Effect on	People	Tasks	Structure	Technology
			How Changes in	
People	—	Educating and training people for new methods.	People need help to learn about their new job duties and responsibilities.	People need extensive training to handle the new technology efficiently. May even require hiring new employees.
Tasks	Need to modify the tasks or goals to rectify and make optimum use of their skills and knowledge	—	Merging two departments into one or splitting one into two, you cannot continue with the same tasks or goals.	Shifting to a newer technology would require making changes to the way things are done.
Structure	For hiring more people, you need more supervisors which requires revamping of the organizational structure	When business processes are reengineered, the organizational structure has to be different	—	Computerization or automation needs a different organizational structure. Interdepartmental communication, too, needs to change
Technology	If you're hiring computer-literate employees, you cannot ask them to work on typewriters.	If the material planning is to be done by the purchasing department, different software would be required.	If you want to cut down staff, you will have to automate some processes, to maintain the same level of production.	—

The mission statement is an extension of the vision statement, which would establish the objectives and help in formulating the strategies. It emphasizes *what* the company is going to do and *why* it's doing it. It concentrates on how you would like to achieve your vision and defines the customers, processes, and the desired levels of performance. It should be concise, outcome-oriented, and indicate the company's key goals.

Features of an effective vision statement include:

- Clarity and lack of ambiguity
- Vivid and clear picture
- Description of a bright future
- Memorable and engaging wording
- Realistic aspirations
- Alignment with organizational values and culture

7.18   CAUTION IN THE APPLICATION OF STRATEGIC PLANNING

1. While it can be appreciated that the managers have to spend considerable time for strategizing, they must be cautioned not to spend too much time in the name of strategic planning, but schedule their overall attention for strategic planning, as well as their routine activities, so as to minimize the negative impact of the former on the latter.
2. The formulators of strategic planning must be intimately involved in the implementation, so as to accept the responsibility for input to the decision process and subsequent actions. In other words, strategic planning must limit its action plan in a way that can be delivered by the decision-makers and their subordinates.
3. The strategic planners must anticipate, minimize, or proactively respond when their team members get frustrated over expectations that are not met.

7.19   CONCLUSION

As cited in the definitions, strategic planning is an organization's process of defining its strategy or direction, and making decisions on allocating its resources to pursue the company's vision and mission. It is a top-down approach concerned with the:

- long-term mission and objectives of an organization,
- the resources used in achieving those objectives,
- the policies and guidelines that govern the acquisition,
- usage and disposition of those resources.

Hence, it forms the very foundation for the success of an organization.

We can finally conclude with Sir Brian Pitman's observation that, *"there is always a better strategy than the one you have. You just haven't thought of it yet."*

On the Lighter Side

1. *Remember the Golden Rule –*
   Whoever has the gold will make the rule

2. Corridor Conference

   After finishing an evening meal and hearty talk at your friends' apartment, you come out to the corridor and are half way down the stairs. But your spouse still continues the chatting at the door steps when you are waiting near the car. This can be called as corridor conference.

FURTHER READING

[1] Kemp RL. America's cities: strategic planning for the future. Danville, IL: The Interstate; 1988.

[2] Kemp RL. Strategic planning in local government: a casebook. Chicago, IL: Planners Press; 1992. American Planning Association (APA).

[3] Kemp RL. Handbook of strategic planning. East Rockaway, NY: Cummings & Hathaway; 1995.

[4] Burkhart PL, Reuss S. Successful strategic planning: a guide for nonprofit agencies and organizations. Newbury Park: Sage Publications; 1993.

[5] Bradford RW, Duncan JP. Simplified strategic planning. Worcester, MA: Chandler House; 2000.

[6] Haines SG. ABCs of strategic management: an executive briefing and plan-to-plan day on strategic management in the 21st century; 2004.

[7] Kono T. Changing a company's strategy and culture. Long Range Plan 1994;27(5):85–97 [October 1994].

[8] Kotler P. Mega marketing. Harvard Business Review; 1986 [March–April 1986].

[9] Naisbitt J. Megatrends: ten new directions transforming our lives. Macdonald; 1982.

[10] Freidman T. The world is flat. New York: Farrous, Strauss and Giroux; 2005.

[11] Levitt T. Marketing myopia. Harvard Business Review; 1960 [July–August 1960].

[12] Lorenzen M. Strategic planning for academic library instructional programming. Illinois Libraries, 86; 2006. p. 22–9 No. 2 (Summer 2006).

[13] Fahey L, Narayman VK. Macro environmental analysis for strategic management and rdquo. St. Paul, MN: West Publishing; 1986.

[14] Lusch RF, Lusch VN. Principles of marketing. Boston: Kent Publishing; 1987.

[15] Tracy B. The 100 absolutely unbreakable laws of business success. Berrett: Koehler Publishers; 2000.

[16] Allison M, Kaye J. Strategic planning for nonprofit organizations. 2nd ed. John Wiley and Sons; 2005.

[17] en.wikipedia.org/wiki/hoshin-kanri.

[18] www.tqe.com/stratplan.html.

[19] www.isisigma.com/me/tqm.

[20] http://www.diffen.com

[21] http://ctb.ku.edu.

Chapter 8

Cost of Quality

Chapter Outline

8.1 INTRODUCTION

Cost of Quality is the Cost of NonQuality

As Philip Crosby said, quality is measured by the cost of quality (COQ) which is a direct result of doing things wrong and by nonconformance. He emphasized that quality costs are those incurred in excess of those that would have been incurred if the product were built or the service performed exactly right the first time.

The traditional perception of the COQ was that higher quality requires higher costs, either by buying better materials, machines, or by hiring more labor.

When faced with mounting numbers of defects, organizations typically react by throwing more and more people into inspection roles. But inspection is never completely effective and appraisal costs stay high as long as the failure costs are significant.

It was during the late 1950s, that the concept of COQ emerged, with the realization that quality can be improved, not by increasing appraisal costs, but by striving to achieve defect-free production. Costs include not only those that

Total Quality Management: Key Concepts and Case Studies. http://dx.doi.org/10.1016/B978-0-12-811035-5.00008-8
Copyright © 2017 BSP Books Pvt. Ltd. Published by Elsevier Inc. All rights reserved.

are direct, but those resulting from lost customers, lost market share, and the many hidden costs and foregone opportunities not identified by modern cost accounting systems. This concept of quality has become increasingly relevant in the debates over quality. This has encouraged companies to identify, evaluate, and quantify the several costs associated with the quality function in order to improve quality at optimal costs.

8.2   FORCES LEADING TO THE CONCEPT

Juran lists the following four forces that led to the above concept.

1. Growth of quality costs due to growth in the volume of complex products, which demanded higher precision, greater reliability, etc.
2. The influence of the great growth in long-life products, resulting in higher costs due to field failures, maintenance labor, spare parts, etc. The costs of keeping such products often exceeded the original purchase/manufacturing costs.
3. The need for quality specialists to express their findings and recommendations in the language of upper management, which is the language of money.
4. The fourth force is the phenomenon of *"life behind the dykes,"* a phrase coined by Juran to compare the dependence on quality to the Dutch way of life, where over a third of the country lies below sea level and is protected from the sea by building dykes. The land surely confers great benefits to the people, but the cost of the massive dykes is high, similar to the effective quality controls in a manufacturing situation.

8.3   THE CATEGORIES OF QUALITY COSTS

Quality costs are defined as the costs associated with falling short of product or service quality as defined by the requirements established the organization, customers, and society.

Juran in his book, *Quality Planning and Analysis*, and subsequently, several authors have classified these quality costs associated with making, finding, repairing, or preventing defects. The summary of these costs is represented in Fig. 8.1. As elaborated in later sections, these three costs are known as Prevention, Appraisal, and Failure (PAF) costs.

Internal failure costs are those that would exist before sorting out and removal of the defects prior to shipment. These are the costs of coping with errors discovered during development and testing and would be comprised of:

1. *Scrap*: The net loss in labor and materials resulting from defective products which cannot economically be repaired or used.
2. *Rework*: The cost of correcting defects to make them fit for use. Sometimes this category is broadened to include extra operations done to rectify them.

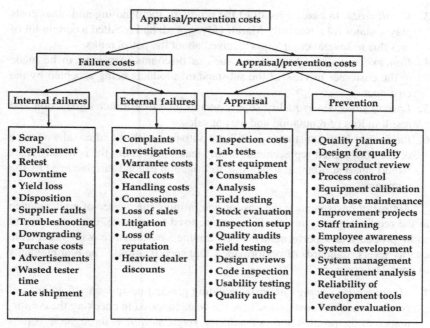

FIG. 8.1 Categories of quality costs.

3. *Retest*: The cost of reinsertion and retest of products which have undergone reworking.

4. *Downtime*: The cost of idle facilities resulting from defects, like a rejected heavy casting being unloaded from the machine and carried to a test rig, as well as the discussions that take place. In some industries, this downtime is large and is quantified. But in most cases, this is ignored, adding to the total quality cost.

5. *Yield losses*: The cost of elemental time added due to ineffective controls of operation. For example, the overfilling of containers such as soft drinks due to variability in the filling and measuring equipment.

6. *Disposition*: The effort required to determine whether the nonconforming products can be used or not. This includes the individuals' time and material review boards.

External failure costs are those that arise from defects that are noticed after the products leave the factory premises. They are distinguished from internal failure costs by the fact they are found after shipment to the customers. These could be comprised of:

1. *Complaints*: All investigations made after receipt of complaints from the customer.

2. *Warrantee costs*: This forms a major cost in the case of defective goods reaching the customer and includes testing and replacement of defective parts.

3. *Recall costs*: In case of recall of the products, the handling and other costs play a major role. Recently Maruti Udyog, Ltd. had recalled a certain lot of cars due to several complaints received about the petrol tanks.
4. *Concessions*: Some concessions such as discounts may have to be made to the customer in view of the substandard products being accepted by the customer.
5. *Loss of sales*: Poor quality supply would affect customer satisfaction and result in loss of reputation and loss of sales.
6. *Litigation*: If the customer is still not satisfied with the after-sales service or the above cited replacement, it may lead to his suing the company. This would entail substantial losses due to litigation expenditure incurred by the company.

Appraisal costs include the cost of determining the degree of conformance to the required quality levels. They are incurred to measure, inspect, test, and audit products and performance to determine conformance with acceptable quality levels, standards of performance, and specifications. The four elements of appraisal costs are

1. *Manufacturing appraisal costs* including product design, qualification, and conformance test costs, which are the costs incurred in checking the conformance of the product during the design stage, as well as throughout its progression in the factory, including the final acceptance and check of packing and shipping.
2. *Purchase appraisal costs* including supplier product inspection cost, incoming inspection, testing costs, etc.
3. *External appraisal costs* when there is a need for field trials of new products and services, including the field setup and checkout before official approval.
4. *Miscellaneous quality evaluation costs* that include the cost of all supports to enable continual customer satisfaction, such as the quality of packing, shipping process, promotions, and audits.

Preventions costs are those that are involved to rectify the processes that lead to the above losses. They include:

1. Quality planning
2. Design for quality
3. New product review
4. Process control
5. Equipment calibration
6. Database maintenance
7. Improvement projects
8. Staff training
9. Employee awareness
10. System development
11. System management

12. Requirement analysis
13. Reliability of development tools
14. Vendor evaluation

8.4   HIDDEN QUALITY COSTS

Among the above, the hidden quality costs and costs of lost opportunities are listed below:

1. Potential loss of sales
2. Cost of redesign
3. Cost due to change in the manufacturing processes
4. Extra manufacturing costs due to defects
5. Cost of software changes
6. Unaccounted scrap

8.5   COST OF LOST OPPORTUNITIES

Among the above, the cost of lost opportunities is generally ignored by nonprofessional management. They include:

1. Revenue loss due to order cancelation
2. Losing to the competitors by not conforming to the customers' needs
3. Loss of customer loyalty
4. Reduced repeat orders

8.6   SERVICE COSTS

In manufacturing, quality costs are basically product-oriented and have a definite relationship to the quantity of output, rejection percentage, etc.—tangible factors. However, in service functions, they are basically labor-oriented and can be as high as 70%, according to Evans and Lindsay.

8.7   TANGIBLE AND INTANGIBLE COSTS

The costs can also be classified into tangible and intangible costs as shown below:

Intangible costs

- Delays and stoppages caused by defectives
- Customer goodwill
- Loss in morale due to friction between departments

Tangible costs—originated at the factory

- Materials scrapped or junked
- Labor and the burden on product scrapped or junked

- Labor, materials, and the burden necessary to effect repairs on salvageable components
- Extra operations added because of the presence of defective products
- Burdens arising from excess production capacity necessitated by defectives
- Excess inspection costs
- Investigation of causes of defects

Tangible costs—originated at the sales

- Discount on seconds
- Customer complaints
- Charges to quality guarantee account

8.8   VISIBLE COSTS AND INVISIBLE COSTS

A majority of the quality costs are hidden and not manifested outside. They can be equated to huge icebergs floating in the ocean. A major portion of it lies under the surface and only a small tip is seen above. The ships which can see the visible portion on the surface can steer clear of the same. However, the invisible portion poses a major threat to their safety unless they become conscious of it. Similarly, the hidden costs form the bulk of the quality costs and we should concentrate more in identifying and reducing them. There is no point in concentrating our efforts to reduce only the visible costs, such as rework and rejection. This is illustrated by Fig. 8.2.

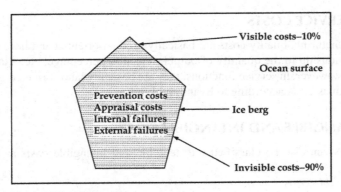

FIG. 8.2   Visible and invisible quality costs.

8.9   QUALITY COST DATA

Organizations need data on quality with respect to origination and information on quality costs because they lead the organization in a particular direction to drive strategies and organizational changes.

Some of the leading practices to collect reliable and appropriate data for driving quality excellence are:

1. To manage resources needed to achieve this road map by evaluating the effectiveness of these action plans
2. To operate the processes that make the organization work and continuously improve
3. To develop a comprehensive set of performance indicators that reflect internal and external customer requirements and the key factors that drive the organization
4. To use comparative information and data to improve the overall performance and competitive position
5. To continually refine information resources and their uses within the organization
6. To use sound analytical methods to when using the results to support strategic planning and daily decision-making
7. To involve everyone in measurement activities and ensure that the performance information is widely visible throughout the organization
8. To ensure the data and information secured is accurate, reliable, timely, and confidential
9. To ensure that the hardware and software systems are reliable, user-friendly and the data is accessible to all related personnel
10. To systematically manage organizational knowledge and to identify and share the best practices.

8.10   CASE STUDIES ON RESEARCH DONE IN THE AREA OF QUALITY COSTING

- In 1997, Schemahl et al. investigated the two areas of internal failure costs and measured their magnitudes. A simulation analysis disclosed the impact of rework inventory level and cycle times on the COQ.
- In 1998, Guruswamy observed that the COQ is considered by management as one of the important techniques of TQM, especially when the organization changes its approach from detection to prevention as a part of its exercises.
- In 1999, Campanella suggested that quality costs allow identifying the soft targets to which we can apply our improvement efforts.
- In 1999, Yasin et al. asserted that one potential critical facet of an organization's TQM is its ability to measure costs related to quality.
- In 2000, Wali reported that to maintain or sustain a competitive edge achieving a low COQ by streamlining of processes, cutting down costs, ability to reduce wastes, and ability to meet customer needs are the most important in their quality improvement journey.
- In 2004, Jaju and Lakhe discussed the relationship of quality costs with several measures of the organization's performance, such as market share, sales, profit, and return on investment.

8.11 SUGGESTED MODEL FOR QUALITY COSTING

This is also called the PAF approach and is illustrated in Fig. 8.3.

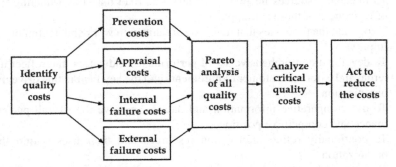

FIG. 8.3 Optimal quality costing model.

8.12 SOURCES FOR COLLECTING QUALITY COST DATA

The main sources for data collection on quality costs can be:

1. Data from normal accounting ledgers
2. Enlarged accounts by obtaining clarifications where needed
3. Company operating systems and procedures
4. Company standards and specifications
5. Data calculated or estimated from specific cases
6. Information from reports
7. Development of data formats to be filled in by concerned personnel to your requirement.

BS 6143 Part II suggests the following source documents for collecting the needed information for calculating the quality costs

1. Product cost information
2. Payroll analysis
3. Manufacturing expense reports
4. Scrap reports
5. Rework or rectifications reports
6. Inspection and test reports
7. Travel expenses claims
8. Field repair, replacement, and warranty cost reports
9. Nonconformance reports

8.13 USES OF QUALITY COST ANALYSIS

The quality cost analysis techniques are used in number of ways as shown below:

1. To identify profit opportunities
2. To make capital budgeting and other investment decisions

3. To improve purchasing and supplier related costs
4. To identify waste in overhead caused by activities not required by the customer
5. To identify a redundant system
6. To determine whether quality costs are properly distributed
7. To establish goals for budgets and profit planning
8. To identify quality problems
9. To act as a management tool for comparative measures of input-output relationships
10. To distinguish between the "vital few" and the "trivial many," using the Pareto analysis
11. To allocate resources for strategic formulation and implementation
12. To act as a performance appraisal measure

8.14   PARETO PRINCIPLE

The Pareto principle (also known as (i) the *80-20 rule*, (ii) the *law of the vital few*, and (iii) the *principle of factor sparsity*, states that, for many events, roughly 80% of the effect comes from 20% of the causes. In other words, in any population, 20% of the people contribute to 80% of a parameter says the GDP. This is similar to the Principle of ABC analysis which states that in an engineering industry, 10% of the production items contribute to 70% of the total annual consumption. When applied to the maintenance situation, we can say that as per the Pareto principle, 20% of the machines cause 80% of the total machine down time. It is also illustrated in Chapter 3 on Quality Gurus.

This principle would help in identifying which of these quality costs the management should pay more attention to in reducing the total cost. Illustrated below are some of the quality costs that are generally found to be critical.

1. Warranty claims
2. Rejection
3. Returned goods
4. Rework
5. Waste and scrap
6. Inspection and test equipment
7. Traveling outside endorsement
8. Calibration cost
9. Lost sales
10. Poor quality administration
11. Lack of quality training
12. Ineffective inspection and testing process
13. Poor quality planning
14. Internal quality audit
15. Extra operations

8.15   QUALITY CONFORMANCE LEVEL

The top management always wishes to have tangible proof in terms of profit improvement, for continuing all efforts, and providing the resources to continue quality improvement programs. Evans and Lindsay observe that the quality cost approach, also called the COQ approach, has numerous objectives, but perhaps the most important one is to translate quality problems into the language of top management—the language of money and profits (Fig. 8.4).

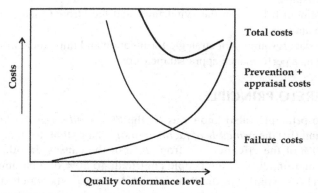

FIG. 8.4   Optimal quality conformance level.

8.16   TOP MANAGEMENT ROLE IN CONTAINING QUALITY COSTS

Crosby emphasizes a program for measuring and publicizing the cost of poor quality to bring them to the attention of management and provide opportunities for corrective action. Such data provide visible proof improvement and recognition of the achievement. Even Juran supported this approach.

This statement obviously accepts quality as a strong business driver and enables understanding the comparison between the quality and cost. We can call this cost-benefit analysis of TQM, in line with such an analysis in providing a safe environment.

In general, all organizations measure and report quality costs as suggested above. Of course, cost accounting has been a traditional tool for reporting rejection-related costs, but has been limited to inspection and testing results. All other costs, such as rework, have been accounted and clubbed as indirect costs or overhead and hence, not analyzed in depth nor could a serious attempt be made to isolate and improve them. Feigenbaum observed that quality and cost are a sum and not a difference; the objectives are complementary and not conflicting.

As managers began to define and isolate the costs as propounded in the first paragraph of this chapter, then the following facts were revealed:

1. The quantity-related costs are much higher than previously conceived, generally to 20–30% of the total sales.
2. The quality-related costs are not only limited to manufacturing operations, but also to ancillary activities like purchasing, inventory management, and customer service.
3. Most of these costs due to poor quality are avoidable, though superficially, they appear unavoidable.
4. Most of these avoidable costs are due to management shortcomings by not defining the clear responsibility or formulating the structural approach for the actions to reduce costs.
5. Another mistaken notion that better quality required higher costs has been exposed. This mistaken notion or myth has prompted several companies to invest heavily on quality-related projects, without realizing that any such investment logically and effectively planned and executed after a good analysis and homework, yields effective results at optimal costs.

8.17   QUALITY AND SAFETY

Until the early 20th century, most consumer products were repetitions or modifications of products made for centuries. Consumers could use their common-sense in identifying the safety of using these products. Consequently, in case of any accidents or injuries caused by the defective product, the manufacturer generally related it to the customer's carelessness in using the product, asserting that "*goods once sold cannot be taken back.*"

But during the last century, new products were introduced, with new designs and processes of manufacture, and together with the faulty manufacturing operations, these presented potential sources of injuries to the user. This increased occurrence of accidents and injuries attracted government and other agencies to protect the customer, not only against products causing injuries, but also against defective products that do not perform the required function. This has given rise to several legal problems with liabilities to the manufacturer. Consumer product safety has hence become a mandatory part of the design (we may call it DFS-Design for safety) and manufacturing process. Besterfield et al. say that in 1977, in a case involving defective latches in minivans, Chrysler Motor Corporation was asked to pay $262.5 million (equal to Rs. 262 crores those days, which is six times as much at today's exchange rates).

8.18   RESPONSIBILITY OF TOP MANAGEMENT FOR PRODUCT SAFETY

Juran and Gryna emphasize that the top management has the responsibility to define a corporate policy with respect to product safety, considering the following elements:

1. A commitment to make and sell only safe products, and to adhere to the published regulations, industry standards, etc.

2. Mandated formal design reviews and product reviews for safety.
3. Requirement that all company functions prepare formal plans defining their roles in carrying out the corporate policy.
4. Broad guidelines for documentation, product identification, and traceability to assure product integrity and to assist in defense against law suits.
5. Guidelines for defense against claims, whether rigid or flexible.
6. Guidelines for evaluation of safety performance and publication of corporate reports on results obtained.
7. Provision of audit to adhere to the policy.

8.19   CASE STUDY ON QUALITY COST

Does minimizing the risk and designing for safety always result in more expensive alternatives?

Spending a long time in design and spending extra rupees for meticulously providing safety features in the design may appear to be a very expensive proposition, especially early in the design cycle of developing the prototype. This is a short-term viewpoint. Unsafe products in the market ultimately result in costly replacements or repair processes, or even expensive lawsuits. Hence, it is absolutely ethical to let the engineer spend as much time as required to achieve a safe design to minimize the future risk of injury or losses.

The United States is known for large-sized cars, including limousines, unlike small and compact foreign cars, especially from Japan and the United Kingdom. In late 1960s, the Ford Motor Company designed a compact car called the Pinto, weighing less than 2000 pounds and costing around US$ 2000, with estimated annual sales of about 11 million cars. Anxious to be in competition, Ford Motors provided a very short time for the design process, due to which styling preceded engineering design. One of the compact features resulted in positioning the fuel tank between the differential and the rear bumper. The possibility of the differential bolts puncturing the gas tank during rear impacts was not given due consideration. After the car had been put on the market, reports poured in about the seriousness of this puncturing and the high number of accidents it caused, and the law suits/claims that were filed. On a review of the design, the cost of providing a safe feature for the car was calculated as US$ 11 per unit for almost 11 million cars.

Comparing the social costs of US$ 200,000 per death, as the claims cost, the management had surprisingly decided that the annual cost of improving the design was more than the social cost and decided to continue the design. However, apart from the death claims, what they had not considered was the loss of reputation. The Pinto had poor sales subsequently, and the company paid a much higher price.

8.20 CONCLUSION

As illustrated in Fig. 8.2, most costs are hidden and not seen or realized during the course of work. More than just causing production losses and delays, poor quality products result in lost business opportunities. Consciousness of product safety, too, is of prime significance in the design and manufacture. Economizing in design features at the cost of safety is not true value engineering, and would lead a company to a steep downfall.

On the Lighter Side

1. If you had purchased a memento for your wife during a trip abroad, better to tell her the price to be 20 to 30% lower than what you paid. The value of a product would be higher when purchased at a lower price and the wife will perceive the value of the gift as higher for the money spent.

—N. Ravichandran

2. I have concluded that wealth is a state of mind and anyone can acquire a wealthy state of mind by thinking rich thoughts.

—Andrew Young

FURTHER READING

[1] Besterfield DH, et al. Total quality management. 3rd ed. New Delhi: Prentice Hall; 2002.
[2] Feigenbaum AV. Total quality control. 3rd ed. New York: McGraw-Hill; 1983.
[3] Juran JM. Quality control handbook. 2nd ed. New York: McGraw-Hill; 1962.
[4] Juran JM, Gryna FM. Quality planning and analysis. 2nd ed. New Delhi: Tata McGraw Hill; 1980.
[5] Kiran DR. Professional ethics and human values. New Delhi: McGraw-Hill; 2013.

Chapter 9

Organization for TQM

Chapter Outline

9.1 WHY ORGANIZATION?

Organization, as defined by the American Society of Mechanical Engineers, is the process of determining the necessary activities and positions within an enterprise, department, or a group, arranging them into the best functional relationships, clearly defining the authorities, responsibilities, and duties of each position and assigning them to individuals, so that the available efforts can be effectively and systematically applied and coordinated.

The concept of organization can be equated to the structure of the human body, which is divided into parts or organs that perform different functions in close cooperation and coordination with each other. The integrating force of the whole system is the brain, which plans and controls the functioning of each and every organ. So is the industrial organization.

As shown by the above definitions and explanations, there are two aspects of the organization. The *static aspect* of the organization defines the relationship between the various subtasks or subsystems. This viewpoint places emphasis on job content, definition, analysis, and job relationships. The *dynamic aspect* of the organization is the pattern in which a number of people relate to each other in a planned systematic accomplishment of the objective. This viewpoint places the emphasis on the human element and personal relationships.

9.2   WHAT NEEDS TO BE ORGANIZED IN THE QUALITY FUNCTION?

1. The traditional quality inspection activities, such as the receiving inspection, process inspection, etc.
2. Supplier quality to interact with the purchasing department and the suppliers.
3. Statistical quality control.
4. New product introduction quality to interact with the design and development department, especially in regards to design for quality and design for six sigma, as discussed in Chapter 32.
5. Field quality team to interface with customers for quality-related tasks.
6. Quality standards unit to interact with the standard bodies, such as Bureau of Indian Standards, as well as the internal standards department, if any.
7. Failure Analysis Laboratory to perform electrical and physical product analysis for identifying the causes of failures.
8. Reliability Laboratory to conduct life and environmental studies to validate long-term reliability.
9. Several other activities involved, each being identified and grouped in accordance with the general company policy.

9.3   PRINCIPLES OF ORGANIZATION

Urwick presents the following ten principles of organization, which are better remembered by their first letters, OSCAR-DCSBC.

1. *Principle of objective*: Every organization and part of the organization must reflect the very purpose of the undertaking or the department, otherwise inclusion of that part is meaningless and therefore, redundant.
2. *Principle of specialization*: The activities of every member of the organization should be specialized for the performance of a single function. This principle put forth by Winslow Taylor, was the most preferred principle during the mid-20th century. However, the latest trend is the principle of delegation where the shop floor operatives are given the overall responsibility of their operations, including machine maintenance, 5S, etc. This is further explained in the chapter on total productive maintenance.
3. *Principle of coordination*: The purpose of organizing the people, as distinct from the purpose of the undertaking, is to facilitate coordination and unity of effort.
4. *Principle of authority*: In any organized group, the supreme authority must rest somewhere. There should be a clear line of authority from the supreme to every individual of the group. This principle has also been called the principle of hierarchy, or the principle of chain of command.

5. *Principle of responsibility*: The responsibility for the functioning of each and every activity should lie somewhere. The superior also should accept the responsibility of the acts of the subordinates. A factor that influences whether the ultimate responsibility for a certain function is to be placed at a higher level reporting directly to the Managing Director, or a lower level, is generally based on its financial importance to the company, as illustrated in Fig. 9.1.

6. *Principle of definition*: The extent of each person or position, with reference to the duty involved, the authority, the responsibilities, and the relationships with other positions should be clearly defined in writing and made known to all concerned.

7. *Principle of correspondence*: In every position, the responsibility should correspond to the authority vested, and vice versa, for the performance of any function.

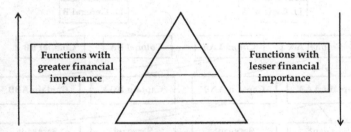

FIG. 9.1   Level of reporting is judged by the functional responsibility.

8. *Principle of span of control*: No person should supervise more than six direct subordinates whose work interlocks.

9. *Principle of balance*: It is essential that the various units of the organization should be kept in balance with respect to their importance to the organization and the level of reporting.

10. *Principle of continuity*: Reorganization is a continuous process. Specific provision must, hence, be made for future modifications or alterations at every stage.

The production Handbook, edited by Carson, adds an eleventh principle, viz., the principle of staff function, which is an extension of the principle of specialization. The staff managers do not interfere in the routine day-to-day production job, but assist in the smooth flow of the production and effective functioning of the production department, with their specialized knowledge and skills in their field. Maintenance, purchasing, etc. functions are typical examples of the staff function. Their organization specialties are best understood from the following section.

9.4   CLASSES OF ORGANIZATIONAL STRUCTURES

For the overall organization for any industrial undertaking, we know there can be basically four types of structures.

The line organization: Each department is a self-sufficient unit and has full authority and responsibility for the performance of its particular product, process, or function. This is equivalent to the decentralized organization for the maintenance function as detailed below. This is also called the military organization (Fig. 9.2).

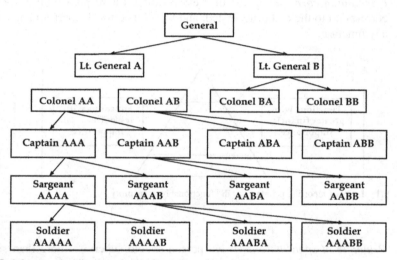

FIG. 9.2   A typical line organization in a military organization.

The functional organization: Conceived by F.W. Taylor, this is the most popular structure in today's industry. It may be noted that the concept of TPM has in recent years favored the decentralized control prevalent in the line organization. The overall responsibilities of the company are divided based in the functions, such as production, maintenance, finance, material control, and marketing. Each of these functional heads controls all aspects of his department and reports to the GM. The GM and in his absence, the functional head, has a centralized control over the functions of all activities of his department. This is equivalent to the centralized organization. This type of organization (Fig. 9.3) based on the principle of specialization takes advantage of the individual's proficiency and specialized knowledge.

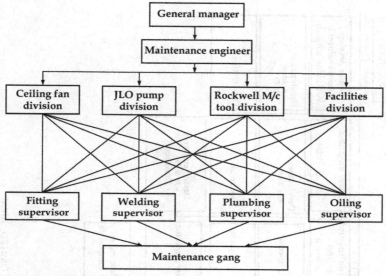

FIG. 9.3   A typical functional type of organization.

Line and functional staff organization: Instead of having a special department for each staff function, some of the major departments, like production, are provided with experts in certain functions, such as maintenance and PP&C given to the production head, inventory control, and purchasing given to the material control head. A typical line and functional staff organization as adapted in Ralli Machine Tools of Surat, India, is illustrated on the next page.

Committee organization: While the organization is headed by the GM or Chief Operating Officer, each of the functions such as material control and maintenance, would be handled by the respective committees. This is prevalent in professional associations and may be in small organizations. This is similar to the functional type of organization as discussed above, or the matrix type of organization for maintenance discussed later.

9.5   ORGANIZATION FOR THE QUALITY FUNCTION

The organization for the quality function, just like any other industrial function, is based on the above principles, and in view of its high importance, not only technically, but also financially, the ultimate responsibility is normally placed at the highest level, except in the case of decentralized type, as seen in the following section.

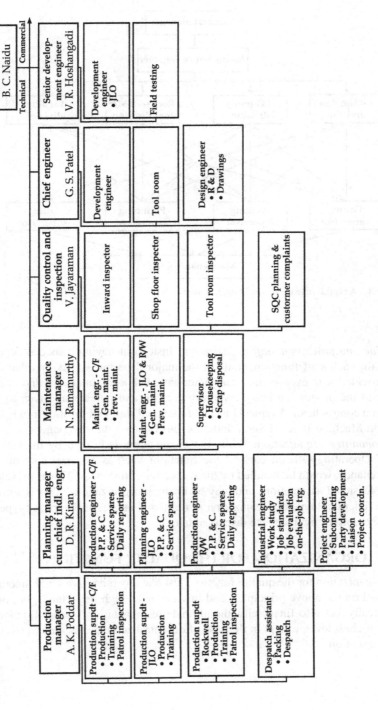

ORGANIZATION CHART—BALLI MACHINES LIMITED, SURAT

There can be three types of organizations suitable for the quality function:

(a) Centralized organization,
(b) decentralized organization, and
(c) matrix type of organization.

9.6 CENTRALIZED ORGANIZATION

This is an organization unit having similar organization principles and having the same status as the other departments, as in the production department. This is equivalent to staff organization as detailed above. This unit is normally headed by a manager, who is responsible for all the quality functions and activities of the company, irrespective of the geographical location, subdivisions of the production units of the company. All quality related staff and facilities, working anywhere in the factory ultimately report to the quality head. We can compare this to the line and staff and functional type of the traditional classification of organizations.

Factors that influence the degree of centralization are:

1. Size of the company, number of the employees, and turnover, etc. Actual plant size sometimes dictates the strength of the maintenance staff, and the amount of supervision needed for this staff. Many more subdivisions in both line and staff may be justified because the overhead can be distributed over more of the department.
2. Geographical distances of the individual units, the maintenance that works in a compact area differs from that dispersed in several buildings over a large area.
3. Technology used, and type of production.
4. Number of the working shifts, for example, during the day shifts, the maintenance function may be more centralized, while the same company's maintenance activities during the night shifts may be more decentralized that is done by the production personnel themselves.
5. Possible use of subcontractors.
6. Internal relationships between each department, as well as with the top management.
7. Level of training and reliability of the workforce. In industries where sophisticated equipment predominates with high wear or failure rate, more skilled mechanics and supervisors would be required, and they generally are under centralized control.
8. Management policy.

Advantages of the centralized organization:

1. Efficient use of technology and opportunity for further development of personnel skills, with good scope for sending employees for higher training. Skill preservation and skill development is possible for different trade personnel.

2. Less total number of departmental personnel required, due to possibility of emergency mobilization of staff from one section to another of the centralized department, as this helps in better redistribution of resources.
3. Better feedback to the management.
4. Less chance of quality inspection standards relaxed or bypassed by production managers, who are reluctant to stop production in their eagerness to have more production time.
5. Better utilization of the specialized quality measuring instruments and tools.
6. Departmental costs can be better isolated and analyzed for the efficient running of the company.
7. Ready availability of data and past quality history.

Disadvantages of the centralized organization:

1. Might develop into constant friction between the production and quality personnel.
2. Tendency for long red tape.

9.7   DECENTRALIZED ORGANIZATION

This is more or less similar to the line organization, as each shop or a production section will have an inspection foreman of its own, reporting to the respective section head in charge of production (Fig. 9.4), day-to-day planning for the section is done independently by the section head, in addition to this production responsibilities. In this organization, the Quality Control Manager is generally responsible only for the central planning. This concept of decentralization has gained so much ground in the past two decades that it is now popularized as "Total Productive Maintenance," which is discussed more elaborately in a later chapter.

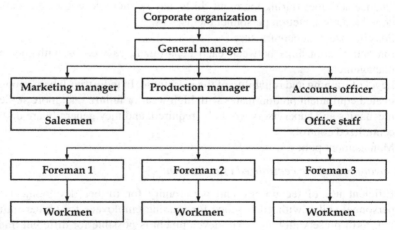

FIG. 9.4   Decentralized organization.

9.8   MATRIX TYPE OF ORGANIZATION

This is partly centralized and partly decentralized, though the overall responsibilities for this function are still vested centrally.

Here people with similar skills are pooled for work assignments, resulting in more than one manager. This is mostly applicable for a large organization having several major products and/or several geographical locations. Each product or location may have its own head and its own production staff, but does not justify having a separate team under each product or location. All the quality team members would be under a central manager, but whenever needed, the team goes to the particular location or the product, and for that job, they fully report to the location manager. Therefore, each engineer may have to work under several managers to get his job done. This is illustrated in Fig. 9.5.

Rallis India group of engineering industries provides a good example for the matrix organization. During the 1960s, it basically had three units, Ralliwolf making portable drilling machines, etc., Rallifan making table and pedestal fans, both located in Bombay, and Ralli Machines, located in Surat, making three distinct products, the ceiling fans under Rallis brand, the table-top machinery under the brand of Rockwell machines, and 25 cc petrol engines under the brand of JLO. While each of the three units has independently centralized control, corporate management activities like audits have the matrix function. In the Ralli machines unit of Surat, while each of the three product units have separate production managers, other functions like quality control, maintenance, industrial engineering, and purchasing are common staff functions and form a matrix type of control.

Kevan Hall, in his book, *Making the Matrix Work*, identifies four C factors for the success of matrix types of organization:

- *Context*: Matrix managers need to make sure that people understand the reasoning behind matrix working and change their behaviors accordingly.
- *Cooperation*: A matrix is intended to improve cooperation across the groups, but shall not lead to more meetings and slower decisions, where too many people are involved.
- *Control*: In a matrix, managers are often dependent on strangers on whom they don't have direct control. Matrix managers need to directly build trust in distributed and diverse teams and to empower people, even though they may rarely get face-to-face.
- *Community*: The formal structure becomes less important than getting things done in a matrix, so managers need to focus on the "soft structure" of networks, communities, teams, and groups that need to be set up and maintained to get things done.

Wikipedia reports that visual representation of matrix charts has been a challenge ever since it was invented. Most organizations use dotted lines to

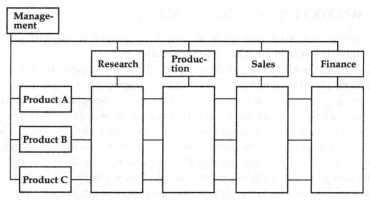

FIG. 9.5   A typical matrix of organization. *Source: Wikipedia*

represent secondary relationships between people, and charting software such as Visio and OrgPlus supports this approach. Until recently, enterprise resource planning and human resource management systems software did not support matrix reporting, but recently SAP software was developed to support matrix reporting, and Oracle e-Business Suite can also be customized to store matrix information.

9.9   FACTORS TO BE CONSIDERED IN DECIDING THE MANPOWER REQUIREMENT

- Complexity of the job
- Periodicity of the activities
- Time required for each of these activities
- Quantum and type of activity
- Layout and location of the activities
- Allowances for leave, holidays, shifts, etc.
- The company policy

9.10   SIZE AND TYPE OF AN ORGANIZATION

Manpower is the main resource for any organization and careful planning is needed to determine the size of the department. The actual type of quality control department depends on the following factors:

- Type of industry, whether process or continuous or batch.
- Size of the industry and volume of its operations.
- Complexity and nature of the business or jobs in that type of industry.
- Availability of skilled labor and inspection personnel employed by the company.

- Availability of external agencies for the required skills and experience needed in the field.
- Extent of automation and built-in testing and monitoring equipment facilities provided.
- Local labor laws, industrial, culture, and practices.
- Policy of the company.

9.11 CONCLUSION

A well-conceived and clear definition of the relationship between the various subtasks or subsystems, not only for the quality department, but holistically among all the departments, as well as other stakeholders like suppliers, shareholders, etc., is an essential aspect factor in developing an organization.

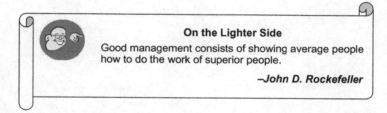

On the Lighter Side

Good management consists of showing average people how to do the work of superior people.

–John D. Rockefeller

FURTHER READING

[1] www.freesale.com.
[2] http://www.pera.net.Tools/Glossary.

Chapter 10

Customer Satisfaction

Chapter Outline

10.1 SELLERS' MARKET VERSUS BUYERS' MARKET

Four decades ago, when you wanted to buy a car, you had to pay the amount in full and wait for long periods, even up to 3 or 4 years to get delivery. If you wanted the color of your choice, or any special feature, you had to wait longer, until the manufacturer programmed that color or feature in lots of 5000 or more. You had to be content with what you got. You would have little choice. Even if the performance was not good, you could not complain. When took it in for repairs, the service personnel would treat you casually, and you had to just leave the car for weeks and months, because the service personnel would say they had too many customers to attend to. This was because the demand was more than the supply, and competition was minimal. It was the sellers' market and the seller could dictate the terms.

Total Quality Management: Key Concepts and Case Studies. http://dx.doi.org/10.1016/B978-0-12-811035-5.00010-6

But today, the situation is not the same. With globalization, the supply and thereby, the competition among producers have increased. The customer is not satisfied with just what he gets. If his requirements are not met, he has other makes or models from which to choose. So the company must improve its quality of production, as well as service to satisfy the customer and to stay in the industry. Today, the customer can dictate his terms and get what he wants. This is called a buyers' market.

10.2   CUSTOMER IS KING

The customer is the most important person in our business.
He does not depend on us. We depend on him.
He is not an interruption to our work. He is the purpose of it.
He is not an outsider. He is a part of our business.
We are not doing him a favor by serving him. He is giving us an opportunity for us to do so.
He is the life and blood of our business. Without him, we will close our shop.
He comes to us with his needs and wants. It is our duty to fulfill them.
He is a human with feelings and emotions just like us.

Adapted from **Mahatma Gandhi's** writings.

We can cite other quotations as below:

All of management efforts for kaizen bolt down to two words, Customer Satisfaction.

–Masaki Imai

A market rarely gets saturated with good products, but it very quickly gets saturated with bad ones.

–Henry Ford

Customers are the most important asset any company has even though they do not show up in the balance sheet.

–Thomas Berry

There is only one boss, the customer. And he can fire everybody in the company from the chairman down, simply by spending his money somewhere else.

–Sam Walton

10.3   POSITION OF THE CUSTOMER IN AN ORGANIZATION

Yasuthoshi Washio, in his address to National Institution for Quality and Reliability, Chennai, in May 2015, stressed that while Crosby's definition of quality as *Conformance to Specification* might be applicable in the past, the better definition applicable today is *Conformance to Requirements*. Even if the quality is acceptable, it depends to what extent it satisfies the customer. It is the customer and not the company's Engineering Department who decides on

FIG. 10.1   Customers' position is at the pinnacle of the organization.

the quality. As the customers' requirements are constantly changing, the quality conformance has an unending journey (Fig. 10.1).

10.4   CUSTOMER'S PERCEPTION OF QUALITY

1. *Performance*: The product's primary functional characteristics. It is the functional ability and reliability to serve the customers' purpose, also called "Fitness for use," as defined by Juran. For example, in an automobile, the characteristics include correctly working accelerator, brake lever, properly closing doors, etc.

2. *Features*: The accessories and add-ons, etc. For automobiles, these may include cassette deck, power steering, burglar alarms, etc. In refrigerators, this may include built-in stabilizer, freeze control, water dispenser, etc. For electrical appliances, this may include electronic speed control, etc.

3. *Aesthetics*: This includes the shape, the external appearance, color, and also how the product looks, feels, smells, etc.

4. *Durability*: This would include the period for which the product would serve without failures, as well as with minor failures that could be adjusted by the customer himself, without having to go to service personnel. In general, this applies to the wear and tear endured by the product due to constant usage. For automobiles, this would also include the mileage between services.

5. *Serviceability*: It is the speed, courtesy, and competence of the after sales service. For an automobile, this also includes the kilometer service between breakdowns, the cost of service, availability of spare parts, etc.

6. *Warranty*: It is the quality assurance, reliability, and durability for a minimum length of time. While the term "guarantee" is used for a verbal assurance as well as the confidence created in the customer due to the reputation of the product, the term "warranty" refers to a written commitment from the supplier for free repair or replacement for a minimum period in case

of failures. During the days of a seller's market, the supplier's liability was minimal, but today with the competition, the customer's awareness of his right to a warranty and guarantee has increased. This has contributed significantly to the adherence to a high quality levels by the manufacturers.

7. *Reputation*: This is the confidence created by the company and the brand name in maintenance of quality, reliability, and aesthetics. In general, customers prefer products from companies like Tata, Godrej, TVS, and Colgate rather than experimenting with new products or brands, even if they are cheaper.

8. *Price*: Last, but not least, and notwithstanding the previous parameters, a more significant perception of the customer is the price of the product without any sacrifice in the quality.

10.5   TYPES OF CUSTOMERS

While the Japanese Professor Noriaki Kano classifies the customer needs as satisfiers and dissatisfiers, further explained in Section 10.9, we can broadly classify the customers into four types.

Type A Customers: Customers who know what they want, but do not express themselves or demand their wants. They are called *normal customers*. They can be awakened and satisfied by customer surveys, advertisements, campaigns, promotions, gift/discount offers, etc., which would have a positive effect on them. Nevertheless, satisfying them fully would be difficult. Their requirements are called *satisfiers*.

Type B Customers: Customers who do not know what they want, but are conscious that they do not know or express themselves. However, any knowledge about the innovations made or improvements made in the product would excite them and keep them satisfied. They are called *active customers* and their requirements are called *exciters*. They are amicable to personal and direct contacts. A little bit of talking to them about the product, explaining positive features, and how to use the product would satisfy them easily.

Type C Customers: Customers who know what they want and expect these wants to be incorporated into the product, even when they neither express, nor ask for them. They are called *demanding customers*. They are not satisfied easily. Even when their requirements are satisfied, they may not express any joy, considering it as a minimum the company can do for them, and they expect even more. But on the other hand, any nonprovision of their expectations would dissatisfy them and drive them away from the product and the company. Their demands are hence called *dissatisfiers*, as per Kano's model. A large percentage of customers fall into this group. In order to maintain a high level of customer satisfaction, it is essential to master the art of estimating the customers' needs by constant interaction with the customers, either directly by surveys, questionnaires, etc., or indirectly through media, etc.

Type D Customers: Customers who do not know what they want, and are neither conscious of this fact, nor care much about the characteristics of the product, as long as they are functional. They are called *passive customers*. Being gentle with them would easily satisfy them.

While Kano classifies the customers based on their satisfaction detailed above, Besterfield classifies information collected based on the voice of the customer as follows.

1. *Solicited, Measurable, Routine*: Obtained through Customer and Market Surveys, trade trials, etc.
2. *Unsolicited, Measurable, Routine*: Obtained through Customer Complaints, Lawsuits, etc.
3. *Solicited, Subjective, Routine*: Obtained through Focus Groups.
4. *Solicited, Subjective, Haphazard*: Obtained through Trade and Customer Visits, Independent Consultants, etc.
5. *Unsolicited, Subjective, Haphazard*: Obtained through conventions, Vendors, Suppliers, etc.

10.6   INTERNAL CUSTOMERS

As seen in earlier chapters, the concept of total quality management (TQM) is essential for all levels and functions across the organization, and it would be necessary to draw all people into the TQM process. This can be achieved through the concept of internal customer within the company. Here, each and every department is treated as an internal customer for the preceding department from which material or services flow. The internal customer-supplier link among individuals, departments, and functions build up the chain of customers throughout an organization that connects every individual and function to the external customers and consumers, thus characterizing the organization's value chain and overall awareness of the quality maintenance through customer satisfaction.

$$\rightarrow \text{Press shop} \rightarrow \text{Machine shop} \rightarrow \text{Welding shop}$$
$$\rightarrow \text{Grinding shop} \rightarrow \text{Assembly} \rightarrow$$

For example, in a bicycle factory, where the general flow of in-process goods are in the machine shop, this can be treated as the internal customer to the press shop so that the former department would ensure quality, service, delivery dates, etc., as much as the company cares for its external customer. This will enable each person in each department to get the motivation that he is a very important cogwheel of the TQM process. As a result of this motivation and the process focus, several companies could continuously improve the quality and simultaneously bring down the cost of quality, which had earlier been the prime cause for the poor performance of the company.

Similarly, the plant engineering department would consider those departments that consume steam or compressed air as their customers. The head of the Human Resources department would consider each trainee as the internal customer for their training programs. Fig. 10.2 illustrates this concept of internal and external customers.

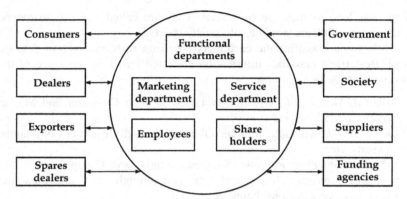

FIG. 10.2   Internal and external customers for an industrial organization.

Similarly, an educational institution can be said to have their internal and external customers as illustrated in Fig. 10.3.

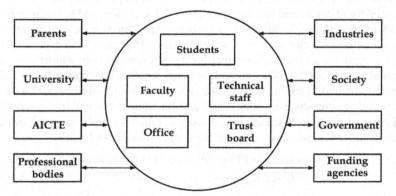

FIG. 10.3   Internal and external customers for an educational institution.

10.7   CUSTOMER SATISFACTION

During the days of a sellers' market, whatever goods were manufactured were sold because the demand is higher than the supply. The customer had to be content with what he received.

But today in the wake of the buyers' market, as illustrated in Section 10.1, the customer is not satisfied unless he gets what he wants. If not, he has many competitors' products from which to choose. Hence, an organization's success depends largely on the customer and how he expresses his satisfaction about the products and services. In other words, customer satisfaction is the goal of an organization. A majority of the repurchases come from customers who are satisfied with the company's products. Customer satisfaction plays an important role in customer retention. There is also an old saying that if a customer likes a company's service, he will tell three people about it and if he does not like it, he will tell eleven people. On the other hand, as another saying goes, the satisfied customers tell their satisfaction or their good experience to twice as many friends than they tell about their bad experience. One study found that companies with 98% customer satisfaction and retention are twice as profitable as those with 90%.

10.8   CUSTOMER DELIGHT

When the organization's products and services exceed the customer's expectations, he is delighted, like getting an unexpected new feature in the product, or the CEO inviting him for a cup of coffee, when he goes to shop. To achieve this, the company should understand what the customer expects and establish effective channels of communication and feedback with the customer.

10.9   KANO MODEL OF CUSTOMER SATISFACTION

Noriaki Kano puts forward three major classifications of the customer requirements as:

- Unspoken, but anticipated by all (basic),
- Spoken (specified or required),
- Unspoken and unanticipated, but delighted if incorporated.

Most hidden facts about the quality features the customer expects are usually unspoken or unclear, but are presumed to be basic. However, the customer expectations that are spoken have only a small impact compared to those unspoken, but are anticipated as basic requirements of a product.

With these points in mind, Kano developed a model as shown in Fig. 10.4, indicating the possible impact on customer satisfaction by fulfilling or not fulfilling the above requirements. The three curves shown (Fig. 10.4) as A, B, and C can further be explained needs are not fully satisfied.

Curve A: Customer requirements that are easily identified both by the manufacturer and the customer himself. Hence, the more these requirements are satisfied, the more the customer is satisfied.

Curve B: Innovations by the manufacturer that are well publicized. The customer gets excited about these and is happy, even if his needs are not fully satisfied.

Curve C: The requirements that are not indicated specifically by the customer, but which he expects to be automatically satisfied. This requires a careful analysis by the manufacturer. Sometimes despite much effort made to incorporate the estimated requirements, the customer still wants something more. This dissatisfaction may not be indicated in written terms, but he may get attracted by the competitor's products, which he thinks would satisfy these requirements.

When there are two customer needs of which only one can be satisfied at the same time, due to technical or financial needs, Kano's model helps to identify which of these two needs is more significant and should be satisfied.

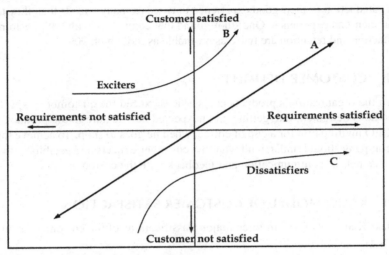

FIG. 10.4   Kano model of customer satisfaction.

In 1996, Sauerwein proposed a questionnaire to analyze the problems that can be anticipated by not fulfilling them as below:

1. Which functions does the customer expect while using the product?
2. Which problems/defects/complaints does the customer associate with the use of the product?
3. Which criteria does the customer take into consideration when buying the product?
4. Which new features or services would better meet the expectations of the customer?

10.10   AMERICAN CUSTOMER SATISFACTION INDEX

The American Customer Satisfaction Index (ACSI) is an economic indicator (Fig. 10.5) that links customer satisfaction to its determinants, such as customer satisfaction, perceived quality, and the value addition to the customer. This index uses tested economic models to produce four levels of indexes drawn on the basis of a survey conducted in a national sample of 46,000 consumers who recently bought, or used products from 40 specific industries from seven industrial groups, and 203 other companies and agencies within that group. ACSI was developed by the American Society of Quality (ASQ) with the help of Michigan Business School, with a primary goal to create public awareness and perception of quality.

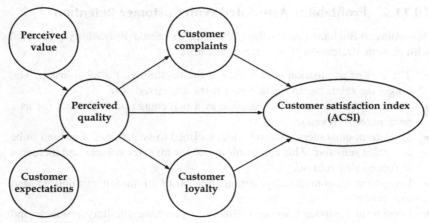

FIG. 10.5 ACSI model.

10.11   CUSTOMER RETENTION

Besides providing products and services that achieve customer satisfaction and delight, it is also essential to achieve customer retention. It is the process of creating customer loyalty, so that he continues to prefer to use these products and services, despite minor variations in the market conditions, such as the competitor offering the same at a marginally lower price. The customer should be made to feel proud of using these products and services. The product quality and the customer relationships maintained by the front line employees play a vital role in this customer retention. If 70% of the customers are retained and efforts are made to improve this to 90%, it would be easier than creating a new customer level of 10%, the customer growth of 10% would reflect on the total business growth.

10.11.1   Tips for Customer Retention

- *Market to your Own Customers*: As explained earlier, once the existing customers are satisfied with the product, they would more easily buy more products, in addition to recommending them to their friends.
- *Use Complaints to Build Business*: Maintain follow-up with the customers and if they have any complaints or suggestions, express your seriousness in examining them. If a prompt follow-up action is made to resolve a complaint, the customer is more likely to do more business than an average customer.
- *Reach Out to Your Customer*: The more the customer sees your staff, the more likely he is to make the next purchase.
- *Loyal Workforce*: Make all out efforts in retaining the front-line staff ,like technical support and customer service staff, who form a vital link between the company and the customers. This is further illustrated in Section 10.18.

10.11.2   Profitability Associated With Customer Retention

According to Buchanan and Gilles (1990), the increased profitability associated with customer retention efforts occurs because:

- The cost of acquisition occurs only at the beginning of a relationship. The longer the relationship is, the lower is the amortized cost.
- Account maintenance costs decline as a percentage of total costs (or as a percentage of revenue).
- Long-term customers tend to be less inclined to switch, and also tend to be less price sensitive. This can result in stable unit sales volume and increases in rupee sales volume.
- Long-term customers may initiate free word of mouth promotions and referrals.
- Long-term customers are more likely to purchase ancillary products and high-margin supplemental products.
- Long-term customers tend to be satisfied with their relationship with the company and are less likely to switch to competitors, making market entry or competitors' market share gains difficult.
- Regular customers tend to be less expensive to service because they are familiar with the processes involved, require less "education," and are consistent in their order placement.

Increased customer retention and loyalty makes the employees' jobs easier and more satisfying. In turn, happy employees feed back into higher customer satisfaction in a virtuous circle.

10.12   CUSTOMER LOYALTY

Customer loyalty is customers adhering to a single company or its products. They become satisfied with those products, either by way of the quality of the product, or the esteem they feel by possessing the product. They not only feel loyal to the product, but also tend to purchase other products from the same company depending on their need, rather than purchasing from the competitor. They make positive referrals. For example, if a customer is satisfied with Godrej steel furniture that has served him for decades, he prefers to purchase Godrej refrigerator in preference to other makes, when he next needs a refrigerator. Even for purchasing soaps and shaving cream, he may prefer to buy those of Godrej make only.

This author would like to distinguish between customer satisfaction and customer loyalty by saying "satisfaction is a behavior, while loyalty is an attitude." (This is different from what Evans & Lindsay say, "satisfaction is an attitude and loyalty is a behavior.") The satisfied customers may often purchase competitors' products due to convenience, or promotion, or the discounts, etc., offered. But loyal customers place priority on purchasing goods from a particular company, even at a higher price than the competitors'.

For example, Kodak is considered as the best for rolls of film for 35 mm cameras and generally receives higher ratings. However, its market share is less than that of Konica. This lack of correlation suggests that the satisfaction of Kodak customers does not necessarily influence their survey opinions. On the other hand, loyal customers stick to Kodak films.

In 1987, Le Boeuf cited the results of a study to establish that companies spend six times more to get new customers than to retain old customers. Similarly, customer loyalty is worth ten times that of marketing of a single product or service. The virtuous circle of customer loyalty can be illustrated by Fig. 10.6.

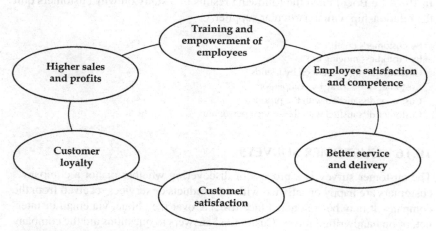

FIG. 10.6   Virtuous circle of customer loyalty.

10.13   FACTORS FOR ESTABLISHING LOYAL CUSTOMERS

The factors that influence the ability to build a strong loyal customer base can be stated as:

1. Products that are highly differentiated from those of the competitors.
2. Higher-end products, where the price is not the primary purchasing factor.
3. Products with a high service component.
4. Multiple products for the same customer.
5. Last but not the least, is the fact that "you cannot create loyal customers without first creating satisfied customers."

10.14   CUSTOMER ATTRITION

Wikipedia introduced another term—Customer attrition, an antonym for customer retention. It can also be described as customer churn, customer turnover, or customer defection, terms used to describe loss of customers.

Banks and mobile phone service companies often use customer attrition analysis and customer attrition rate as one of their key business indices. The customer churn can either be voluntary or involuntary, the former occurring due to a decision by the customer to switch to products of a competitor company, and the latter occuring due to circumstances, such as the customer's death, or relocation to a distant city, etc.

10.15   HOW COMPANIES LOSE THEIR CUSTOMERS

In 1987, Le Boeuf cited the following results of a study on why customers quit the relationship with a particular supplier.

By customer's death	1%
By customers moving out	3%
Customer being influenced by friends	5%
Customer lured away by competitors	9%
Customer dissatisfied with the product	14%
Customer dissatisfied with the service personnel	68%

10.16   CUSTOMER SURVEYS

The customer survey is a process of discovering whether or not a company's customers are happy or satisfied with the products or services received from the company. It may be conducted face-to-face, over the phone, via email or internet, or on handwritten forms. Customers' answers to questions aid the company to analyze whether or not changes need to be made in business operations to increase overall customer satisfaction.

Business directory defines customer survey as, "Customer polling to identify their level of satisfaction with an existing product, and to discover their express and hidden needs and expectations for new or proposed product(s)."

The Qualifications and Credit Framework (*QCF*) suggests the following guidelines for conducting customer surveys.

1. Data collected through survey should not only be rich, but also informative to take appropriate action.
2. Distinguish between clients and customers.
3. Through surveys customer expectations are realized.
4. The answers from a questionnaire will be directly in tune with the way a question is asked. If we want a better answer, the questions asked should be specific.
5. The time for getting responses to a questionnaire is very short, maybe only 10 to 12 minutes.
6. A thoughtfully, well-planned questionnaire will help in easy data analysis and its interpretation. How you ask and what you ask are as important as whom you ask the *questions*.

7. Before collecting the data, plan first how the data is proposed to be analyzed and used, and also decide on the nature of clients who will be surveyed.

10.17   CUSTOMER AND QUALITY SERVICE

What all we discussed in the previous section holds true for the quality of service provided to the customer by the organization. In fact, service quality has a direct and immediate impact on customer satisfaction and hence, is discussed more in detail below.

Service can be defined as any primary or complementary activity that does not directly produce a physical product. It is the nongoods part of the transaction between the customer and the service provider. The service can be as complex as a repair of a car, or as simple as handling a complaint.

When the customer is happy with service and the product, and interacts with enthusiastic, knowledgeable, and committed service personnel who are anxious to help, the company would continue to enjoy the patronage of these customers for a long time, as illustrated in Section 10.12. As per an old saying that if a customer likes the service, he will tell three people. If he does not like it, he will tell eleven people.

10.18   THE KEY ELEMENTS OF SERVICE QUALITY

1. *Reliability*: It is the ability to provide what was promised, as inmaking the repair correctly the first time, following the customers' instructions.
2. *Assurance*: It includes not only the courtesy extended by the employee, but also the knowledge possessed by him, so as to convey trust and confidence among the customers.
3. *Empathy and responsiveness*: The willingness to help customers and the degree of individual attention given to provide prompt service is the basic element of the survive quality. This will make the customer feel valued.
4. *Communication*: The number of contact points shall be minimized and the documents shall be in customer-friendly language.
5. *Front line people*: Frontline people are those service personnel who keep in direct contact with the customers. They should be very customer-friendly, treat the customers with courtesy, and have considerable patience to listen to them.
6. *Organization*: The customer reception area shall be clean and comfortable. The full details of the services provided by the office shall be displayed in the office. It is also a good idea to display customer-oriented slogans in the reception area to attract and give a positive impression to the customers.
7. *Tangibles*: Tangibles are the physical facilities, equipment for service, and also the appropriate appearance of the front line people, etc., which would be witnessed by customers who come to the service shop. Availability of good tangibles with the company will instill confidence in the customers.

10.19   CUSTOMER RETENTION VERSUS EMPLOYEE MORALE

See Fig. 10.7.

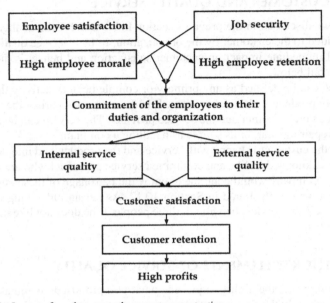

FIG. 10.7   Impact of employee morale on customer retention.

10.20   ACTION TO BE TAKEN TO HANDLE CUSTOMER COMPLAINTS

1. Investigate customer complaints by proactively soliciting feedback, both positive and negative.
2. Develop procedures for complaint resolution that include empowering the front-line personnel.
3. Analyze the complaints, fitting them into clear and definable categories.
4. Work to identify the process and material variations and eliminate the root cause.
5. Remember that subjecting the defects to more inspection is not the right solution.
6. When the survey response or the feedback is received, the senior manager should contact the customer and strive to resolve the issue.
7. Establish customer satisfaction measures and constantly monitor them.
8. Communicate complaint information and the results of the investigation to all related persons in the organization.

9. Provide a monthly complaints log report to the quality council or the process improvement teams for their evaluation.
10. As far as possible, strive to identify the customers' expectations beforehand, rather than as a postmortem.

10.21   HEALTHY PRACTICES BY CUSTOMER FOCUSED ORGANIZATIONS

Evans and Lindsay emphasize the following healthy practices adopted by customer-focused organizations.

1. They clearly define the key customer groups and markets, considering competitors and other potential customers, and then segment their customers accordingly.
2. They understand both near-term and long-term customer needs and expectations, and employ systematic procedures for listening and learning from the customer. This is called the voice of the customer by Dr. Mizuno, who developed quality function deployment, an effective technique to incorporate customer requirements in the design process. More of this is discussed in later chapters.
3. They understand the linkage between the voice of the customer and design, production, and delivery process.
4. They build relationships with customers through commitments that promote trust and confidence, provide easy accessibility to people, and informs them, set effective service standards, train customer contact employees, and effectively follow up on products, services, and transactions.
5. They have effective complaint management systems by which customers can easily comment, complain, and receive prompt resolution of their concerns.
6. They measure customer satisfaction, compare the results in relation to competitors and use the information to evaluate and improve upon the internal processes.

10.22   CUSTOMER CODE OF ETHICS TO BE FOLLOWED

1. Keep your promises made to the customer.
2. Return calls to the customer in a prompt manner.
3. Patiently listen to the customers about their concerns, referring to the appropriate staff member for problem-solving.
4. Treat the customers with respect, courtesy, and professionalism at all times.
5. Give them personal attention.
6. Be aware and evaluate the customer satisfaction regularly.
7. Continuously search for customer-related improvements.
8. Deliver services and products quickly and effectively.
9. Maintain a clean and neat environment and personal appearance.

10. Review and implement customer feedback and suggestions into current procedures when possible.
11. Engage in any training or education that will enhance the job performance and the commitment to the customer.
12. Finally, treat every customer just as you would like to be treated by others.

10.23 RECENTLY HELD INTERNATIONAL QUALITY SYMPOSIA

In October 2013, the second IAQ/ANQ Symposium was held in Bangkok. In his keynote address at this symposium, Prof. Noriaki Kano emphasized that the basic step in increasing the customer satisfaction is to eliminate customer problems. He summarized that sales depended on demand and marketing share. While demand is affected by change of market structure and also the appropriateness of the demand prediction, market share is affected by what are called "business approach ratio" and "successful ratio," which again depended upon the quality, price, delivery, and sales activities.

On Nov. 18, 2014, the second World Quality Day Celebrations were held at Zayed University in Dubai, the theme being "Culture of quality—accelerating growth and performance in the emerging world economy." The keynote address emphasized that "In the future, there will be no such people as quality professional as we know today. Everyone in every function is responsible for the quality of output."

10.24 CONCLUSION

As we have seen in this chapter, in a competitive marketplace, customer satisfaction plays a key role in business strategy. It is therefore essential for companies to effectively manage customer satisfaction by taking reliable and representative measures for this satisfaction.

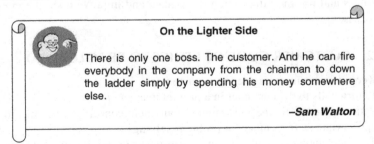

On the Lighter Side

There is only one boss. The customer. And he can fire everybody in the company from the chairman to down the ladder simply by spending his money somewhere else.

–Sam Walton

FURTHER READING

[1] Goodman J, DePalma D, Breetzmann S. Maximizing the value of customer feedback. Qual Prog 1996;29:35–9.
[2] Godfrey B. Beyond satisfaction. Quality Digest; 1996.

[3] Buchanan R, Gilles C. Value managed relationship: the key to customer retention and profitability. Eur Manag J 1990;8(4):523–6.

[4] Finch B. A new way to listen to the customer. Qual Prog 1997;30:73–6.

[5] Afors C, et al. A quick, accurate way to determine customer needs. Qual Prog 2001;34:82–7.

[6] Cochran C. Customer satisfaction, the elusive quarry. Quality Digest; 2001.

[7] Labowitz G. Keeping your Internal Customers satisfied. Wall Street Journal; 1987.

[8] Clarke G, editor. Managing service quality. UK: IFS Publ; 1990.

[9] http://www.businessdictionary.com/definition.

[10] Jeffrey J. Preparing the front line. Qual Prog 1995;28:79–84.

[11] Horowitz J. Putting service quality into gear. Qual Prog 1991;24:54–8.

[12] Harrington J. Looking down at the customer. Quality Digest; 2001.

[13] Rosenberg J. The five myths about customer satisfaction. Qual Prog 1966;29:57–60.

[14] Brecka J. The American Customer Satisfaction Index (ACSI). Qual Prog 1994;27:41–4.

[15] Kiran DR. Customer satisfaction. In: Proceedings of the NIQR/IIPE seminar on TQM; 2006.

[16] Moloney CX. Winning your customer's loyalty: the best tools, techniques and practices. In: Proceedings of AMA workshop, San Diego; 2006.

[17] Duray R, et al. Improving customer satisfaction through mass customization. Qual Prog 1999;32:60–6.

[18] Gardner B. What do customers value? Qual Prog 2001;34:41–8.

[19] Patton SM. Unhappy employees and unhappy customers. Quality Digest; 1999.

[20] Aman S. The essence of TQM—customer satisfaction. J Ind Technol 1994;10:2–4.

[21] Vavra T. Is your satisfactory service creating dissatisfied customers? Qual Prog 1997;30:51–7.

[22] Top providers of telephone customer service. Business line 1995.

[23] Hic VB. Technology is redefining the meaning of customer service. St. Louis Post Dispenser; 1999.

Chapter 11

Total Employee Involvement

Chapter Outline

11.1   WHAT IS TOTAL EMPLOYEE INVOLVEMENT?

Total employee involvement refers to an approach by which all the employees participate in work-related decisions and improvement projects. The principle behind this approach is the fact that *the man on the job is the best person to spot and pinpoint areas for improvement.* This approach comes from the Japanese philosophy of Poka Yoke. As a matter of fact, the term *total* is not used in the same sense as the term *total* in total quality management (TQM). The more commonly accepted term is employee involvement, but in view of the popularity of the term *total* in TQM, total productive maintenance, etc., it is used here also by some authors and has come to stay. In this context, the term *total* is an attribute of *involvement,* rather than of *employees.*

Total employee involvement approaches can range from simple sharing of information or providing input, or suggestions on work-related issues, to self-directed responsibilities, such as setting goals, making business decisions, and solving problems, often as a part of cross-functional teams.

The major components pertaining to total employee involvement are:

- Motivation
- Teamwork
- Participation
- Performance appraisal
- Recognition and rewarding.

11.2   MOTIVATION

Motivation is the basic psychological factor in that makes an employee feel he is part and participle of the company, and getting him to work with his full heart and soul on the job. This has been a subject of interest since man started living as a social animal. Several theories have been developed to understand this psychological factor. The following are some of the theories propounded on motivation.

11.2.1   Theory X and Theory Y

These two theories are based on the general attitude of individual employees with regards to their work.

The former, X theory, propounded by Sigmund Freud, considers the negative aspects of human nature and presumes that a worker, by nature, is sluggish and avoids working. This theory propounds that:

1. The average worker has an inherent dislike of work and tries to avoid it when possible.
2. The average worker lacks ambition, shows no initiative, and accepts no responsibility.
3. The average worker desires job security and economic rewards above all other things.

4. The average worker prefers to be closely directed. To make him work, the management has to resort to rewarding, coercing, and sometimes punishing.
5. The average worker is self-centered and indifferent to overall organizational goal achievement.
6. The average worker is resistant to change.

The latter, Y theory, propounded by McGregor, on the other hand, considers the positive aspects of human nature and presumes that the worker, by nature, wants to learn and develop himself. This theory propounds that:

1. For the average worker, working is a natural and necessary activity, and he subjects himself to self-discipline.
2. For the average worker, rewards and job security are only two of the economic benefits desired. He learns under intrinsic rewarding conditions to seek and accept more job responsibilities.
3. The average worker is committed to goals that enable him to self-actualize.
4. The average worker seeks freedom to work on difficult and more challenging jobs all by himself.
5. The average worker has a lot of initiative and creativity to aid his growth and the accomplishment of the goals.

11.2.2   Theory Z

The Japanese have added one more theory based on employee relations developed by William Ouchi —that workers like to build relationships with other workers and management, to feel secure in their jobs, develop skills through training, and value their family life and traditions.

The existence of these three theories helps the management in realizing the capabilities and limitations of workers and to adjust their leadership approaches as demanded by the situation.

11.2.3   Maslow's Theory of the Hierarchy of Basic Needs

Abraham Maslow propounded that man is motivated mostly by fulfillment of his basic needs, which vary as time passes, and as he achieves higher levels during his career. Once the lower level of needs is satisfied, he is no longer motivated by that need. He wants to go a step higher. As illustrated in Fig. 11.1, as a man starts his life, his basic need would be food, shelter, and clothing for survival. That means he just wants to earn sufficient money to feed himself and his family. Once this is satisfied, his need would then be to have a secure job and life. This level includes a satisfying job and a safe working environment. At the next level, he needs to satisfy his ego or social needs, that is, the need to be part of a society around him. Within the organization, he would like be a part of the group of workers with whom he can have day-to-day free exchange of thoughts. Once this is satisfied, his ego or esteem needs surface. He wants to be recognized in the society as someone important. At the self-actualization level, he wants to be appreciated as a man of achievement. He wants to be given opportunities to go as far as his abilities can take him.

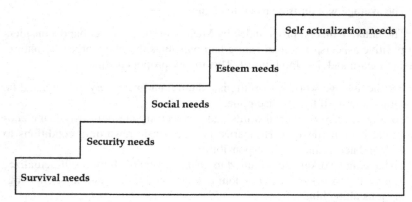

FIG. 11.1   Maslow's hierarchy of needs.

11.2.4   Herzberg's Two Factor Theory

Frederick Herzberg propounded the two factor theory. He classified the motivating factors into two broad classifications as motivators and dissatisfiers.

The need for recognition of achievement, advancement, or accepting responsibility, which are basically intrinsic in nature, are called *motivators,* whereas low salary, an unsecured job, poor working conditions, mediocre management policies, or snoopervision, which are basically extrinsic in nature, are called *dissatisfiers.* The latter are also called hygiene factors, as they can be minimized by effective management, thereby increasing the motivation factors. Absence of motivators does not necessarily dissatisfy the employee, but their presence would provide strong levels of motivation.

11.2.5   Achieving a Motivated Workforce

Besterfield et al. indicated the following eight requirements of management in creating an environment of motivation among their subordinates, as propounded by Theodore Kinni.

1. *Know thyself:* First the managers must understand their own strengths, weaknesses, and motivations and conduct strengths, weaknesses, opportunities, and threats (SWOT) analysis. They should realize that their success largely depends upon their capacity to influence and guide the subordinates in their progress toward the goal.
2. *Know your subordinates*: Managers should have the full knowledge of the interests and emotions of the subordinates. They would thereby assist them

in directing their effort toward the achievement of their goals by utilizing their strengths.

3. *Establish a proper attitude*: Managers must treat their ideas and suggestions in a positive and constructive attitude, and implement them as and when they become appropriate. The feedback, too, shall be positive and constructive.

4. *Share the goals*: A well-established team shall have well-established and clearly understood goals, and know their individual role in the team. All the team members, including the leaders, shall strive equally for achieving the goal.

5. *Monitor progress*: All the activities involved in the efforts of achieving the goal should be well-documented and all periodic (daily or weekly) targets and time schedules shall be indicated and clearly explained to each member. This is like preparing and explaining the road map for achieving the goal, indicating details of the journey milestones and individual assignments. The progress shall be monitored by the leader periodically with proper feedback to the members, as well as the top management.

6. *Develop interesting work*: The management should endeavor to make the job interesting to the employee, instead of being monotonous, repetitive, and creating mental fatigue. As a matter of fact, man by nature is prepared to exert himself for physical fatigue, but abhors being loaded with jobs that cause mental fatigue. Job rotation, job enlargement, and job enrichment are the latest methods for creating job interest. *Job rotation* makes the employee work on different jobs, one after the other. *Job enlargement* ensures that the employees perform a number of jobs sequentially. This is in typical contrast with the job specialization propounded by F.W. Taylor that had been long accepted as the basic scientific management principle. *Job enrichment* adds planning, inspection, scheduling, etc., decision-making elements to the job, so as to increase the sense of responsibility to the employees and satisfy their esteem needs.

7. *Communicate effectively*: The importance of proper communication needs no emphasis. Providing the employees with full knowledge about their work eliminates the unwanted grapevine information. A road map as indicated in Fig. 11.4 shall be planned and explained in detail.

8. *Celebrate success*: Always be free and frank in complimenting the employees for their contribution to the successful achievements. This will motivate them to contribute further.

11.3   EMPLOYEE INVOLVEMENT STRATEGIES

The following are some of the strategies to ensure the involvement of the employees.

Suggestion Schemes: Programs that encourage individual employees to put forth their ideas of improvement in the process, or any operation or work environment.

Survey Feedback: The use of employee surveys as a part of the larger problem-solving process, in which the survey data is used to encourage, structure, and measure the effectiveness of employee participation.

Quality Circles: Groups at the shop floor-level, meeting voluntarily and regularly in a structured environment under the guidance of the supervisor to identify and suggest work related improvements. In view of the significance of this concept in the recent decades, this is discussed further in detail in later chapters.

Job Redesign: This involves the three elements discussed earlier—job rotation, job enlargement, and job enrichment.

Self-Management Teams: Similar to the quality circles, but with a definite assignment like safety incorporation in certain processes. This may also include certain support functions, like hiring of personnel for that assignment.

Employee Participation Group: This involves all levels of employees, convened by the management for a particular task.

11.4 TEAMWORK

A team is a group of interdependent individuals who have complementary skills and are committed to a shared, meaningful purpose and specific goals with a common collaborative approach. They have a clear understanding of their responsibilities and hold themselves mutually accountable for the performance of the team. Effective teams display confidence, enthusiasm, and continuously seek to improve their performance.

11.4.1 The Three Elements of Teamwork

(a) *Respect*: Respect means giving due value to other team members for their professional expertise and their commitment to the purpose of the team.

(b) *Commitment*: It implies sharing of loyalty and devotion to the principles of engineering practice. The engineers must share their ideas freely and frankly with one another, despite the fact that there may be competition among engineers in their profession. This is possible by their commitment to their profession getting the better of their need for individual recognition.

(c) *Connectedness*: This is where the team spirit lies. To be working closely with other team members, one requires the spirit of cooperation and mutual understanding. Each should know the emotional attitudes of the other and be ready to share his personal feelings with them. This element leads to the success of many endeavors.

The teamwork is also called synergy, a term indicating synchronized application of individual energies (Fig. 11.2).

```
Team means

Together,
Everyone
Achieves
More
```

```
Coming together is beginning,
Keeping together is progress, and
Working together is success
```

FIG. 11.2   Two illustrative slogans for Teamwork.

11.4.2   Categories of Teams Based on Natural Work Units

Often corporations are not systematically organized, and work is allocated based on current workload and/or past experience. By this, work appears to move randomly around the organization. On this basis of work allocation, work creates little or no meaning or value to the employees, and they feel little sense of ownership and responsibility. On the other hand, creating natural work groups depending upon what is to be achieved and allocating work on the basis of these groups, make the employees identify more with the work and to take ownership and pride of achievement. This will often lead to improved quality through the sense of responsibility of the individual. Table 11.1 below indicates some of the categories and features of these natural work groups.

TABLE 11.1 Natural Work Units

S.No.	Category	Features
1	Geographical	Assigning work based on a particular location, state, or country branch to which the employee is attached. For example, the Chennai unit, or Andhra Pradesh Unit, or the US branch of an Indian Company.
2	Organizational	Allocating work according to the division or department, like manufacturing or marketing.
3	Product-wise	In cases where a corporation manufactures a variety of goods and has separate production lines or departments for them, this category allocates work based on the product group.
4	Alphabetical	Customer processing work may be assigned according the alphabetical order of the customer's name, for example their initials can be grouped as A–E, F–L, M–R, S–T, and U–Z.

Continued

TABLE 11.1 Natural Work Units—cont'd

S.No.	Category	Features
5	Numerical	Allocating work in supply depots, etc., which involves processing a large number of components numbering over 10,000, then the work allocation can be as per the component number or the bin location.
6	Customer-wise	Allocating work as per the customer size or type, like large corporations, medium industries, small industries, high net-worth individuals, etc.
7	Industry sector	In order to make the optimal use of an individual's experience in specific industry sectors, the work can be allocated based on the company's dealings with industry sectors, such as linesmen, automobile, property, medical, educational institutions, etc.

11.4.3 The Basic Functions of the Team

The basic functions of the team are better understood by the word "SREDDIM."

Select   the job to be studied.
Record   all the facts and factors about the present method of operation in sufficient detail by using easily communicable representations, such as charts and diagrams.
Examine   these factors in the original process, by critical questioning, using techniques best suited for this purposes.
Develop   the most practical, economic, and effective methods with due regard to all the factors.
Define   the new method so that it can always be identified and interpreted correctly by anyone, anytime.
Install   the new method as a standard practice, and
Maintain the standard practice by regular routine checkups.

The same procedure is identical to that specified in Chapter 22 and more or less the same as plan-do-study-act, developed by Shewhart and Deming and discussed in more detail in later chapters.

11.4.4 Characteristics of Successful Teams

1. Sponsor
2. Team charter
3. Team composition
4. Appropriate leadership
5. Balanced participation
6. Cohesiveness

7. Clear cut ideas
8. Training
9. Rules of functioning and conduct of the teams
10. Accountability
11. Well-designed decision processes
12. Resources
13. Open communications
14. Effective problem-solving and
15. Trust, not only the mutual trust among the team members, but also from the top management which sponsors this, as well as by the workforce which implements the decisions.

11.4.5   Some Nicknames for the Nonconducive Team Members

Quite often we come across some team members who are not conducive to the smooth progress of teamwork. Wikipedia has compared their behavior with those of some animals as follows;

1.	The Lion	Keeps overpowering other members and fights whenever others disagree with their ideas.
2.	The Giraffe	Looks down on others and the program in general, with an attitude *"I am above all this childish nonsense."*
3.	The Elephant	Blocks the way and prevents others from continuing in a fruitful discussion.
4.	The Donkey	Very stubborn and will not change his point of view.
5.	The Ostrich	Buries his head, refuses to face reality, and does not admit that the problem exists.
6.	The Monkey	Fools around and distracts others from serious discussions.
7.	The Peacock	Always shows off, competing for attention.
8.	The Rhino	Suddenly keeps changing his point of view, upsetting his supporters unnecessarily.
9.	The Hippo	Sleeps or dozes all thorough the meeting. Never puts his head up except for a yawn.
10.	The Frog	Croaks on and on about the same subject in a monotonous voice.
11.	The Tortoise	Is too slow in his understanding of the discussion points or in putting forward his ideas.
12.	The Mouse	Is too timid to speak on any subject.
13.	The Cat	Always looks for sympathy, *"It is so difficult for me to do this!"*
14.	The Rabbit	Runs away or quickly changes the topic as soon a tension or an unpleasant job is sensed.
15.	The Owl	Looks solemn and pretends to be very wise. Generally talks in long and complicated sentences.

11.5   EMPOWERMENT

Empowerment is an environment in which people have the ability, the confidence, and the commitment to take the responsibility and ownership to improve the process and initiate the necessary steps to satisfy the customer requirements, within well-defined boundaries in order to achieve organizational values and goals. It is a philosophy that bestows on the employees the authority and responsibility for making decisions which affect their jobs. It also signifies the process of employee participation in company working practice and organization performance.

As indicated above, the conditions needed to be satisfied for empowerment are:

- The capacity for the employee to make the right decision,
- the commitment to take responsibility, and
- The confidence and trust bestowed upon him by the management.

Since empowerment largely involves people making decisions, reference may be made to Chapter 6, which elaborates on the process of decision-making.

11.5.1   Types of Supervisors as per Harvard Business School Study

A study of first line supervisors' responses to employee involvement programs by Harvard Business School showed that 72% of the supervisors saw the program as good for the company, 60% saw them as good for the employees, and only 31% saw them as good for themselves. This study also classified the supervisors into five types in this connection:

1. *Proponents of Theory X,* who think the workers would take advantage of the program,
2. *Status seekers*, who do not want to relinquish their authority,
3. *Skeptics*, who question the ability of the organization to change,
4. *Equality seekers*, who think they too, along with the workers, should be included in the programs, and
5. *Deal-makers*, who prefer one-to-one interactions with the employees, through which they can strike deals.

11.6   PARTICIPATIVE MANAGEMENT

11.6.1   Resistance to Change

This is related to the third basic component of total employee involvement as cited in Section 11.1. The man on the job is the best person to spot and pinpoint areas for improvement, and is also the right person to implement this improvement.

Empowering the employee in making the suggestions and implementing after due analysis would make him happier and more responsible and to achieve higher results because the suggestion is his own.

In the day-to-day shop meetings at the factory manager's level, if the workers or their representatives are invited to participate along with the departmental heads, the employees would feel responsible for the job in view of their participation. This is called participative management. On the other hand, if the management decides to implement certain changes or improvements without consulting the workers or their representatives, there would be a certain amount of resistance from the workers, even though those improvements would ultimately benefit the worker. This is called *Resistance to change* and is experienced by a majority of the industrial engineers working in an industry. In a majority of the cases where a resistance is encountered to a change, the reason is more psychological than the worker being against the change itself. Hence, the manner in which the implementation is planned and introduced, with due regard to the ego of the concerned worker, and the manner in which he is taken into confidence in assisting the change, and in the process of implementation, is the very key to the success for any continuous improvement project.

The following paragraphs highlight the significance of this factor and how a manager or the supervisor should strive in overcoming this resistance. A case study from the author's experience in a medium-scale industry during the early 1970s is given at the end of the chapter to illustrate this point.

11.6.2 Types of Changes That Usually Meet Resistance

1. Changes that threaten to lower the prestige of the work of a group or of an individual, like an office clerk being asked to work on the shop floor.
2. Changes that highlight the inefficiency in the present procedures, resisted mostly by the higher-ups.
3. Changes that reduce the authority and scope of decision. If the personnel department is asked to scrutinize or to take control over the constant hiring and firing of the casual labor, the foreman may feel loss of his importance.
4. Changes that interrupt the routine work, such as the re-layout project.
5. Changes of processes where the skilled workers apprehend loss of their jobs or reduction in their earnings.
6. Changes that create fear of loss of employment or reduction in wages, as in unannounced changes in the processes or systems.
7. Changes involving development of skills in new directions.
8. Changes involving transfer to new environments, like moving from a cozy room to large hall, or moving into a new set of workers.

11.6.3   Reasons for Resistance

When viewed from the workers' point of view, they do have their own valid reasons to resist a change. A few commonly cited reasons put forth by workers or their representatives are cited below:

1. New methods may result in retrenchment, and he may be among those affected.
2. He may not be proficient in the new job, resulting in reduction of wages earned.
3. The experience gained in the old job over years would be reduced to a waste.
4. He has to create a new circle of friends under a new supervisor, which may or may not suit him.
5. Psychological attachment to the place, or machine, or the group with which he has worked for years.
6. Fear of de-recognition caused by insufficient information received by him about the purpose of the change, resulting in distorted rumors.
7. Though he may not be against the change personally, he may be bound by the decision of the union or group to which he belongs.
8. Last, but not the least, his ego might have been hurt for not being consulted before the change. This fact could be understood clearly in Fig. 11.3.
9. Of the above, the last cited reason created a revolution in the management thinking, appreciating the need for involving the workforce, not only in the improvement projects, but also in the day-to-day management, suggestions, etc., which concept is generally known as *Employee participation* or sometimes *Workforce Empowerment*.

11.6.4   Some Criticisms Encountered by Industrial Engineers From Higher-Ups

- Why change it? It is still working alright.
- We already thought of it and we know it cannot be done.
- We already tried this and it was a flop.
- We don't have the time.
- Our place is different.
- Good thought, but not practical.
- Not practical for the operating staff.
- Let us get back to realities.
- Don't you realize that the present one was developed after so much research and practices, considering all the factors you are pointing out now?
- Who are you to tell me? I know my job.
- Don't you think we too get ideas?

11.7   EFFECT OF WORKER REPRESENTATION ON PRODUCTIVITY

Fig. 11.3 illustrates the effect of worker representation on productivity.

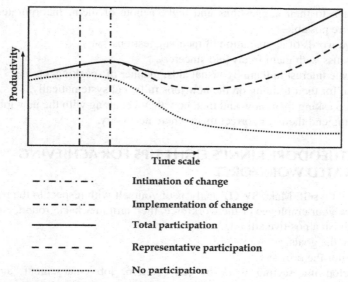

FIG. 11.3   Effect of worker representation on productivity.

11.8   HOW TO SUCCESSFULLY IMPLEMENT A CHANGE

In a majority of the cases where a resistance is encountered to a change, the reason is more psychological than the worker being against the change itself. Hence, the manner in which the implementation is planned and conducted, with due regards to the ego of the concerned worker, and the manner in which he is taken into full confidence in assisting the change, and in the process of implementation, is the very key to the success of any continuous improvement project

One of the seminars on participative management summarizes the following tips on how to successfully implement a change.

1. Understand the change itself, its benefits to the worker, its benefits to the company, etc., and enlist them. You personally need to be convinced of these benefits.
2. Forecast the reasons for the possible resistance from the workers' point of view. Be prepared with your solutions for these factors and other criticisms.
3. Explain to the workers fully and convince them about the benefits of the change for themselves. Allay their fears.
4. Do not forget the channels of authority. First explain to the concerned supervisor or the head of the department.
5. Review the proposal and make a list of the reasons for possible resistance from the worker's point of view and other possible criticisms from different levels and personnel. Be prepared with your solutions.

6. Listen to their suggestions and make a note of them. Incorporate them where possible.
7. Give words of appreciation of their suggestions.
8. Discuss with them freely and sincerely.
9. Create interest in them by visual aids, manner of speaking, etc.
10. Plan for their training on the new jobs fully and systematically.
11. Keep asking them now and then how they get along with the new job.
12. Commend them, or correct them where necessary.

11.9 THEODORE KINNI'S EIGHT TIPS FOR ACHIEVING MOTIVATED WORKFORCE

1. Know thyself. Make SWOT analysis of yourself with respect to the issue.
2. Know your employee or the workforce, their attitudes background, etc.
3. Establish a positive attitude.
4. Share the goals.
5. Monitor the progress.
6. Develop intersecting work by job rotation, job enlargement, and job enrichment.
7. Communicate effectively.
8. Celebrate success.

11.10 BENEFITS OF EMPLOYEE INVOLVEMENT

Besterfield cites the following benefits of employee involvement and participative management:

1. Employees make better decisions using their expert knowledge and skills in the process.
2. Employees are more likely to implement and support decisions if they had a part in making them.
3. Employees are better able to spot and pinpoint areas for improvements.
4. Employees are better able to make immediate corrective action.
5. Employee involvement reduces the labor/management friction by encouraging more effective communication and cooperation.
6. Employee involvement increases the morale by creating a feeling of belonging to the organization.
7. Employees are better able to accept change because they control the work environment.
8. Employees have an increased commitment to unit goals because they are involved.

Peter Grazier puts the following points in favor of total employee involvement.

1. Everyone has something to contribute and will, if the environment is right.
2. The human element performance is more important than the technical element.
3. Most decisions can be significantly improved through collaboration.

4. People need leaders. Good leaders build trust and a higher sense of mission and source of worth.
5. Employee involvement is not a program. It is a corporate philosophy.
6. Continuous improvement is beautifully simple. As improvements begin to flow, confidence will build and the progress will feed on itself.

11.11   ROLE OF SENIOR MANAGEMENT IN EMPLOYEE INVOLVEMENT

In Chapter 4, we have seen several types and qualities of senior management. They must commit themselves to the philosophy of TQM. They must understand the present internal culture, environmental features, and plan all the infrastructure that is required to implement TQM culture in their organization by fully involving the employees in all the decision-making. The following are some of the guidelines that the top management must follow for planning the road map.

FIG. 11.4   Top management role in employee involvement strategy.

1. Conduct an assessment of the organization's existing attitudes, structures, culture, systems, and barriers to the desired implementation of the TQM process, or any other change.
2. Develop a vision statement for the future.

3. Develop a formal management policy in the development of employee involvement strategy.
4. Inculcate a team spirit both in the employees and the management staff.
5. Publish the employee involvement policy, the goals, and targets.
6. Communicate the above to employees at all organizational levels.
7. Implement the strategies that will fit the abilities of the employees and management to adapt them.
8. Provide facilitators, advisors, and other change agents.
9. Train employees in the new methods, problem solving, group skills, and in decision-making.
10. Evaluate the employee involvement strategies to ensure that in the program, the proposed methods are implemented as intended, monitor the progress, and that the method is producing the desired results.

This can be represented by Fig. 11.4.

11.12   RECOGNITION AND REWARDS

This is the fourth component of employee involvement. Recognition is a process by which management acknowledges the good performance of an employee. This is based on the esteem need as indicated by Maslow in his hierarchy of human needs. We have already seen that recognition and satisfying the esteem need of the worker is a major contributor for his motivation. This sustains the employee's interest and commitment in moving towards the common goal. Employee recognition is not just a nice thing to do for people. Employee recognition is a communication tool that reinforces and rewards the most important outcomes people create for your business. The rewards can either be intrinsic like the nonmonetary rewards, and extrinsic like monetary rewards.

11.13   FORMS OF RECOGNITION AND REWARDS

The following can be some of the methods of recognition and rewards.

A. *Intrinsic Rewards*:
- *Verbal appreciation*: Supervisors can give on-the-spot appreciation in front of other employees.
- Certificates or plaques.
- Letters of appreciation from the Chief Executive Officer, Chairman, etc.
- Displaying their names on the notice boards.
- Other nonmonetary rewards, such as inviting them for a get-together or family dinner.
- Group incentives like departmental picnics, or departmental annual holidays. Khatau Mills has the convention of annually sending their senior officers, along with their families to holiday resorts like Goa, Mahabaleshwar, etc.

B. *Extrinsic Rewards*:
- Profit-sharing
- Gain-sharing
- Employment security
  Benefits of recognition and rewarding systems:
- Monetary rewards like cash rewards for the suggestion schemes
- Job-related incentives, such as double increments, promotions, cash bonus, gain-sharing, etc.
- Productivity based incentives.
- Quality based performance bonuses

11.14 CRITERIA FOR EFFECTIVE RECOGNITION OF EMPLOYEES

The website *humanresources.about.com/od/rewardrecognition/a/recognition_tip*.cites the following tips for effective recognition of employees for establishing criteria for what performance or contribution constitutes rewardable behavior or actions.

- All employees must be eligible for the recognition.
- The recognition must supply the employer and employee with specific information about what behaviors or actions are being rewarded and recognized.
- Anyone who then performs at the level or standard stated in the criteria receives the reward.
- The recognition should occur as close to the performance of the actions as possible, so the recognition reinforces behavior the employer wants to encourage.

11.15 ADVANTAGES OF EFFECTIVE REWARDING SYSTEMS

1. It is an effective employee motivator by letting them know that they are valuable members of the company.
2. Better and committed involvement of the employees can be ensured.
3. It also motivates other employees.
4. It creates a healthy competition among individuals and teams.
5. It increases morale in the company.
6. It provides a specific goal to the employees.
7. It provides the organization an opportunity to thank high achieving people.
8. It gives publicity to the company that they value quality and productivity.

11.16 CONCLUSION

As Taoism, the Chinese philosophy says that the real value of employee empowerment lies in not unleashing the power, but in sharing the power. It professes giving employees a certain degree of autonomy and responsibility for decision-making regarding their specific organizational tasks, which are better understood by them than by their superiors.

APPENDIX   A CASE STUDY ON WORKER INVOLVEMENT

The Works Manager of M/s XYZ & Co, a medium-scale industry manufacturing engineering products called the supervisors of all the departments for a meeting. An MBA from Harvard University has a high reputation of being a dynamic leader, having experience in industrial engineering. He opened the meeting with an appreciation for the good performance the company made in the past 2 years and commended the role of the workmen, especially the supervisors, in achieving this high performance level.

He recalled how the company started its operation 10 years back in an old barracks type of building. The office was located in the front room, while each of the seven rooms housed stores and machinery, as and when they are procured. A small extension for this building was made with a semiopen asbestos roofing, housing maintenance, fettling, bench drills, etc., a large backyard is left unused.

Due to the good performance of the company, expansion of activities became inevitable and it was contemplated to extend the building to cover a major portion of the backyard. The existing layout had obvious inefficiencies causing obstructions to the free movement of the men and materials. A comprehensive study was made with specific reference to the movement. The improved layout, the works manager explained, took care of all these factors providing more facilities for the workmen. The basic changes, according to him were insignificant, except movement of certain machinery, removal of certain partition walls, and relocation of the workmen.

All the supervisors stared blankly at the works Manager and showed no signs of happiness. They were apprehensive that their workmen would resist the change.

In the existing system of working, senior workers of the machine shop were closeted in a cozy room, but would have to be shifted to a large hall, together with other semiskilled operatives working on smaller machines. In the assembly shop instead of clustering around a larger work table in quiet conditions and chit-chatting, they would now have to work in a larger hall with individual smaller tables, one behind the other.

All this meant too much for senior workers, since skilled workers are equated with semiskilled workers and are made to move out of the cozy rooms. They decided to approach their union, which in turn could force the management not to undertake the re-layout work.

The above is just one case in point to illustrate how resistance to changes occur. We all know that most of the management techniques to raise productivity aim at methods improvement or systems improvement. But the very word improvement is linked with the action of change and the consequent resistance to change is more psychological than the workers being against the change itself. While in a majority of cases, the resistance is from the operatives, many times the resistance comes from personnel higher up in the ladder, even from the departmental heads.

A sincere attempt to understand the possible motive for such a resistance, as well as the past history responsible for the development of such a resistance, together with complete analysis of the situation, would possibly enable the management to plan a rational course of action for successful implementation of the change.

Quiz: How Would You Have Handled the Situation?

The layout improvement project sited in our case study was planned very meticulously, providing several new facilities to the workers, elimination of crowded work areas, etc. This project was worker-friendly and was expected to be implemented without resistance. But it was not so. If you are the Works Manager, how would you have handled this situation?

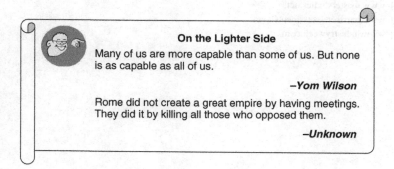

On the Lighter Side

Many of us are more capable than some of us. But none is as capable as all of us.

—Yom Wilson

Rome did not create a great empire by having meetings. They did it by killing all those who opposed them.

—Unknown

FURTHER READING

[1] Besterfield DH, et al. Total quality management. Upper Saddle River, NJ: Prentice Hall; 2005.

[2] James E, Williams L. The management and control of quality. Mason, OH: Thompson's SW; 2005.

[3] Ireson WG, Grant EL. Handbook of industrial engineering and management. Englewood Cliffs, NJ: Prentice Hall; 1971.

[4] Fegenbaun AV. Total quality control. 3rd ed. New York, NY: McGraw Hill; 1983.

[5] Maynard HB. Industrial engineering handbook. 3rd ed. New York: McGraw Hill; 1971.

[6] Snape E, et al. Managing human resources for TQM—possibilities and pitfalls. Employee Relat 1995;17(3):42–51.

[7] Surman D, et al. Integrating TQM and HRM. Employee Relat 1995;17(3):75–86.

[8] Rubinsteen SP. QC circles and U.S. participative management. In: ASQC technical conference transactions, Washington, DC; 1972. p. 391–6.

[9] Scholtels PR, et al. The team handbook—how to use teams to improve quality. Madison, WI: Joint Associates; 1988.

[10] Hackman JR, Oldham GR. Work redesign. Reading, MA: Addison-Wesley; 1980.

[11] Kiran DR. Resistance to change. Excell Superv 1985;3: [NPC].

[12] Kiran DR. Participative management. Manufacturing Technology & Management (of IIPE) 1988.

[13] Drucker P. Management challenges for the 21st century. New York, NY: Harpers Business Publ; 1989.

[14] Juran JM. Juran on leadership for quality, an extensive handbook. New York, NY: The Free Press; 1989.

[15] Joseph P, Furr D. Total quality in managing human resources. Delray Beach, FL: St. Louis Press; 1995.

[16] Kondo Y. Human motivation a key factor for management. Tokyo: 3A Corporation; 1989.

[17] Filbeck G, Preece D. Fortune's best 100 companies to work for in America: do they work for shareholders? J Bus Finan Account 2003;30:771–97.

[18] How companies satisfy employees. Fortune 2003. www.fortune.com.

[19] www.acas.org.uk.

[20] humanresouces.about.com.

[21] www.isynergy.idoscovery.com.

[22] www.crativeevents.ie.

[23] www.timesjobs.com.

[24] www.masterstudies.net.

[25] www.harrisonassessments.com.

[26] www.industryweek.com.

Chapter 12

Supplier Partnership

Chapter Outline

12.1 INTRODUCTION

The quality of product depends on the quality of the individual components that go into it. Because most components are outsourced, it is essential to sustain the quality of the incoming goods. It is not sufficient to subject them to meticulous inward inspection, but to ensure that the supplier maintains high quality standards at his premises and is motivated to be committed towards this goal. This is possible only by mutual trust, commitment, understanding, and communication between the supplier and buyer. The former shall be made accountable, morally at least, for any rejections or difficulties, either during or after assembly of these components. This extended relationship between the supplier and purchaser has been a major business strategy, and has come to be known as the supplier partnership, as explained in the following section.

This partnership is more effective if the supply sources are limited to one or two. Dr. Edward Deming recognized that the cost of production is associated with the poor quality of materials, both from the production shops and the purchased components, and recommended limiting the supply base and forming single-source relationships to reduce the sources of variance in incoming

Total Quality Management: Key Concepts and Case Studies. http://dx.doi.org/10.1016/B978-0-12-811035-5.00012-X
163

product quality. He specifically observed that the supplier and manufacturer must be considered as a macro organization.

The Malcolm Baldrige National Quality Award, which includes supply quality criteria, clearly has increased awareness of the importance of supply management.

12.2 TRADITIONAL VERSUS TQM ORIENTED VENDOR RELATIONS

Juran, in his Paper *Vendor Relations, an Overview*, published by Quality Progress indicated the following contrast between the two concepts of purchasing systems as shown in Table 12.1.

TABLE 12.1 TQM Oriented Vendor Relations

	Traditional	TQM Oriented
1	Natural or semi-processed materials	Purchase of designs, plans, and technical services, in addition to materials
2	Wide tolerances and variable quality standards	High precision and high reliability
3	Rudimentary specifications	Sophisticated designs and quantified design parameters
4	Exclusively designed product with low interchangeability	High interchangeability with usage flexibility
5	Incoming inspection being meticulous	Incoming inspection not undertaken, putting trust in the supplier
6	Limited subcontracting	More *buy* than *make*
7	Geographic proximity of supplier	Geographic proximity only secondary factor
8	Secrecy concept by both parties	Mutual disclosure essential
9	Single line of communication	Multiple lines of communication
10	Vendor supplies goods only	Vendor supplies goods along with proof of compliance

12.3 PARTNERSHIP DEFINITION

In general, any business partnership is that type of business strategy in which two or more individuals pool money, skills, and other resources, and share profit and loss in accordance with terms of the partnership agreement. In the absence of such agreement, a partnership is assumed to exist where the participants in an enterprise agree to proportionately share the associated risks and rewards.

With the development of ancillary industries, especially in the automobile industry, the practice of increasing the proportion of purchased components has increased from about 20% in the 1970s to 60% by the turn of the century. In recent years, a growing number of manufacturing firms have established supplier partnerships and supplier development programs. The partnering concept has received attention for a variety of reasons, most significant of which is TQM.

12.4   STRATEGIC PARTNERSHIP

A strategic partnership is a formal alliance between two commercial enterprises, usually formalized by one or more business contracts, but falls short of forming a legal partnership or, agency, or corporate affiliate relationship. One common strategic partnership involves one company providing engineering, manufacturing, or product development services, partnering with a smaller, entrepreneurial firm or inventor to create a specialized new product. Typically, the larger firm supplies capital, and the necessary product development, marketing, manufacturing, and distribution capabilities, while the smaller firm supplies specialized technical or creative expertise.

Business Dictionary (www.businessdictionary.com) defines customer-supplier partnership as

In quality control, extended relationships between buyers and sellers based on confidence, credibility, and mutual benefit. The buyer, on its part, provides long-term contracts and assurance of only a small number of competing suppliers. In reciprocation, the seller implements customers' suggestions and commits to continuous improvement in quality of product and delivery.

Kaizen Institute (www.kaizen.com/glossary.html) offers the following definition:

It is a long-term relationship between a buyer and supplier characterized by teamwork, mutual confidence, and common goals regarding customer satisfaction. The supplier is considered an extension of the buyer's organization, based on several commitments. The buyer provides long-term contracts and uses fewer suppliers. The supplier implements quality assurance processes to limit or eliminate incoming inspection by the buyer. The supplier also helps the buyer reduce costs and improve product and process designs.

Partnership Sourcing Ltd., a body created by England's government and various industries defines partnership sourcing as:

Here customers and suppliers develop such a close and long-term relationship that the two work together as partners. It isn't philanthropy: the aim is to secure the best possible commercial advantage. The principle is that teamwork is better than combat. If the end customer is to be best served, then the parties must work together – and both must win. Partnership sourcing works because both parties have an interest in each other's success.

The following are the other explanatory definitions offered by several sources as indicated:

The process by which partners adopt a high level of purposeful cooperation to maintain a trading relationship over time. The relationship is bilateral; both parties have the power to shape its nature and future direction over time. Mutual commitment to the future and a balanced power relationship are essential to the process. While collaborative relationships are not devoid of conflict, they include mechanisms for managing conflict built into the relationship.

Monczka, Trent, and Handfield

A bilateral effort by both the buying and supplying organizations to jointly improve the supplier's performance and/or capabilities in one or more of the following areas: cost, quality, delivery, time-to-market, technology, environmental responsibility, and managerial capability, and financial viability.

Krause and Handfield

12.5   PRINCIPLES OF CUSTOMER/SUPPLIER RELATIONS

Kaoru Ishikawa in his book, *What is Total Quality Control?*, has suggested ten principles to ensure quality products and services, and to eliminate any issues due to mistaken interpretations or otherwise, between the supplier and customer.

1. Both the customer and the supplier are fully responsible for the control of quality.
2. Both the customer and supplier should be independent of each other and respect each other's independence.
3. The customer is responsible for providing the supplier with clear and sufficient requirements so that the supplier knows precisely what, when, and how much to produce.
4. Both the customer and the supplier should enter into a non-adversarial contract with respect to quality, quantity, price, delivery methods, and terms of payment.
5. The supplier is responsible for providing the quality that will satisfy the customer and submit the necessary data upon the customer's request.
6. Both the customer and the supplier should decide the method to evaluate the quality of the product or service to the satisfaction of other the parties.
7. Both the customer and the supplier should establish in the contract the method by which they can reach an amicable settlement of any disputes that may arise.
8. Both the supplier and customer should continuously exchange information, sometimes using multifunctional teams, in order to improve the product or service quality.
9. Both the supplier and customer should perform business activities such as procurement, production and inventory planning, clerical work and systems so that an amicable and satisfactory relationship can be established.
10. When dealing with business transactions, both the parties should always have the best interest of the end-user in mind.

12.6 THE THREE PRIMARY AND NECESSARY REQUIREMENTS FOR PARTNERING

The three key elements to a successful partnership are long-term commitment, mutual trust, and shared vision.

(a) *Long-term commitment*: Benefits of partnership cannot be achieved in a short time. Only when the supplier is sure that the partnership will continue for a long time, then he will be motivated to undergo all the responsibilities imposed on him, apart from just producing to the specifications. The long-term commitment provides the needed environment for both parties to work toward continuous improvements. Every concerned person from the CEO to the worker from both sides, should feel fully involved. Each party has its own strengths and weaknesses and they should pool them together. If any new equipment or more labor force is required to undertake a new order, the supplier may not be able to take this risk unless a long time commitment is assured. In addition, there should not be any hesitation from either parties to share or integrate resources like training activities, administrative systems, and equipment. A striking example for this is the training project funded by M/s Ashok Leyland for the cluster training to their ancillary units by National Institution for Quality and Reliability (NIQR).

(b) *Mutual trust*: Mutual trust is required between the two partnering parties. They should share technical information and accept reduced cross monitoring and control. This mutual trust forms the basis for a strong working relationship. Besterfield et al. emphasized the fact that the purchasing function of the organization should be treated only as a subsidiary function of the overall relationship goals and objectives.

(c) *Shared vision*: Each party must understand the other's business and business practices. They must have an open and candid exchange of the needs and expectations with the requirements of the end-user in mind. These shared goals and objectives must ensure a common direction and align with each party's vision and mission. This element must percolate thorough all the employees on both sides. They should have no hesitation in sharing their business planning to aid a mutual strategic planning.

12.7 MULTIPLE SUPPLIER PARTNERSHIP

Despite Deming's recommendation to have a single supplier source, companies today prefer to have partnership deals with at least two suppliers for each component, more as a factor of safety. Nevertheless, both suppliers shall be treated in unison, encouraging mutual trust and understanding between them, and making them perform not in competition, but to supplement each other's efforts.

12.8 ADVANTAGES OF SUPPLIER PARTNERSHIP

1. Achieves the highest standard of safety and health in all the activities in which the purchasing organization is engaged;
2. Identifies and works with best practice suppliers;
3. Ensures that suppliers understand the business objectives of the purchaser and vice versa;
4. Ensures that the structure of relationship best suited to the achievement of those objectives are established;
5. Encourages efficiency, continuous improvement, and innovation among the suppliers;
6. Gives and receives open and honest feedback on performance to and from the suppliers.

12.9 SUPPLIER SELECTION

Before outsourcing subassembly or assembly, it must be decided whether nor not to outsource them. This is traditionally called the Make or Buy Decision (MBD). While the factors considered in the traditional MBD were the costs and the technical conformance, in today's TQM scenario the reliability of the supplier with respect to the quality and delivery is the most important criterion. Kaoru Ishikawa also suggested the following 12 criteria in the process of selecting the right supplier before entering into partnership deals.

1. The supplier shall understand and appreciate the management philosophy of the buyer organization.
2. The supplier shall have a stable management system. In determining this condition, several questions should be asked:
   - Is there a quality policy statement that includes objectives for quality and its commitment to quality?
   - Is the policy implemented and understood at all levels of the organization? Is there documentation that indicates who is in charge and responsible for quality in the organization?
   - Is there a member of top management with the authority to execute a quality system?
   - Does the management have scheduled reviews of its quality system to determine its effectiveness?
3. The supplier shall maintain high technical standards and have the capability of dealing with future technological innovations.
4. The supplier shall provide those raw materials and parts required by the purchaser, and those supplied shall meet the quality specifications.
5. The supplier shall have the capability to produce the amount of production needed and can attain that capability.
6. There shall be less probability of the supplier breaching corporate secrets.

7. The price shall be right and the delivery dates could be met.
8. The supplier shall be easily accessible in terms of transportation and communication.
9. There must be a system to trace the product or lot from receipt and all changes of production delivery.
10. The supplier shall be sincere in implementing the contract provisions. Check the following:
    • Does the supplier have a system for contract review, and does that system include a contract review of requirements?
    • How should differences between the contract and/or accepted order requirements be resolved?
    • Does the system allow the inclusion of amendments?
    • Does the system include maintaining records of reviewed contracts?
11. The supplier shall have an effective quality system and improvement program such as ISO/QS 9000.
12. The supplier shall have a track record of customer satisfaction and organization credibility.

12.10 VENDOR RATING

A supplier rating (or vendor rating, as it sometimes referred) is a business term used to describe the process of measuring an organization's supplier capabilities and performance. A major part of the effort of the management of manufacturers or service providers is to ensure that the desired characteristics of a purchased product is built in the system so that it does not depend upon after-the-event indicator. Thus, the vendors or suppliers are rated according to their attainment of some level of performance. Vendor rating is the result of a formal vendor evaluation system. Vendors or suppliers are given standing, status, or title, according to their attainment of some level of performance, such as delivery, lead time, quality, price, or some combination of variables.

Supplier rating often forms part of an organization's supplier relationship management program. Such systems can vary in the criteria that are assessed, which broadly fall into quantitative and qualitative types, and can be used for vendor rating. Yet, the criteria or the process may vary from one organization to another.

12.11 CRITERIA FOR EVALUATION

Most commonly, the vendor rating is a result of formal and objective rating system of chosen parameters. Some of the factors that determine the rating are:

1. Quality factors
2. Pricing factors
3. Delivery factors
4. Service factors

Quality factors include:

- *Compliance with purchase order* and showing an understanding of the customer's expectations.
- *Conformity to specifications*.
- *State-of-the-art product/service*. The vendor should consistently refresh product life by adding enhancements.
- *Reliability* of both the product and after sales services.
- *Durability* of product that is the time for replacement or mean time between failures being as high as possible.
- Provision of effective warranty.

Pricing factors include:

- Competitive pricing that is comparable to those of vendors providing similar product and services.
- Price stability over time.
- Price accuracy with a low number of variances.
- Advance notice of price changes if the price has to be increased.
- Sensitivity to purchaser's costs and showing an understanding of their needs. The vendor should exhibit knowledge of the market and share this insight with the buying firm and suggest possible cost savings.
- Billing shall be accurate, clear, and timely. Final invoice should not vary significantly from the estimates.

Delivery factors include:

- Time, not only for delivery, but in answering requests for information, proposals, and quotes.
- Quantity, delivering the correct items or services as per the contracted quantity.
- Lead time.
- Packaging which should be sturdy, suitable, damage-proof, and properly marked.
- Extra effort demonstrated to meet requirements when an emergency delivery is requested.

Service factors include:

- Prompt information with up-to-date catalogs, price information, and technical information.
- Knowledge of buying firm's needs and being helpful with customer inquiries, regarding order confirmation, shipping schedules, shipping discrepancies, and invoice errors.
- Technical support, such as providing training on the effective use of its products or services.

- Warranty support and *warranty* protection offered and problems are resolved in a timely manner.
- Problem resolution response is timely.

Wikipedia has given the following factors for the objective vendor rating.

1. Quality—for example, number of incorrect first-time deliveries
2. Delivery schedule adherence
3. Cost/Price
4. Capability
5. Service

Results of each variable are then weighted into a final score—usually a percentage, allowing suppliers to be ranked. Fig. 12.1 illustrates a model to indicate several factors to be considered in supplier rating and performance evaluation.

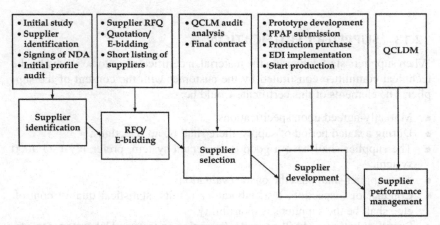

FIG. 12.1   Supplier performance model. *NDA*, non-disclosure agreement; *RFQ*, request for quotation; *QCLM*, quality cost logistics management; *PPAP*, production part approval process; *EDI*, electronic data interchange; *QCLDM*, quality cost logistics delivery management.

12.12   THE PARTNERSHIP INDICES

The effectiveness of a supplier partnership system can be quantified by the following four types of distinctive indices, which measure the extent to which both the customer requirements and the supplier capabilities match or mismatch. The bidding party or parties with the most promising indices are considered further for awarding the contract.

- *Satisfaction Index (SI)* is the measure of the extent to which a customer requirement is satisfied by a supplier capability. The larger the value of SI, the greater the potential that this pair of customer and supplier become partners. SI is based on the overlap between the customer requirements and supplier capabilities.

- *Flexibility Index (FI)* is the measure of the extent to which a supplier capability exceeds a customer requirement. The larger the value of FI is, the more flexible the supplier to satisfy the changing customer requirement. FI is based on the surplus of the supplier capabilities.
- *Risk Index (RI)* is the measure of the extent to which a supplier capability fails to meet a customer requirement. The larger the value of RI is, the more risky the partnership between them. RI is evaluated through the shortage of the supplier capabilities.
- *Confidence Index (CI)* is the measure of trustworthiness of the supplier meeting the customer requirements over a period of specified time. The higher the value of CI for a longer time, the more reliable the supplier is. Longer-term partnership may be considered, instead of using short-term competitive tendering. CI is evaluated through historical records of supplier performance as measured in selected indicators or inquiries.

12.13   SUPPLIER CERTIFICATION

When suppliers start delivering the material, a certificate has to be issued by the technical committee constituted by the customer with the consent of the supplier. The elements of this certificate would be:

- Mutually-agreed upon specifications.
- During a stated period of supply, there shall be no rejections.
- The supplier shall have a good documented system, preferably ISO 9000 system.
- The supplier shall pass the onsite evaluation.
- Conduct of inspection, test, laboratory results, statistical quality control, etc., shall be the supplier's responsibility.
- Supplier shall provide all needed information on tests and laboratory reports.

12.14   BENEFITS OF SUPPLIER RATING

1. Helps minimize subjectivity in judgment and makes it possible to consider all relevant criteria in assessing suppliers.
2. Provides feedback from all areas in one package.
3. Facilitates better communication with vendors.
4. Provides overall control of the vendor base.
5. Requires specific action to correct identified performance weaknesses.
6. Establishes continuous review standards for vendors, thus ensuring continuous improvement of vendor performance.
7. Builds vendor partnerships, especially with suppliers having strategic links.
8. Develops a performance-based culture.
9. Ensures complete and well-interpreted communication with suppliers.

10. Provides the supplier with a detailed and factual record of problems for corrective action.

11. Enhances the relationship between the customer and supplier.

12.15   LEAN INSPECTION THROUGH SUPPLIER PARTNERSHIP

Incoming inspection is a non-value adding activity. A properly planned and executed supplier partnership moves away from the traditional sample inspection to a system where the supplier bears the responsibility of inspecting in his premises. Here the materials move directly from the supplier to the final assembly section of the purchaser. Section 12.18 illustrates how this has reduced the inventories of the purchasing company.

The author was initially surprised to see that during his few visits to one of the TVS group of companies in Chennai during the early 1990s, the supplier's goods would go directly to the shop floor without undergoing inward inspection. The reason lies in the partnership deal entered into and the resulting confidence in the quality of the goods.

12.16   VENDOR MANAGED INVENTORY

Vendor Managed Inventory or supplier-managed inventory is a system in which the purchaser of manufactured components, usually the larger company, places a blanket order on the vendor, but specifies the delivery schedule mostly on a daily basis, so that the purchaser does not carry inventory not more than 1 day. The responsibility of maintaining the regularity of the supplies lies with the vendor. This can be achieved through information sharing and removing many of the adverse factors. Mark Bowles calls this RSP (Retailer Supplier Partnership), with specific reference to a supermarket scenario, while Lee calls this VMI (vendor managed inventory). The financial responsibility associated with the inventory is also often retained by the supplier or vendor, thus reducing the retailer's risk. The net effect on the supply chain partners is reduced cost and improved service level.

As explained above, this system requires a significant confidence bonding between them so that the supplier would not fail in his daily supplies even for a single day. While this was made popular by American supermarket chains, such as Walmart and P & G in the late 1980s, it soon caught up in India and today, all the major Indian industries like that of the TVS group follow this system.

12.17   RETAILER SUPPLIER PARTNERSHIP

When such partnership occurs in retail marketing business like that of supermarkets, it is called Retailer Supplier Partnership (RSP). The level of partnership and the scope of cooperation may vary under different agreements, but the principles, in general, are the same as detailed above for the industrial situation between the component supplier and the final product manufacturer. The supermarkets sometimes assess the suppliers and award them periodically, or annually based on their performance. This is called Supplier Performance Awards by Retail Category (SPARC). Some of the other awards are:

- the Specialty Retailer of the Year award added in the mid-1990s, by Walmart,
- the Merchandising Innovation award by 3M,
- the Food Retailer of the Year award,
- the Supplier Continuous Quality Improvement Award (SCQI) by Intel,
- the Gold Pentastar Award and the Platinum Pentastar Award, both by DaimlerChrysler.

This author is not aware of any such award in India, proposed by, and if not, he sincerely wishes that leading Indian industries promote such awards to encourage the quality culture in the supply chain management.

12.18   IMPACT OF SUPPLIER PARTNERSHIP ON INVENTORY NORMS

- During the 1960s, the norm for average production inventory (excluding finished products) in the engineering industry was 25% of the total annual turnover.
- During the early 1970s, this author undertook a comprehensive material control project that includes Codification, ABC Analysis, and Inventory parameters by which project, the total inventory level fell down to 15% of the turnover, which was said to be a big achievement.
- By the 1990s, due to additional and more stringent controls, this norm was lowered to 5%, especially in the TVS group companies of Chennai.
- The zero inventory drive started around the turn of the century and companies like Brakes India devised systems so that the receiving inspection is conducted at the vendor's place under the supervision of the buyer's staff, and on the specified date, the goods equal to 1 or 2 days requirement go directly to the shop floor for assembly, eliminating the internal store inventories.
- The vendor place inspection, a major element of supplier partnership, enables achievement of the zero inventory principle with an inventory level of less than 1%.

12.19   CONCLUSION

Buyer-Vendor partnership has now come to stay as one of the most significant features of total quality management. During the 1980s, this author was surprised to find in a TVS group company, that the bought-out components from a vendor going directly to the assembly shop were bypassing the inward inspection. This was possible due to the established mutual trust and accountability as discussed earlier in this chapter.

On the Lighter Side

During the sixties, pumpkins were lot cheaper in Madras (now Chennai) than in Bangalore. Once, when we were visiting our uncle in Bangalore, he wanted us to bring a pumpkin. As it was too heavy to carry in the train, we went straight from Bangalore Station to City Market, bought a pumpkin and gave it to him. He was happy that we brought him something from Madras and we were happy to have a lighter journey. This is the win-win situation of Supply Management.

FURTHER READING

[1] Measure for Measure. Supply Manag 2001;39.

[2] Muralidharan C, Anantharaman N, Deshmukh SG. Vendor rating in purchasing scenario: a confidence interval approach. Int J Oper Prod Manag 2001;21(9/10):1305–25.

[3] Trent RJ, Monczka RM. Purchasing and supply management: trends and changes throughout the 1990s. Int J Purch Mater Manag 1998;34:2–11. fall.

[4] Walton SV, Handfield RB, Melnyk SA. The green supply: integrating suppliers into environmental management processes. Int J Purch Mater Manag 1998;34:2–11. Spring.

[5] Lee HL. Creating value through supply chain integration. Supply Chain Manag Rev 2000;4:30–6.

[6] Waller M, Johnson ME, Davis T. Vendor-managed inventory in the retail supply chain. J Bus Logist 1999;20(1):183–203.

[7] Roylance D. Purchasing performance: measuring, marketing, and selling the purchasing function. Aldershot: Gower; 2006.

[8] Pyzdek T. What every engineer should know about quality control. Boca Raton, FL: CRC Press; 1990.

[9] Lock D. Gower handbook of management. Aldershot: Gower Publishing Company; 1998, p. 726.

[10] www.marcbowles.com.

[11] www.businessdictionary.com.

[12] www.kaizen.com/glossary.htm.

[13] www.supplierratingsystemssupplychainmechanic.com.

[14] Web site of Supply Chain Resource Cooperative (SCRC).

[15] www.historeum.org/wiki/Supplier_rating.

Chapter 13

Total Productive Maintenance

Chapter Outline

13.1 INTRODUCTION

During the early days of the industrial working scenario, the routine maintenance of machines was carried out by the operator himself, while major repairs were done by the supervisor, or an external specialist. As we have seen in Chapter 5, the development of Taylors' principle of specialization resulted in the creation of a separate specially trained maintenance team to do all maintenance work, including lubrication by the patrol maintenance gang, together with all minor repairs, such as screw tightening. The industry did benefit a lot by such specialization in light of the centralized maintenance planning and control, the benefits of which we had seen in the previous chapters.

However, it turned out that whenever small machine hold-ups need minor works, such as lubrication or screw tightening, the operator simply waits for the maintenance man to come and do that job, thereby losing production time. Hence, the logical outcome is the idea that the operator himself can trained to do such minor adjustments. This thinking lead to the development of the TPM philosophy, which has become a part and parcel of Total Quality Management (TQM) philosophy.

Total Quality Management: Key Concepts and Case Studies. http://dx.doi.org/10.1016/B978-0-12-811035-5.00013-1
177

In other words, TPM philosophy emphasizes auto-maintenance, meaning every worker is responsible for the proper working of his machine, apart from being responsible for the quality. The production operators share the preventive maintenance efforts, assist the mechanics with repairs when the equipment is down, and together they work on equipment and process improvements on a team activity basis. TPM aims to use all equipment at its maximum effectiveness by eliminating waste and loss incurred by failure of the equipment, increased set-up time, reduced speeds, and processed effects, etc., which finally lead to reduced output.

The objective of TPM is to maintain the plant or equipment in good condition without interfering in the daily process. To achieve this objective, preventive and predictive maintenance is required. By following the philosophy of TPM, we can minimize the unexpected failure of the equipment.

To distinguish TPM from the quality circles, we can say that quality circles are formed by the workmen of a particular activity location, while TPM is formed by senior managers, supervisors, and workmen who carry out similar exercises on a company-wide basis.

13.2   THE MEANING OF TPM

T represents *Total* employee involvement, indicating the teamwork with well-coordinated work between the production and maintenance workers. This term "Total" also means total equipment effectiveness.

P represents *Productive*, indicating the production of goods and services that meet the customers' expectation by maintaining the reliability of the outputs, which would only be possible if the reliability of the respective machines is kept at a high level.

M represents *Maintenance*, keeping the equipment and plant in good working condition at all times.

13.3   EVOLUTION OF TPM

TPM is an evolving process, starting from a Japanese idea that can be traced back to 1951, when preventive maintenance was introduced into Japan from the United States (Deming). Nippondenso, an ancillary unit of Toyota, was the first company in Japan to introduce plant-wide preventive maintenance in 1960. While, as stated above, the main principle behind TPM, that is, the production operator taking care of routine maintenance of his machine is traced back to the prewar industrial scenario, the development of present TPM philosophy is traced back to the early 1950s, when the Toyota group of companies in Japan made headway introducing this practice. Because of the high level of automation, the equipment maintenance became complex, needing more specialized maintenance personnel that the management lead by Seiichi Nakajima, aptly called the father of TPM, and decided to pass on the routine maintenance jobs to the production operator, focusing the maintenance department's attention more towards the major maintenance jobs. This practice refined the concept of TPM.

The evolution of TPM is summarized as:

- Pre-Industrial Revolution — Operator is responsible for operation and maintenance
- Pre-1950s — Maintenance department's gang for cleaning and lubricating all machines, plus specialized preventive maintenance, etc
- 1960s — Productive maintenance
- 1970s — TPM
- 1980s — Predicative maintenance by operatives and foremen
- 1990s — Maintenance preventive design— Design out maintenance
- 2000s — TPM— the concept of the factory of the future

13.4   DEFINITIONS OF TPM

- TPM is a proactive approach that essentially aims to identify issues as soon as possible, and plan to prevent any issues before occurrence. One motto is "zero error, zero work-related accident, and zero loss."
- Seiichi Nakajima, the Japanese originator of the TPM concept, had defined it as a process to continuously improve all operational conditions within a production system, by stimulating the daily awareness.
- Business Dictionary defines TPM as a methodology designed to ensure that every machine in a production process always performs its required task and its output rate is never disrupted.
- QCFI gives an explanatory definition to TPM as a manufacturing philosophy that pursues production efficiency to its ultimate limits of comprehensive efficiency by,
  - Putting together a practical shop floor system to prevent losses before they occur throughout the entire production system's life cycle,
  - Involving all functions like production, development, sales, and management,
  - Having employee participation from top executives to front line workers,
  - Achieving zero losses through overlapping small groups.

There are several other definitions of TPM and most of them imply the following:

1. TPM aims to maximize the equipment effectiveness.
2. TPM establishes a thorough system of preventive maintenance for the equipments' entire life span.
3. TPM is implemented by various departments, such as line operations, maintenance, engineering, etc.
4. TPM involves every single employee from the top management to the shop floor workers.
5. TPM is based on the promotion of preventive maintenance through motivation.

13.5   TPM IS AN EXTENSION OF TQM

In manufacturing and service industries, improved quality of products and services depend very much on the features and conditions of the company's equipment and facilities. TPM is a maintenance process developed for improving productivity by making processes more reliable and less wasteful. This is where TPM plays a major supplementary role for TQM.

TPM focuses primarily on manufacturing and is the first methodology Toyota used to improve its global position during the 1950s. After integrating TPM with it, the focus was stretched to include various other aspects of TQM, like Supply Chain involving suppliers and customers. The next methodology was called lean manufacturing, where TPM plays a critical role. If machine uptime is not predictable and if process capability is not sustained, the process must keep extra stocks to buffer against this uncertainty and flow through the process will be interrupted. This would deteriorate the product quality, and also delay the delivery to the customer. In this context, TPM contributes a lot to TQM and hence, we can say TPM is an extension of TQM.

While we have stated above that TPM is an extension of TQM, the following distinguishing features of TQM and TPM would enable us to understand the statement better.

1. TPM focuses on the reliability of the equipment, whereas TQM focuses on the reliability and quality of the products and services, which, however, cannot be achieved without achieving TPM's focus on equipment.
2. TPM focuses on improvement of the performance of the equipment, that is, the hardware, while TQM focuses on the improvement of systems and standards that is the software.
3. In TPM, personnel training centers on maintenance technology and skills specific to maintenance, whereas TQM centers on the personnel training in management technologies, such as SC, Kaizan, quality circles, etc.
4. Like quality circles popular with TQM, TPM also advises weekly meetings among small groups of production and maintenance operators, which some companies prefer to call "productivity circles."

13.6   TPM STARTS WITH CLEANING

Fig. 13.1 illustrates how the simple job of cleaning transforms into high quality standards of a company.

Operation and maintenance are like the wheels on the both sides of a car. If one is turned, the other turns, too.

TPM starts with cleaning

- *Cleaning* gives the opportunity for *inspection* of the machine
- *Inspection* reveals *abnormalities*
- *Abnormality* identification enables *rectification* steps
- *Rectification* steps result in *improvement*
- *Improvement* brings *positive results*
- *Positive results* increase *high quality standards*
- High quality standards give *PRIDE* to the work environment.

FIG. 13.1   Metamorphosis of cleaning to high quality standards.

Another allusion is how we take daily care of our body by brushing, cleaning ourselves. We go to the doctor only in case of major illnesses. Like the machine cleaning, lubrication, etc. should be done by the operator himself. Only the major preventive or breakdown maintenance should be assigned to specialists, or the maintenance department.

13.7   THE SEVEN TYPES OF ABNORMALITIES

We can categorize the abnormalities as follows:

1. *Minor Flaws*
(a) Contamination — dust, dirt, oil, grease, and rust
(b) Damages — cracking, crushing, chipping, bending, and deformation
(c) Play — shaking, falling out, run out, eccentricity, and wear
(d) Slackness — in belts and chains
(e) Abnormal phenomena — unusual noise, overheating, vibration smell, and discoloration
(f) Adhesion — hardening, obstructing, accumulation of debris, and malfunction

2. *Unfulfilled basic condition such as*
(a) Poor lubrication — insufficient, dirty, and leaking lubricant
(b) Lubricant supply — dirty, unsuitable deformed oil inlets, faulty lubricant channels
(c) Tightening — nuts and bolts being loose, cross-threaded, and unsuitable washers

3. *Inaccessible places*
(a) Cleaning — covers, plant and machine layout, lack of design out maintenance, and available space
(b) Checking — instrument position and orientation
(c) Lubricating — position of lubricant inlets, height, levers
(d) Tightening — covers, layout, and leverages
(e) Operation — like machine layout, valve position, switches and levers
(f) Adjustment — the position of the pressure and other gages, adjusting instructions

4. *Contamination sources*
(a) Product — leaks spills and scatter
(b) Raw materials — leaks scatter, overflow
(c) Lubricant — leaking, wet, and contaminated with fuels and other fluids
(d) Gases — leaking compressed air, vapors, and exhaust fumes
(e) Liquids — leakages, half-finished products, and wastewater
(f) Scrap — flashes, packaging, and nonconforming products

5. *Defective sources*
(a) Foreign matter — inclusions, insects, and rust
(b) Shock — jolting, collision, and vibration
(c) Moisture — infiltration and wetness
(d) Grain size — punctured mesh-screen, separators, unwanted grain sizes

| (e) Concentration | inadequate warming, mixing, and evaporation |
| (f) Viscosity | on-homogeneous mixture, evaporation, and presence of water droplets |

6. *Unnecessary and nonperforming items*

(a) Machinery	pumps, fans, and tanks
(b) Piping equipment	unwanted pipe connections, dampers, and valves
(c) Measuring equipment	pyrometers and gages that are not referred to
(d) Electrical equipment	wiring, switches plugs, and power leads
(e) Jigs and tools	general fixtures, cutting tools, molds, and frames
(f) Spare parts	standby equipment, and auxiliary materials
(g) Makeshift	tapes, wire, and metal plates repair items

7. *Unsafe places*

(a) Floors	uneven, slippery, projections, and cracking floors
(b) Steps	irregular, too steep, and missing handrails
(c) Lights	dim, dirty, fused, and with broken covers
(d) Rotating machinery	displaced covers, unguarded, and without emergency stops
(e) Lifting gear	cables, hooks, and other parts of cranes
(f) Others	special substances, toxic gases, and lack of protective clothing

13.8   THE EIGHT PILLARS OF TPM

The eight pillars of TPM as emphasized by the Japanese are:

1. *Focused improvement* (Kobetsu Kaizen)—Continuous improvement, even though small steps.
2. *Planned Maintenance*—It focuses on increasing availability of equipment and reducing breakdown of machines.
3. *Initial Control*—To establish the system to launch the production of a new product and new equipment in a minimum run-up time.
4. *Education and Training*—Formation of autonomous workers who have skill and techniques for autonomous maintenance.
5. *Autonomous Maintenance* (Jishu Hozen)—meaning "Maintaining one's equipment by oneself."
6. *Quality Maintenance* (Hinshitsu Hozen)—Quality Maintenance is establishment of machine conditions that will not allow the occurrence of defects, and control of such conditions is required to sustain Zero Defect.
7. *Office TPM*—To make an efficient working office that eliminates losses.
8. *Safety, Hygiene, and Environment (SHE)*—The main role of SHE is to create a safe and healthy workplace where accidents do not occur, uncover and improve hazardous areas, and do activities that preserve the environment.

Fig. 13.2 illustrates the pillars that support the structure of TPM, each pillar representing each of the above concepts.

13.9 THE FIVE ZEROS OF TPM

In any industrial situation, the major effort of the management is to minimize the factors that increase production hold ups and cause losses to the profitability of the company. To create the awareness and commitment among all employees, companies successful in TQM practices publicize the following through posters, etc., so as to create a mindset among employees.

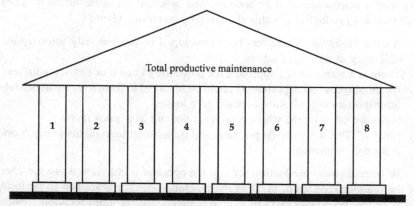

FIG. 13.2   TPM structure.

- Zero breakdowns
- Zero defects
- Zero accidents
- Zero pollution
- Zero inventory

Actually, the last one, viz. zero inventory, is applied in general to the production situation wherein the inventory plays a major cost-adding factor, more than for the maintenance function, whose inventory requirements are different. Normally TPM books cite four zeroes. Nevertheless, since some of the syllabi specify five zero's for Maintenance Management, this concept is explained as above, together with the last one.

13.10 WHY OPERATIVES FAIL TO ADAPT TPM AS A WAY OF LIFE?

- They do not know what and where regular checks are to be made on the equipment
- When equipment is to be lubricated and the lubricating points are not relayed to him

- When an abnormality is routinely found and the operative is not trained to address it
- The operative himself is not very conscious of the losses created by the machine downtime

13.11 WHAT CAN TPM ACHIEVE?

Wikipedia emphasizes that an accurate and practical implementation of TPM will increase productivity within the total organization, where:

1. A clear business culture has been developed to continuously improve the efficiency of the organization.
2. There is a standardized approach for preventing known or unknown losses.
3. All employees and departments could be involved to form a transparent multidisciplinary organization to reach zero losses.
4. Steps are taken on the whole road map, and not as a quick menu.
5. Finally, TPM will provide practical and transparent ingredients to reach operational excellence.

Whereas in most production settings the operator is not viewed as a member of the maintenance team, in TPM the machine operator is trained to perform many of the day-to-day tasks of simple maintenance and fault-location. Teams are created that include a technical expert (often an engineer or maintenance technician) as well as operators. In this setting, the operator is enabled to understand the machinery and identify potential problems, righting them before they can impact production and by so doing, decrease downtime, and reduce costs of production. He attends to the daily cleaning of the machine bed and slide ways, and lubricates the lubrication points, including the work surfaces. Where an oil well is provided, he checks the oil level, starts the machine and then checks the oil flow, etc., before commencing the production work. Any defect is reported to the supervisor to ensure his immediate attention. To assist the operative lubrication, points should be painted in distinctive colors. These maintenance jobs should be incorporated as suitable allowances in the standard output rates.

13.12 OVERALL EQUIPMENT EFFECTIVENESS (OEE)

Overall equipment effectiveness quantifies how well a manufacturing unit performs relative to its designed capacity, during the periods when it is scheduled to run. OEE breaks the performance of a manufacturing unit into three separate, but measurable components: Availability, Performance, and Quality.

Availability is	Operating time/planned production time
Performance is	Net operating time/operating time
Quality is	Rate or percentage of good parts out of total production

$$OEE = Availability \times Performance \times Quality$$

13.13　THE SIX LOSSES FROM POOR OEE

The website of Optimum FX Consulting lists six major losses that result from poor OEE

- PDT or external unplanned event
- Breakdowns (>5 min)
- Minor stops (<5 min)
- Speed loss
- Production rejects
- Start-up rejects

These losses are explained in Table 13.1 and further illustrated in Fig. 13.3

13.14　THE THREE LEVELS OF AUTONOMOUS MAINTENANCE IN TPM

1. Routine lubrication and minor repairs, such as screw tightening or belt tightening, if necessary under instructions from the supervisor. This is called the *repair level of TPM.*
2. In case of any abnormal sound, like vibration or bearing notice, identify the root cause or notify the supervisor, as a part of the condition monitoring. This is called the *prevention level of TPM.*
3. Not only taking corrective action as above, but also take it up during quality circle meetings, departmental meetings, etc., and discuss improvements. This is called *improvement level of TPM.*

13.15　THE FIVE GOALS OF TPM

1. TPM's chief goal is to *improve system effectiveness.* It identifies and examines all losses that occur, whether it is downtime losses or speed losses or defect losses.
2. TPM achieves *autonomous maintenance* by motivating the operators to take responsibility for routine maintenance tasks as explained in the previous paragraph.
3. TPM adopts a *systematic approach* to all the maintenance activities. The level and nature of preventive maintenance for machine and equipment is identified and standards developed for condition monitoring. While operators are considered as owners of the machines taking their general care, the maintenance staff is considered as specialists providing supportive role for preventive and corrective maintenance activities.
4. TPM defines the *responsibilities* of the operators and maintenance staff, and that each has all the needed skills to carry out their roles. TPM emphasizes appropriate and continuous training and the maintenance department is given the responsibility of training the operators in routine and minor maintenance.

TABLE 13.1 The Six Losses due to Poor OEE

No.	OEE Measure	Six Loss Category	Reason for Loss	Countermeasures
1	Availability	Planned downtime or external unplanned event	1. Changeovers 2. Planned maintenance 3. Material shortages 4. Labor shortages	• Planned Downtime Management • 5S Workplace Organization • ABC Planning
2	Availability	Breakdowns	1. Equipment failure 2. Major component failure 3. Unplanned maintenance	• Kaizen Blitz • ProACT • Root cause analysis • Asset Care
3	Performance	Minor stops	1. Fallen product 2. Obstruction 3. Blockage 4. Misalignment	• Opportunity Analysis • 5S Workplace Organization • Management Routines • Line Minor stop audits
4	Performance	Speed loss	1. Running lower than rated speed 2. Untrained operator not able to run at nominal speed 3. Misalignment	• IFA Opportunity Analysis • Line Balance Optimization • Management Routines
5	Quality	Production rejects	1. Product out of specification 2. Damaged product 3. Scrap	• IFA Opportunity Analysis • Six Sigma • Error proofing
6	Quality	Rejects on startup	1. Product out of specification at start of run 2. Scrap created before nominal running after changeover 3. Damaged product after planned maintenance activity	• Planned Downtime Management • 5S Workplace Organization • Standard Operating Procedures • Precision settings

5. TPM strives to *attain the early design out maintenance* aspects for equipment. Its aim is to move towards zero maintenance through a maintenance prevention program (MP). This involves considering and analyzing failure causes and maintainability of the equipment during every stage, whether during design, manufacture, installation, or commissioning of the equipment.

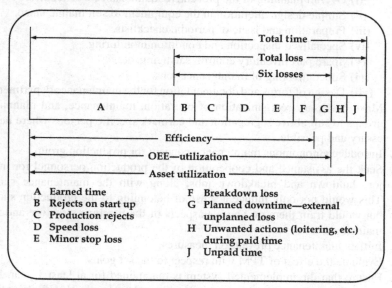

FIG. 13.3   Illustration of the six OEE losses.

13.16 PROCEDURE FOR THE IMPLEMENTATION OF TPM

1. Study the existing equipment history, maintenance records, etc., for all equipment and make a preliminary report of the need for TPM.
2. Obtain the consent of top management for the introduction of TPM in the organization.
3. Discuss with the concerned department heads, supervisors, and the unions.
4. Set goals and norms for TPM parameters such as the equipment's effectiveness and availability standards.
5. Segregate the maintenance jobs into three classes
   (a) *Routine maintenance jobs that could be done by the operator*:
      (i) Machine lubrication and oil levels monitoring.
      (ii) Cleaning and upkeep of the equipment.
      (iii) Adjustments, set up, and minor repairs like screw tightening.
      (iv) On-line condition monitoring with the help of production supervisor

   (b) *Jobs essentially to be done by the maintenance staff*:
      (i) Major breakdown jobs.
      (ii) Related works, like work order, material indents.

(iii) Plant shutdowns for preventive maintenance.
(iv) Major checking and off-line condition monitoring.
(v) Major shutdowns and capital repair jobs.

(c) *Jobs that are to be done by the planning and technical service groups*:
(i) Overall planning of the preventive maintenance e-schedules.
(ii) Simple design alteration in the equipment to suit maintenance.
(iii) Preparation and issue of periodic checklists.
(iv) Specialized inspection and condition monitoring.
(v) Spare part inventory control, salvaging, etc.
(vi) Subcontracting of maintenance jobs.
(vii) Design of forms and documentation for the maintenance department.

6. Make necessary reorganization of operation, maintenance, and planning groups to suit above segregation and earmark specific persons where necessary and possible.
7. Introduce autonomous maintenance concept for production group.
8. Seek the assistance and cooperation of the production personnel for major shutdown and breakdown jobs, along with the maintenance staff. This would not only induce a sense of belonging to the production staff, but would train them on various aspects of the machine structure and its maintenance.
9. Initiate maintenance prevention measures.
10. Evaluate the effect of TPM with respect to the set goals.
11. Ensure that the implemented system is maintained for at least 1 year. Be present to answer and solve any queries from the concerned staff about the system.

13.17 MAINTENANCE WORK SAMPLING

Work sampling is a proven measurement technique of industrial engineering, based on a random sampling of the workforce to determine what types of activities they are performing over the course of the day. It identifies and quantifies maintenance workforce efficiency opportunities. World-class maintenance operations can potentially perform the same amount of work with half as many workers. By identifying and eliminating barriers to productivity, the value-added contribution of existing maintenance resources can be significantly increased.

13.18 CONCLUSION

As we had seen, preventing equipment breakdowns and standardizing the equipment result in less variance and the quality of the products increases. Thus, as explained in Section 13.4, TPM is, and continues to be, one of the cornerstones in the quality movement and forms an extension of TQM.

APPENDIX   (SOURCE: TOTAL QUALITY MANAGEMENT BY SAMUEL K. HO)

CHECKLIST FOR JIPE'S PRODUCTIVE MAINTENANCE EXCELLENCE AWARD

Japan Institute of Plant Engineers (JIPE) sponsored the *Productive Maintenance Excellence Award* recognizing outstanding achievements in TPM field for Japanese industries as indicated earlier. JIPE developed a checklist as indicated below, to decide on the award. This checklist also gives the full insight into the salient aspects of TPM and helps us better understand the several factors that contribute for the success of TPM.

1. *Policy and objectives*
   - **1.1** How is equipment management integrated into company policy?
   - **1.2** Are equipment management policy and objectives set and prioritized correctly?
   - **1.3** Are the management guidelines and evaluation criteria good?
   - **1.4** Are long-term and annual plans integrated?
   - **1.5** Are policy and objectives thoroughly understood and implemented?
   - **1.6** Are accurate checks done to make sure objectives are being met?
   - **1.7** Are the year's results considered in formulating goals, objectives, and plans for the next year?

2. *Organization and operation*
   - **2.1** Are the organization and personnel assignments right for managing the equipment?
   - **2.2** Is a good organization in place for promoting TPM?
   - **2.3** Is the TPM promotion organization in close contact with production lines?
   - **2.4** Are all the necessary departments participating fully in TPM?
   - **2.5** Is effective coordination being done between the head office and the factories?
   - **2.6** Are there any impediments to the exchange and use of information?
   - **2.7** Are relations good with any subcontractors responsible for equipment, dies, tools, and maintenance work?

3. *Small-group activities and autonomous maintenance*
   - **3.1** Is the organization of the small groups done right?
   - **3.2** How are small-group activity objectives set?
   - **3.3** Do the groups meet frequently and are their meetings lively?
   - **3.4** Are there lots of good suggestions, and are they handled properly?
   - **3.5** How is goal-attainment ascertained?
   - **3.6** Do operators take the initiative in maintaining their equipment?

4. *Training*
   - **4.1** Do the different departments understand TPM?
   - **4.2** Are the training programs broad enough and their curricula right?

    4.3 Do the training programs follow the curricula?

    4.4 Do people take part in outside training programs?

    4.5 How many people have technical or expert certifications?

    4.6 How knowledgeable and skilled are people in maintenance techniques?

    4.7 How is skill-assessment done?

    4.8 Is there any way to make sure the training is having an effect?

5. *Equipment maintenance*

    5.1 Are the 5-S being implemented?

- Is the machinery clean and free of dust, filings, oil, and other waste material?
- Have policies been instituted to deal with dirt, places that cause foiling, and places that are difficult to clean, inspect, and lubricate?
- Are lubrication labels, gauge limitations, bolt tighten-to-her marks, and other visible indicators used?
- Are all of the tools, materials, gauges, and other things stored neatly and kept clean?

    5.2 Are equipment diagnosis techniques used in these cases?

- Cracking, corrosion, and loosening?
- Abnormal vibration, heating, and noise?
- Water, oil gas, and air leakage?

    5.3 Are power lines, water lines, hydraulic lines, and other lines neatly and properly handled?

    5.4 Are oils properly selected and properly replaced or filtered at the appropriate intervals?

6. *Planning and management*

    6.1 Are the appropriate efforts being made to improve maintenance techniques and efficiently?

    6.2 Are equipment standards properly set and enforced in a planned manner?

    6.3 Are monthly and annual maintenance plans drawn up and implemented properly?

    6.4 Are the purchasing plans for spare parts and other maintenance equipment properly drawn up (eg, how much of what to buy from where) and are such things cared for?

    6.5 Are equipment blueprints well cared for?

    6.6 Are the dies, tools, and gauges properly cared for?

    6.7 Are good records being kept on equipment wear and equipment failures that mandate stoppages or other maintenance efforts?

    6.8 Are maintenance records used to improved processes?

    6.9 Are the right maintenance techniques properly applied?

7. *Equipment investment plans and maintenance planning*

    7.1 Is equipment investment coordinated with the development of new products and new processes?

7.2 Is equipment investment cost-effective?

7.3 How is the plant investment budget drawn up and controlled?

7.4 Are maintenance planning suggestions duly reflected in equipment investment standards?

7.5 Are reliability and maintenance duly considered in selecting, designing, and placing equipment?

7.6 Is equipment closely monitored in the start-up stages?

7.7 Is the company good at developing its own dies, tools, and equipment?

7.8 Are policies promptly instituted to keep major problems from recurring?

7.9 Are plant assets properly managed?

8. *Production volume, scheduling, quality, and cost*

8.1 Is equipment control closely integrated with production volume and scheduling?

8.2 Is equipment control closely integrated with quality control?

8.3 Are maintenance budgets drawn up and managed properly?

8.4 Is energy and other resource conservation practice?

9. *Safety, sanitation, and environmental conservation*

9.1 Are sound policies in place for safety, sanitation, and environmental conservation?

9.2 Are the right organizations in place for safety, sanitation, and environmental conservation?

9.3 Are safety, sanitation, and environmental conservation methods known and practiced?

9.4 Is equipment investment integrated with safety, sanitation, and environmental conservation considerations?

9.5 Are safety and sanitation polices paying off?

9.6 Do environmental policies meet all of the legal requirements?

10. *Results and assessments*

10.1 Are results properly measured?

10.2 Are policies being implemented and objectives met?

10.3 Is maintenance paying off in terms of enhanced productivity and other management aims?

10.4 Does the company make an effort to publicize its activities and its successes?

10.5 Does the company know where the problems are?

10.6 Have plans been drawn up for future progress?

On the Lighter Side

The secret of managing is to keep the guys who hate you away from those who are undecided

—Casey Stengel

FURTHER READING

[1] Seiichi N. Introduction to TPM. Cambridge, MA: Productivity Press; 1989.

[2] Hartman EG. Successfully installing TPM in a non-Japanese plant: total productive mainte- nance. Cambridge, MA: Productivity Press; 1992.

[3] Seiichi N. TPM development program and implementing TPM. Cambridge, MA: Productivity Press; 1989.

[4] Japanese Institute of Plant Maintenance. TPM for every operator (Shop Floor Series). Taylor & Francis; 1996.

[5] Leflar JA. Practical TPM: successful equipment management at Agilent Technologies. Portland, OR: Productivity Press; 2001.

[6] Campbell JD, Reyes-Picknell JV. Uptime: strategies for excellence in maintenance manage- ment. 3rd ed. Cambridge, MA: Productivity Press; 2015.

[7] Borris S. Total productive maintenance. 1st ed. New York: McGraw Hill; 2006.

[8] http://reliabilityweb.com/index.php/maintenance_tips.

[9] http://www.optimumfx.com/overall-equipment-effectiveness-oee.

Chapter 14

Quality Awards

Chapter Outline

14.1 WHY QUALITY AWARDS?

In order to encourage industry and other corporates to make significant improvements in quality for maximizing consumer satisfaction and for successfully facing competition in the global market, several government and nongovernment organizations promoting quality have instituted awards to promote this quality movement and motivate the corporate in the interest of the development of a nation's economy and trade.

The most significant of them are the Quality Award Trio, namely, the Deming prizes of Japan, Malcom Baldrige Award of the United States, and the European Quality Awards. These three awards are discussed more in detail in the later paragraphs.

Apart from these, many other awards instituted by the governments of several countries, NGO's, and Corporates can be cited here. These awards are uniquely drafted to be the milestones in the journey to excellence and are coveted as symbols of recognition for organizations which have achieved certain quality levels in their pursuit of excellence.

14.2 INTERNATIONAL QUALITY AWARDS

Some of the Quality Awards apart from the Quality Award Trio are indicated below. It may be noted that many industrialized countries, professional associations, and corporates have instituted quality awards in some form or other. Only a few of them are cited here.

A. *National Awards*
- Australian Business Excellence Award
- Canadian Award for Business Excellence
- Rajiv Gandhi National Awards (India) instituted by the Bureau of Indian Standards in 1991
- China Quality Award
- Annual Dutch Quality Prize and Quality Award

Some more national awards are indicated in the website of Centre for Organizational Excellence Research, whose criteria for selection are indicated in Table 14.1.

B. *Quality Awards Issued by Leading Professional Associations*
- Akao Prize, awarded to individuals around the world for excellence in Quality Function Deployment (QFD).
- American Society for Quality's Awards and Medals, to individuals for superior achievements in the development, promotion, and communication of quality.
- National Institution for Quality and Reliability (NIQR), a premier professional association of India has instituted Quality awards given away during its biennial Quality Conventions as listed below.
  - NIQR-GKD Award for Outstanding Organization, instituted in 1970.
  - NIQR-Bazaz Auto Award for Outstanding Quality man, instituted in 1977.
  - NIQR-TS Krishnan Student Award for Winner of Essay Competition, instituted in 1982.

- NIQR-Susira Award for Outstanding Small Scale Industry, instituted in 1997.
- NIQR-Lucas TVS Award for Outstanding Service Organization, instituted in 2002.
- NIQR-India Pistons Award for Outstanding Individual Life Time Achievement, instituted in 2009.

The recipients of these awards for the year 2014 are given in the Appendix 14.4. NIQR has nearly 15 student chapters, the number increasing year by year. There is a proposal to institute another NIQR award for the best Student Chapter.

- FFI and Delhi State Govt.'s Bharat Excellence Award.
- Shingo Prize for Excellence in Manufacturing, referred to by Business Week as the Nobel Prize in manufacturing. This was instituted by Utah State University, in 1988, recognizing Dr. Shigeo Shingo for his lifetime work in promoting world-class manufacturing and recognizes companies that achieve superior customer satisfaction and other positive business results.
- *George M. Low Award*: NASA's Quality and Excellence Award: Recognizes large and small businesses that demonstrate excellence and outstanding technical and managerial achievements in quality and performance on NASA-related contracts or subcontracts.
- *QASAR Award*: The NASA Quality and Safety Achievement Recognition Award (QASAR) promotes quality, safety, leadership, productivity, performance, teamwork, and improvement throughout NASA. The award recognizes individual government and contractor employees at NASA Headquarters and Centers who have demonstrated exemplary performance in contributing to the quality and/or safety of products, services, processes, or management programs and activities.
- *Saturn's Quality Recognition Award*: Given annually to suppliers that exceed Saturn's stringent quality criteria by having less than 25 ppm nonconformance, maintaining a shipping performance of 95% or better, and are proactive in developing cost improvements.

C. *Quality Awards Instituted by Leading Corporates*
- *JPMorgan Chase Quality Recognition Award*: The award is given each year following a continual review process of industry suppliers undertaken by JPMorgan Chase Bank. It is designed to identify and reward companies that have demonstrated specific levels of operating excellence.
- *Citigroup Quality Recognition Award*: A prestigious annual award developed by Citigroup Global Transaction Services and is presented to leading financial institutions around the world, which meet the award's strict correspondent banking and processing criteria.
- Chrysler Quality Award
- Honda Award
- NUMMI Excellence Award for Quality

- Nissan Master Quality Award & Nissan Zero Defect Award
- Toyota Certificate of Achievement Award—Quality
- Memphis Regional Chamber Quality Cup Award

14.3   INTERNATIONAL QUALITY AWARD TRIO

As cited in Section 14.1, the three Quality Awards, namely, the Deming Application Prize of Japan, Malcom Baldrige Award of the United States, and the European Quality Award occupy the pinnacle of the quality awards and are aptly referred to as the Quality Award Trio. The detailed explanations of these awards are given in the following paragraphs.

14.4   DEMING APPLICATION PRIZE

The Deming Prize is a global quality award that recognizes both individuals for their contributions to, and corporates for their outstanding achievement in the field of Total Quality Management (TQM). It was established in Dec. 1950 in Japan to honor W. Edwards Deming, who literally was the chief architect of Japan's quality movement. Initially it was awarded by Japanese Union of Scientists and Engineers (JUSE) for Japanese companies for major quality achievements and from 1989, it has been elevated to international level considering corporates all over the world for the award.

14.4.1   Qualifications and Criteria Specified by JUSE for the Deming Prize

JUSE has specified that the following TQM qualifications should be confirmed for being considered for the prize.

(a) Customer-oriented business objectives and strategies are established in a positive manner, according to the management philosophy, type of policy industry, business scale, and business environment with the clear management belief.
(b) TQM has been implemented properly to achieve business objectives and strategies as mentioned Item (a) above.
(c) The business objectives and strategies in the Item (a) above have been achieving effects as an outcome of the Item (b) above.

JUSE website broadly specifies the following criteria for the award of the Deming Application Prize:

1. The emphasis of examination is on the implementation of TQM.
2. The actual implementation of TQM practices is appreciated.
3. It is not the usage of advanced statistical methods, but the appreciation and implementation of these methods that is taken as the basis for success.

4. Similar patterns of evaluation are adopted for both manufacturing and non-manufacturing companies.
5. The check-list given below is not for assessment, but helps the company to provide an overall picture of TQM.
6. Examiners judge the features that have been applied by the company.
7. The Examination Viewpoint includes:
   • Top Management Leadership, Vision, Strategies
   • TQM Frameworks
   • Quality Assurance Systems
   • Management Systems for Business Elements
   • Human Resource Development
   • Effective Utilization of Information
   • TQM Concepts and Values
   • Scientific Methods
   • Organizational Powers (Core Technology, Speed, Vitality)
   • Contribution to Realization of Corporate Objectives

14.4.2   Check List for Deming Application Prize

1. *Policy*
   1. Policies pursued for management quality and quality control
   2. Method of establishing policies
   3. Justifiability and consistency of policies
   4. Utilization of statistical methods
   5. Transmission and diffusion of policies
   6. Review of policies and the results achieved
   7. Relationship between policies and long and short-term planning
2. *Organization and its Management*
   1. Explicitness of the scopes of authority and responsibility
   2. Appropriateness of delegations of authority
   3. Interdivisional cooperation
   4. Committees and their activities
   5. Utilization of staff
   6. Utilization of QC Circle activities
   7. Quality-control diagnosis
3. *Education and Dissemination*
   1. Education programs and results
   2. Quality- and control-consciousness, degrees of understanding of quality control
   3. Teaching of statistical concepts and methods, and the extent of their dissemination
   4. Grasp of the effectiveness of quality control
   5. Education of related company (particularly those in the same group, sub-contractors, consignees, and distributors)

    6. QC circle activities

    7. System of suggesting ways of improvements and its actual conditions

4. *Collection, Dissemination, and Use of Information of Quality*

    1. Collection of external information

    2. Transmission of information between divisions

    3. Speed of information transmission (use of computers)

    4. Data processing statistical analysis of information and utilization of the results

5. *Analysis*

    1. Selection of key problems and themes

    2. Propriety of the analytical approach

    3. Utilization of statistical methods

    4. Linkage with proper technology

    5. Quality analysis, process analysis

    6. Utilization of analytical results

    7. Assertiveness of improvement suggestions

6. *Standardization*

    1. Systematization of standards

    2. Method of establishing, revising, and abolishing standards

    3. Outcome of the establishment, revision, or abolition of standards

    4. Contents of the standards

    5. Utilization of statistical methods

    6. Accumulation of technology

    7. Utilization of standards

7. *Control*

    1. Systems for the control of quality and such related matters as cost and quantity

    2. Control items and control points

    3. Utilization of such statistical control methods as control charts and other statistical concepts

    4. Contribution to performance of QC circle activities

    5. Actual conditions of control activities

    6. State of maters under control

8. *Quality Assurance*

    1. Procedure for the development of new products and services (analysis and upgrading of quality, checking of design, reliability, and other properties)

    2. Safety and immunity from product liability

    3. Customer satisfaction

    4. Process design, process analysis, and process control and improvement

    5. Process capability

    6. Instrumentation, gauging, testing, and inspecting

   7. Equipment maintenance, and control of subcontracting, purchasing, and services

   8. Quality assurance system and its audit

   9. Utilization of statistical methods

  10. Evaluation and audit of quality

  11. Actual state of quality assurance

9. *Results*

   1. Measurements of results

   2. Substantive results in quality, services, delivery time, cost, profits, safety, environments, etc.

   3. Intangible results

   4. Measures for overcoming defects

10. *Planning for the Future*

   1. Better understanding of the present state of affairs and the concreteness of the plan

   2. Measures for overcoming defects

   3. Plans for further advances

   4. Linkage with the long-term plans

Initially *two* categories, the Deming Prize for Individuals and the Deming Application Prize for corporates were given. The Deming Grand Prize, the most coveted of the Deming Prize group, was instituted in 1969, during the first International Conference on Quality Control (ICQC), in Tokyo. Appendix 14.1 lists the winners of the Application Prize and since 1989 hence, it became international, it is noteworthy that most of the winners are Indian corporates.

14.5   MALCOLM BALDRIGE NATIONAL QUALITY AWARD

The Malcolm Baldrige National Quality Award is the national quality award that recognizes US organizations in the business, health care, education, and nonprofit sectors for performance excellence and in applying the principles of TQM as embodied in the Baldrige Criteria for Performance Excellence as illustrated in Fig. 14.1. This was named in recognition of the contribution of Howard Malcolm Baldrige, Jr. (1922–87), who was the Secretary of Commerce and who played a major role during the presidency of Ronald Reagan in identifying the unfair trade practices and passing the Export Trading Company Act of 1982, which later became the Quality Improvement Act of 1987.

14.5.1   Criteria for the Performance Excellence Framework

The seven criteria listed below reflect the critical aspects of managing and performing as an organization and the individual weightage points in the assessment:

TABLE 14.1 Criteria for Performance Excellence Framework

Criteria	Weightage	Points
1. Leadership		110
a. Leadership system	80	
b. Company responsibility and citizenship	30	
2. Strategic planning		80
a. Strategy development process	40	
b. Company strategy	40	
3. Customer focus		80
a. Customer and market knowledge	40	
b. Customer satisfaction and relationship enhancement	40	
4. Measurement, analysis, and knowledge management		80
a. Selection and use of information and data	25	
b. Selection and use of comparative information and data	15	
c. Analysis and review of company performance	40	
5. Human resource focus		100
a. Work systems	40	
b. Employee education, training, and development	30	
c. Employee well-being and satisfaction	30	
6. Process management		100
a. Management of product and service process	60	
b. Management of support process		20
c. Management of supplier and partnering process	20	
7. Business results		450
a. Customer satisfaction results	125	
b. Financial and market results	125	
c. Human resource results	50	
d. Supplier and partner results	25	
e. Company specific results	125	
Total points		1000

These can also be illustrated by Fig. 14.1.

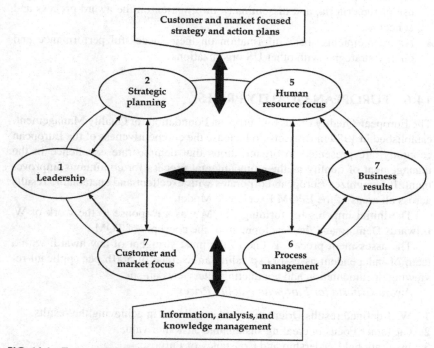

FIG. 14.1   Framework for the Malcolm Baldrige award criteria.

14.5.2   Organizations/Individuals Involved in the Awarding Process

- The Foundation for the Malcolm Baldrige National Quality Award raises funds for the award program.
- The National Institute of Standards and Technology (NIST), an agency of the US Department of Commerce, manages the Baldrige Program.
- The American Society for Quality (ASQ) assists in administering the award program under contract to NIST.
- The Board of Overseers advises the Department of Commerce on the Baldrige Program.
- Members of the Board of Examiners—evaluate award applications and prepare feedback reports.
- The Panel of Judges, part of the Board of Examiners, makes award recommendations to the director of NIST.

- Alliance for Performance Excellence, a network of state, regional, and local bodies provides potential award applicants and examiners, promotes the use of the criteria, and disseminates information on the award process and concepts.
- Award recipients share information on their successful performance and quality strategies with other US organizations.

14.6   EUROPEAN QUALITY PRIZES

The European Quality Prizes by European Foundation for Quality Management, established in 1989 in Brussels, to increase the competitiveness of the European economy, are presented to organizations that demonstrate excellence in the management of quality as their fundamental process for continuous improvement. It recognizes European corporates with excellent and sustainable results across all areas of the EFQM Excellence Model.

The initial impetus for forming EFQM was a response to the work of W. Edwards Deming and the development of the concepts of TQM.

The assessment process is one of the most stringent of any award, with a team of independent assessors spending an average of 500 h per applicant reviewing documentation and conducting interviews on-site.

Award Criteria for European Quality Prizes

1. Well-defined results Orientation and excellence in achieving the results.
2. Customer Focus in creating sustainable customer value.
3. Inspirational Leadership and Constancy of Purpose.
4. Managing the organization through a set of interdependent and interrelated systems, processes, and facts.
5. People Development and Involvement to maximize the contribution of employee.
6. Continuous Learning, Innovation, and Improvement to challenge the status quo and in using learning to create innovation and improvement opportunities.
7. Partnership Development by maintaining value-adding partnerships.
8. Corporate Social Responsibility by serving the society in a way to exceed the minimum regulatory framework and to strive to understand and respond to the expectations of their stakeholders in society (Fig. 14.2).

14.6.1   Categories of the Award

Awards are received by four categories

1. Large Private Sector (over 1000 employees)
2. Small/Medium Private Sector (less than 1000 employees)
3. Large Public Sector (over 1000 employees)
4. Small/Medium Public Sector (less than 1000 employees)

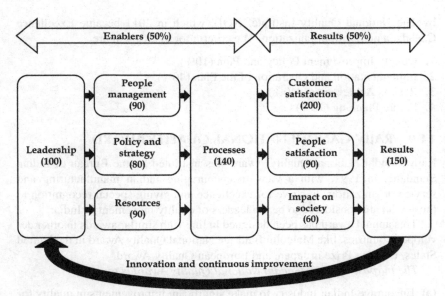

FIG. 14.2 Criteria for European Quality prizes (figures within brackets indicate the points in a total of 1000).

Awardees of the European Quality Prize

- 1992   The very first award given to Rank Xerox
- 1993   Milliken Europe
- 1994   D2D (Design to Distribution)
- 1995   Texas Instruments
- 1996   Brisa (Bridgestone)
- 1997   SGS-Thomson
- 1998   TNT UK
- 1999   Yellow Pages
- 2000   Nokia

14.7   AUSTRALIAN BUSINESS EXCELLENCE AWARD

The Australian Quality Award was first established by Australian Quality Council (AQC) in 1988 and since 2003, managed by SAI Global (Standards Australia International Limited). The Australian Business Excellence Awards are recognized as Australia's premier business awards and have been promoting business success in Australia for nearly 20 years.

14.8   CANADIAN AWARD FOR BUSINESS EXCELLENCE (CABE)

Business excellence in quality was instituted in 1984, to recognize Canadian businesses for excellence in applying the principles of TQM. It is administered by The National Quality Institute (NQI), which in 2011 became Excellence Canada, a nonprofit organization. The criteria for assessment are

1. Quality Improvement Policy and Plan (10%)
2. Implementation and Operation of the Plan (40%)
3. Results Achievement (40%)
4. Future Planning (10%)

14.9   RAJIV GANDHI NATIONAL QUALITY AWARD

Rajiv Gandhi National Quality Award was instituted by the Bureau of Indian Standards in 1991, with a view to encouraging Indian manufacturing and service organizations to strive for excellence and giving special recognition to those who are considered to be the leaders of quality movement in India.

This annual award has been designed in line with similar awards in other developed countries, like Malcolm Baldrige National Quality Award in the United States, Deming Prize in Japan, and European Quality Award.

The purpose of Rajiv Gandhi National Quality Award is to:

(a) Encourage Indian Industry to make significant improvements in quality for maximizing consumer satisfaction and for successfully facing competition in the global market as well;
(b) Recognize the achievements of those organizations which have improved the quality of their products and services and thereby set an example for others;
(c) Establish guidelines and criteria that can be used by industry in evaluating their own quality improvement efforts, and
(d) Provide specific guidance to other organizations that wish to learn how to achieve excellence in quality, by making available detailed information on the "Quality Management Approach" adopted by award winning organizations to change their culture and achieve eminence.

14.9.1   Assessment Criteria

(a) For large scale organizations, the following nine parameters would be assessed,
1. Leadership
2. Policies
3. Objectives and strategies
4. Human resource management
5. Resources and Processes
6. Customer focused results
7. Employees' satisfaction
8. Impact on environment and society and
9. Business results.
(b) For small scale organization six parameters would be assessed namely,
1. Leadership
2. Human resource management
3. Processes

4. Customer-focused results
5. Impact on environment and society and
6. Business results.

Emphasis will be placed on quality achievement and quality improvement as demonstrated through the information provided by applicant organization.

14.9.2   Eligibility of Organizations for This Award

1. Must be located in India.
2. Must have been in existence for at least 3 years as of last date of the application.
3. Applicant organization must be situated at one place or a unit of an organization housed at one location.
4. Never has been convicted by any court for deficiency in product or service and/or found guilty of financial irregularities by any regulatory authority or court. An undertaking to this effect will have to be submitted.
5. Not manufacturing products like tobacco and liquor, etc., which are injurious to health.

14.10   GOLDEN PEACOCK NATIONAL QUALITY AWARD

The annual Golden Peacock National Quality (GPNQ) Award, consisting of a trophy and a certificate, is awarded to encourage total quality improvements in manufacturing and service sectors to the Winner and the Runner up in each category of

- Large Enterprises (LE)   251 and above Employees
- Medium Enterprises (ME)   51 to 250 Employees
- Small Enterprises (SE)   up to 50 employees

The different Golden Peacock awards are:

1. National Quality Award,
2. Environment Management Award,
3. National Training Award,
4. Innovative management Award,
5. Innovative Product/Service Award,
6. Excellence in Corporate Governance,
7. Award for excellence in Corporate Social Responsibility,
8. Eco-innovation Award.

14.11   IMC-RAMAKRISHNA BAJAJ NATIONAL QUALITY AWARD (IMCRBNQA)

IMC (Indian Merchants' Chamber) Ramakrishna Bajaj National Quality Award was instituted in 1996 with an aim to match the stature of international awards, such as Malcolm Baldrige, the European Award, and the Deming Prize in terms of rigorous criteria and standards of excellence. It emphasizes three factors:

- The openness and transparency in governance and ethics,
- The need to create value to customers and business, and
- The challenges of rapid innovation and capitalizing on the knowledge assets.

Reference may be made to http://imcrbnqa.com/Past%20Winners for more details of past winners.

14.12   CHINA QUALITY AWARD

China Quality Award is the top award on the quality of product and the quality of management in China, presented to the enterprise with effective quality management practices. Its purpose is to meet the situation of market competitiveness of economic globalization, stimulate and direct excellent quality business of domestic enterprises, and accelerate the development of domestic enterprises with international competitiveness.

14.13   NATIONAL QUALITY/BUSINESS EXCELLENCE AWARDS IN DIFFERENT COUNTRIES

The Centre for Organizational Excellence Research, in their website www. coer.org.nz had published the details of major national awards of the world in Quality/Business Excellence, based on a research done in Jan. 2010. It lists as many as 96 awards from over 50 countries, indicating

- the country
- the name of award
- the model used
- the name of the organization that administers the award, and
- the relevant website

To know more about these awards, the details of which are beyond the scope of this book, the above website may be referred to. However, it may be mentioned here that during ANQ 2013 conference held in Bangkok in Sep. 2013, Thailand's Ishikawa-Kano Award (IKA) was given to Rit Thirakomen, Chairman of a Bangkok Restaurant Group.

14.14   BASIC DIFFERENCES AMONG THE AWARD TRIO

By critically going through the focus of each of the above three awards, we can say that the basic differences between the Deming, Baldrige, and Europe Quality Awards is that,

- The Deming Prize of Japan focuses in the application of statistical techniques that control the processes,
- The Malcolm Baldrige National Quality Awards of the United States has a broad emphasis on customer satisfaction through the implementation of TQM, while,
- The European Quality Award in the Western Europe is broader still, because it examines the impact of quality not only on the company itself but also in that company's social and environmental community.

14.15   CONCLUSION

Almost every industrialized country has instituted awards to encourage Industry and other corporates to make significant improvements in quality for successfully facing competition in the global market, which are as discussed in this chapter. Of these, the three awards of Deming, Baldrige, and Europe Quality Awards are of prime importance and have become yardsticks to measure the quality performance of each corporation. Winning these awards has become a pride, not only for the corporation, but also to the country.

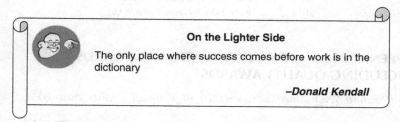

On the Lighter Side

The only place where success comes before work is in the dictionary

–Donald Kendall

APPENDIX 14.1   RECIPIENTS OF DEMING APPLICATION PRIZES FROM 1998

(source: http://en.wikipedia.org/wiki/Deming_Prize)

1951	Fuji Iron & Steel Co., Ltd. (now part of Nippon Steel), Showa Denko K.K., Tanabe Seiyaku Co., Ltd. and Yawata Iron & Steel Co., Ltd (now part of Nippon Steel)
1953	Sumitomo Metals
1989	Florida Power & Light (first non-Japanese winner of award)
1998	Sundram Clayton Limited, Brakes Division
2001	Sundaram Brake Linings, the world's first friction material company to win
2002	TVS Motor Company and Hi-Tech Carbon GMPD
2003	Brakes India Ltd. (Foundry Division), Mahindra & Mahindra Ltd. (world's first tractor company to win), Rane Brake Lining Ltd., Sona Koyo Steering Systems Ltd., and Grashim Industries (Birla Cellousic Division)
2004	Indo Gulf Fertilisers Ltd., Lucas TVS, and SRF limited
2005	Rane Engine Valve Ltd., Rane TRW Steering Systems Ltd. (SGD), and Krishna Maruti Ltd., Seat Division
2006	Sanden International (Singapore) Pte Ltd. (SIS), the first Singapore-based company to win
2007	Rane (Madras) Ltd., Asahi India Glass Ltd., and Relianxe Industries Ltd. (Hazira Division)
2008	Tata Steel, the first integrated steel plant in Asia to win Deming award in 2008
2010	National Engineering Industries Ltd. (CK Birla GrouP)
2011	Sanden Vikas (India) Limited (India)

2012	SRF Limited, Chemicals Business (India), Mahindra & Mahindra Limited, and Farm Equipment Sector, Swaraj Division (India)
	The Deming Grand Prize, the most coveted of the Deming Prize group, institute in 1969, was won by Lucas TVS Ltd., Tata Steel Ltd., and Rane Madras Ltd
2013	RSB Transmissions (I) Limited, Auto Division

Besides the following are the Japan Quality medal Winners from India

2002	Sundaram Clayton Ltd.
2007	Mahindra and Mahindra Ltd.
2011	Rane TRW Steering Systems Ltd.

APPENDIX 14.2   SOME INTERNATIONAL AWARDS INCLUDING QUALITY AWARDS

(source: http://en.wikipedia.org/wiki/list_of_national_quality_awards)

Sl. No.	Name of Award	Instituted by	Year	Some of the Recipients
1	Deming Prize for Organizations	JUSE, Japan	1951	1953— Sumitomo Metals
2	Edwards Medal	ASQ	1960	1975—Golonski 1995—Pyzdek
3	Malcolm Baldridge National Quality Award for Manufacturing Industries	United States	1987	
4	Malcolm Baldridge National Quality Award for Service Organizations	United States	1988	
5	Malcolm Baldridge National Quality Award for Small Business	United States	1988	
6	Rajiv Gandhi National Quality Award	Indian Govt.	1991	2006—Dr. N. Ravichandran
7	European Quality Awards	14 Western European countries	1992	
8	Bharat Excellence Award	IIFS—Delhi State Govt.	1996	2006—D.R. Kiran
10	Harrington/Neron Medal	Quebec Soc. for Quality, Canada	1997	
11	Sri Lanka National Quality Award	Sri Lanka	2000	

Sl. No.	Name of Award	Instituted by	Year	Some of the Recipients
12	Dorian Shainon Award	ASQ	2005	
13	Deming Prize for Individuals			
14	NIQR-GKD Outstanding Organization Award	NIQR, India	1970	2007—Mahindra & Mahindra
15	NIQR-Bajaj Outstanding Quality Man Award	NIQR, India	1977	2007—N. Ravichandran
16	NIQR-Susira Outstanding Small Scale Industry Award	NIQR, India	1982	2007—Aditya Auto Products
17	NIQR-TS Krishnan Student Award	NIQR, India	1982	2007—Aditya Auto Products
18	NIQR-Lucas TVS Outstanding Educational Institution Award	NIQR, India	2002	2007—The Banyan
19	NIQR-Kidao Outstanding Service Organization Award	NIQR, India	2004	2007—Kongu Engineering College
20	NIQR-India Pistons Outstanding Individual Life Achievement Award	NIQR, India	2009	2012—V. Krishnamurthy
21	US Air Force Quality Awards	US Air Force		
22	Shewhart Medal			1949—Harold Dodge 1988—Ishikawa
23	Harrington / Ishikawa Medal	APQO		
24	Grant Medal			1972—Harold Dodge 1990—Golomski
25	Arthur Fleming Award			1986—Hertz
26	Eugene Grant Award	ASQ		1972—Ishikawa
27	Shigo Shimgeo Prize Excellence in Manufacturing			
28	Henry Ford II Distinguished Award for Excellence	SAE		
29	Simon Collier Quality Award			2004—Pyzdek
30	American Deming Medal			—Golomski
31	Frank & Lilian Gilbreth Industrial Engineering Award	AIIE		

Note: The spaces are left blank where full information is not available.

APPENDIX 14.3   RECIPIENTS OF RAJIV GANDHI NATIONAL QUALITY AWARD

Year	Recipient
1991–92	Kirloskar Cummins Limited, Pune
1993	Steel Authority of India Limited, Bhilai Steel Plant, Bhilai
1994	ITC Limited ILTD Division Chirala (A.P.)
1995	ITC Limited ILTD Division, Anaparti, Andhra Pradesh
1996	Tata Bearings (A Division of TISCO), Kharagpur, West Bengal
1997	Larsen & Toubro Limited, Bangalore Works, Bangalore (Karnataka) Ammunition Factory, Khadki Pune, Maharashtra
1998	Mathura Refinery of Indian Corporation Limited, Mathura
1999	Gujarat Co-operative Milk Marketing Federation Limited, Anand, Tata Cummins Limited, Jamshedpur
2000	Tata International Limited, Dewas
2001	Birla Cellulosic, Bharuch
2002	No award
2003	IOC Ltd (Gujarat Refinery), Vadodara, Grasim Industries Ltd (Chemical Division), Nagda
2005	Moser Baer India Limited, Greater Noida
2006	Steel Authority of India Limited Bhilai Steel Plant, Bhilai
2007	Steel Authority of India Limited Bokaro Steel Plant, Bokaro
2008	Satluj Jal Vidyut Nigam Limited, Shimla, Himachal Pradesh
2009	Tata Motors Limited, Lucknow, Uttar Pradesh
2010	Vikram Cement Works, Khor, Madhya Pradesh
2011	DAV ACC Senior Secondary Public School Barmana, Himachal Pradesh
2012	No award
2013	No award

APPENDIX 14.4   RECIPIENTS OF NIQR AWARDS IN 2014

No.	Award	For	Recipient
1	NIQR-GKD Award	Organization	Maruti Suzuki India Ltd.
2	NIQR-Bazaz Auto Award	Quality Man	Er. L. Ganesh, Chairman, Rane Group
3	NIQR-TS Krishnan Student Award	Essay Competition Winner	Joseph Rajan, Minaxi Sundararajan Engineering College, Chennai
4	NIQR-Susira Award	Small Scale Industry	Anusham Industries, Chennai
5	NIQR-Lucas TVS Award	Service Organization	Self Employed Women's Association (SEWA), Gujarat
6	NIQR-India Pistons Award	Individual Life Time Achievement	Prof. Yasuthoshi Washio
7	NIQR-TVN Kidao Award	Educational Institution	Sasi Inst. of Tech. and Engg. (SITE)

APPENDIX 14.5   RECIPIENTS OF GOLDEN PEACOCK AWARDS

Year	Large MFG.	SME MFG	Large Service	Small Service	Other
1991	Kirloskar Brothers				
1992	Kirloskar Electric,				
1993	TELCO, Jamshedpur				
1994	Philips, Kolkata				
1995	Kirloskar Oil, Bhilai Steel Plant Bharat Electronics				
1996	10 companies, including Vizag Steel plant				
1997	L&T Mathura Refinery, Vysya bank	Bharti Enterp., Gabriel	CPTI of SAIL	Bax Global	Defense Electronic Appln. Lab.
1998	TI Diamond Chain, Bhial Steel plant	Perfect Circe Vector, Samore Nutrition's		Bax Global, Military College of Electronics & Mechanical Engg.	Ordinance Equipment Factory
1999	Reliance Inds. BHEL	Wipro InfoTech	BPC, Aviation Business.	Honeywell Software, Malavya Inst. of Petrol Exploration	Ordinance Factory
2000	BPC, IOC	Rainbow Denim, Bharti Teletech	BHEL, Noida	Bharti Cellular, Inst. of Mgmt. Devpt, ONGC	
2001	Indian Rayons, VIP Inds.	Tata Metalinks, Pepsico	Ford Services, IOC Pipelines		

Continued

Year	Large MFG.	SME MFG	Large Service	Small Service	Other
2002	Grasim Ind., Asian paints. Jindal Steels	PepsiCo, Newage Electrical	Mgmt. Devpt. Inst., ONGC		
2003	Tinplate, Asian paints, Gujarat Refineries, HPC	Maini Precisions, Siemens, Coca-Cola	I-flex Solutions	Bharti Health care	RAPP, Ordinance Factory, Acrylic Fibers
2004	Mahindra & Mahindra, Grasim Cements	Moser Baer, Ucal Fuels, Coca-Cola,	Infosys, Power Grid Corpn.		Hindustan Latex, Rail Wheel Factory
2005	BPC, NTPC, ITC	Coach w/ shop. WR & CR.	Manipal Hospitals		
2007	8 companies		AIMA		

Note: The spaces are left blank where full information is not available.

FURTHER READING

[1] en.wikipedia.org/wiki/Deming_Prize.
[2] en.wikipedia.org/wiki/Malcolm_Baldrige_National_Quality_Award.
[3] en.wikipedia.org/wiki/List_of_national_quality_award.
[4] en.wikipedia.org/wiki/Rajiv_Gandhi_National_Quality_Award.
[5] www.bis.org.in/other/rgnqa_geninfo.htm.
[6] https://deming.org/content.
[7] isqnet.org/demingprzwinners.html.
[8] www.gwu.edu/~umpleby/mgt201/201-14.
[9] www.wikiwand.com/en/List_of_national_quality_awards.
[10] http://www.scm.ethz.ch/teaching/Courses/FS14/QM/The_Deming_Prize.
[11] www.caq.org.cn/htm/about/en/China_Quality_Award.shtml.
[12] http://imcrbnqa.com/Past%20Winners.

Chapter 15

Quality Circles

Chapter Outline

15.1 WHAT IS A QUALITY CIRCLE?

Quality circle is defined as a small group of 6–12 employees doing similar work who voluntarily meet together on a regular basis to identify improvements in their respective work areas. By using proven techniques for analyzing and solving work-related problems that are preventing them from achieving and sustaining excellence, the groups work toward mutually uplifting employees, as well as the organization. It is "a way of capturing the creative and innovative power that lies within the work force."

(a) Quality circle is a form of participation management.
(b) Quality circle is a human resource development technique.
(c) Quality circle is a problem-solving technique.

15.2 ORIGIN OF QUALITY CIRCLES

During his visits to Japanese industry, Deming observed that the Japanese work atmosphere is oriented to the operators who feel more involved in the production, unlike in the United States. The voice of the workers in offering any spontaneous suggestions to improve productivity and cost reduction is given serious consideration by the Japanese managers. He argued that American management had typically given line managers and engineers about 85% of the responsibility for quality control and line workers only about 15%, and wanted these roles to be reversed.

Total Quality Management: Key Concepts and Case Studies. http://dx.doi.org/10.1016/B978-0-12-811035-5.00015-5

This fact made Deming encourage the formation of quality circles during the1950s where the bottom-line working force meets regularly to discuss shop problems. Originally formed for controlling defects at the shop floor level, its scope was later widened by Kaoru Ishikawa, the initiator of Cause and Effect diagram, to include method improvements, maintenance problems, etc. Thus, though the concept was introduced by Deming, the quality circles took shape in 1962 due to the efforts of Kaoru Ishikawa, who translated, integrated, and expanded the management concepts of W. Edwards Deming and Joseph M. Juran into the Japanese system of Quality Control.

In 1962, the first QC from Nippon Telephone and Telegraph was registered in Japan. It was also called Quality Control Circles (QCC), and also Small Group Activity (SGA).

15.3   THE AMERICAN SCENARIO

This concept of quality circles spread to the United States after the U.S. aerospace manufacturer Lockheed organized a tour of Japanese industrial plants, and planned to implement in their company. Thereafter, quality circles spread rapidly and by 1980, more than one-half of firms in the Fortune 500 had implemented or were planning to implement quality circles.

- In 1978, 1981, and 1985: three international quality circle conventions were held.
- By 1988, more than one million Circles, with over ten million members were formed in Japan (https://en.wikipedia.org/wiki/Quality_circle).
- Even the U.S. National Labor Relations Board (NLRB) in 1990, which prohibited company unions and management-dominated labor organizations, later relaxed the ruling in the case of quality circles to make them more popular in companies in the interest of productivity.

15.4   THE INDIAN SCENARIO

- In India, the quality circles concept was first introduced by BHEL, Ramachandrapuram, Hyderabad in the year 1980, which grew to 1411 Circles in all BHEL offices covering around 13,362 members by 1985.
- As a direct effect of the success of BHEL's quality circles, the Quality Circle Forum of India came into existence in Apr. 1982, as a nonprofit, nonpolitical, national professional body with the purpose of creating an environment for active involvement and participation of employees and college students in every area of human endeavor.
- Tata Motors (formerly Telco) started quality circles in 1983, and by 1985, they had more than 400 Circles.
- Quality Circle Forum of India (QCFI) represents India in the 13-nation International Committee that has been set up for organizing International conventions on Quality Concept Circle.

- After attending a QC Conference in Hong Kong in 1993, the City Montessori School Lucknow was the first educational institution to set up the student quality circle.
- Since then several engineering colleges and other educational institutions have set up quality circles and QCFI gives the best QC Project award in its annual meets.

15.5   SIGNIFICANCE OF QUALITY CIRCLES

Quality circles are formed of employees working together in an operation who meet at intervals to discuss problems of quality and to devise solutions for improvements, and are led by a supervisor or a senior worker. They usually receive training in formal problem-solving methods, such as brain-storming, Pareto analysis, and cause-and-effect diagrams—and are then encouraged to apply these methods, either to specific or general company problems.

Quality circles played a key role in rejuvenating Japanese industries and economy. This practice spread from Japan to countries all over the world, including India.

15.6   OBJECTIVES OF QUALITY CIRCLES

1. Opportunity to identify and solve problems in their work area.
2. Improve job satisfaction.
3. Create and enhance problem-solving capacity resulting in enhancing competence.
4. Build an attitude of problem prevention and problem-solving.
5. Promote personal and leadership development.
6. Better interpersonal relationship.
7. Recognition leading to motivation.
8. Promote creativity.

15.7   NATURE OF PROBLEMS THAT CAN BE SOLVED BY QUALITY CIRCLES

1. Customer service.
2. Streamlining of factory functioning.
3. Business growth and profitability.
4. Optimum utilization of working space.
5. Optimum unitization of manpower.
6. Reduction of human errors.
7. Reduction in the defects produced.
   (In fact, this was the original objective of quality circles.)
8. Training and knowledge development.
9. Cleanliness of the work area and factory premises.

15.8 TEN CONDITIONS FOR SUCCESSFUL QUALITY CIRCLES

Ron Basu and J. Nevan Wright, in their book *Quality Beyond Six Sigma* specified seven conditions for successful implementation of quality circles. These are summarized below after adding three more:

1. Quality circles must be staffed entirely by volunteers.
2. Each participant should be representative of a different functional activity.
3. The problem to be addressed by the QC should be chosen by the *circle*, not by management, and the choice honored even if it does not visibly lead to a management goal.
4. Management must be supportive of the circle and fund it appropriately, even when requests are trivial and the expenditure is difficult to envision as helping toward real solutions.
5. The members should not be suppressed in giving their opinions. There should ideally be brain-storming, allowing them to "let loose their brains."
6. There should be prior suggestions to the members, to let them think before coming for the meeting.
7. The members should only seek clarification, but no criticism.
8. Circle members must receive appropriate training in problem-solving.
9. The circle must choose its own leader from within its members.
10. Management should appoint a manager as the mentor of the team, charged with helping members of the circle achieve their objectives; but this person must not manage the QC.

15.9 ROAD MAP TO BE FOLLOWED IN A QUALITY CIRCLE MEETING

1. Identify work-related problems
2. Prioritize problems by applying ABC analysis, etc.
3. Define the problem
4. Analyze the problem
5. Identify the root causes by cause and effect principle
6. Develop solutions
7. Foresee the probable resistances
8. Conduct trial implementation and check on the performance
9. Implement
10. Maintain.

15.10 CHARACTERISTICS OF AN EFFECTIVE QUALITY CIRCLE MEETING

1. Commitment to task
2. Informality in the meeting

3. Openness
4. Conflicts on ideas, and not between persons
5. Constructive criticism
6. Agreement by consensus
7. Action plan after the meeting.

15.11   STRUCTURE OF A QUALITY CIRCLE

Any corporate employee irrespective of his cadre, except the head of the unit, can join the circle. The structure of a quality circle consists of the following elements, as illustrated in the following Fig. 15.1.

- **(i)** *A steering committee*: This is at the top of the structure. It is headed by a senior executive and includes representatives from the top management personnel and human resources development people. It establishes policy, plans, and directs the program and meets usually once a month.
- **(ii)** *Coordinator*: He may be a Personnel or Administrative officer who coordinates and supervises the work of the facilitators and administers the program.
- **(iii)** *Facilitator*: He may be a senior supervisory officer. He coordinates the work of several quality circles through the Circle leaders.
- **(iv)** *Circle leader*: Leaders may be from the lowest level workers or supervisors. A Circle leader organizes and conducts Circle activities.
- **(v)** *Circle members*: They may be staff workers. Without circle members, the program cannot exist. They are the lifeblood of quality circles. They should attend all meetings as far as possible, offer suggestions and ideas, participate actively the in group process, and take training seriously with a receptive attitude. The roles of Steering Committee, Coordinator, Facilitator, Circle leader, and Circle members are well defined.

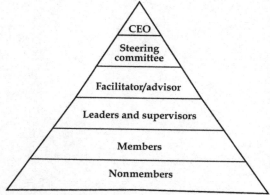

FIG. 15.1   Structure of a quality circle.

15.12   CONCLUSION

Quality circles are not limited to manufacturing firms only. They are applicable for variety of organizations where there is scope for a group-based solution of work-related problems. Quality circles are relevant for factories, firms, schools, hospitals, universities, research institutes, banks, government offices, etc.

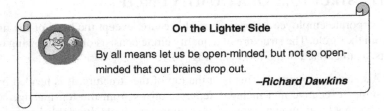

On the Lighter Side

By all means let us be open-minded, but not so open-minded that our brains drop out.
                                                            –Richard Dawkins

Chapter 16

Fundamentals of Statistics— Part I

Chapter Outline

Total Quality Management: Key Concepts and Case Studies. http://dx.doi.org/10.1016/B978-0-12-811035-5.00016-7

16.1 DEFINITION OF STATISTICS

Originally the word "STATISTICS" was derived from the word "STATUS" or "STATE" ie, a science of dealing with the affairs of administration of a state. Today, statistics refers to the scientific approach to the collection, presentation, analysis, and interpretation of numerical data or information by presenting it in a way to understand the information.

Different individuals react differently under the same set of circumstances. But the ability to predict with reliable confidence how the entire group is likely to react under the given circumstances is very important. Hence, it is necessary to evaluate a certain statistical parameter, which would represent the characteristics of the entire group. This is called an average. The method of statistics actually deals with the evaluation of such characteristics and hence it is also defined as the "Science of Averages."

16.2 ROLE OF STATISTICS IN ANALYSIS

Today statistics is indispensable for a clear appreciation of any problem affecting human welfare, whether it is industry, transport, business, medicine, and more significantly, in quality control analysis.

Statistics is an aid to the economic measure. It is a technique of analyzing the data obtained to conclude on the economic progress and forecast the future trend.

Statistics is an aid to the manager particularly in these days of impersonal relationships between the employer and employee. With this, he can estimate the demand more accurately then by guesswork. Statistics help in recording the past knowledge and experience, drawing out standards whose results can be compared from time to time. And also the expected changes, the reasons for these changes, and the effects of these changes on the nation's economy can be estimated.

We can summarize the role of statistics as follows:

1. Helps in simplifying complexity.
2. Keeps a control on the disturbing factors affecting the data and helps in detecting and eliminating them.
3. Studies the relationship between connected facts and in measuring their degree or level of significance.
4. Helps in forecasting the future happening of events based on the present situation.

16.3 LIMITATION OF STATISTICS

1. It is restricted only to the study of quantitative phenomena.
2. It is cognizant of individual items.
3. It does not reveal the entire problem.
4. Its laws are true only on an average.
5. It is liable to be mishandled and misused.

16.4 ELEMENTS OF STATISTICAL TECHNIQUES

In general, the elements of statistical techniques include:

1. Collection of data.
2. Assembling of data.
3. Classification and summarization of data.
4. Presentation of data.
   - in textual form
   - in tabular form
   - in graphic form
5. Analysis of data.

16.5 METHODS OF COLLECTING DATA

The first step in statistical analysis is to collect data, which can be in any of the following methods:

A. *By direct observation* like counting the number of buses passing a particular junction during peak hours, or taking the time study of an operation.
   Advantages: It reduces the chance of incorrect data or guess works being recorded.
   Disadvantages: Many times this method is costly, or not possible, such as observing the various activities done by a housewife during the course of a month.
B. *Interviewing* asking concerned people personally for the required information, such as the market research people going around to the customers' houses to collect information for the census data.

Advantages: Easy and more accurate replies.

Disadvantages:
- **(a)** Sometimes deliberately wrong data may be given due to shyness or forgetfulness.
- **(b)** Wrong understanding of the questions or wrong interpretation of the answer.

C. *Referring to published records:* They can be of two types:
- **(a)** The *primary data*, consisting of raw and unprocessed data collected at the point of generation; and
- **(b)** The *secondary data*, consisting of processed summarized primary data as found in reports and publications, either government or private.

D. *Questionnaires:* Sending questionnaires by post and asking people to complete it and send it back.

Advantages: Reduces wrong interpretation and giving all unnecessary information.

Disadvantages: Poor response. Just out of laziness, namely a maximum number of 15% respond. Also by the time you get back the answers, it might be too late.

Factors for the Design of a Questionnaire:

- **(a)** Questions should be simple.
- **(b)** Questions should not be vague or ambiguous.
- **(c)** The questionnaire should be as short as possible.
- **(d)** The questions should not be irrelevant or pursuant.
- **(e)** Leading questions should not be asked. For example "Don't you think all the sensible people use XYZ Soap"?
- **(f)** Questions in a questionnaire should fall into a logical sequence.
- **(g)** The best kinds of questions are those which allow preprinted multiple choice answers, so that the respondent has just to tick his answer.

16.6   DATA CLASSIFICATION

To make the data comprehensive, it is necessary to classify it into homogeneous groups, subgroups, etc., as per their respective characteristics, into useful and logical categories.

The main objects of classification are:

1. To simplify the understanding of such huge data.
2. To enable specifying the main objective of collecting such figures.
3. To trace out certain characteristics in the order of their importance.
4. To facilitate comparison.

16.7   DATA PRESENTATION

The next step after data classification is to arrange them in an orderly manner, so that they can be readily understood. In general, the following are the different methods of data presentations.

1. Narrative form.
2. Tabular presentations.
3. Single dimensional diagrams, such as bar charts.
4. Two-dimensional presentations.
   - Squares and rectangles.
   - Graphs.
   - Binominal curves.
   - "Z" charts.
   - Lorenz curves.
5. Pictorial presentations.
   - Pictograms.
   - Pie charts.
   - Statistical maps.
6. 3-Dimensional presentations.

16.8   POPULATION VERSUS SAMPLE

Before understanding the elements of statistical techniques, the first step in statistical analysis is to understand the basic data groups like population, sample, attributes, variables, etc.

16.8.1   Population

Population is the entire body of items about which we want to obtain information. That is when all the items, values, or attributes are taken into consideration for any statistical enquiry, and will be called a population or universe. There are many types of populations.

1. If we take the values of variables, then, the set of all these values will be a population of values.
2. If we take all the forms of a certain attribute, then we get a population of attributes.
3. A population which consists of an infinite number of items is called an infinite population.
4. A population of finite members is called a finite population.
5. A population defined under certain law is called hypothetical population. For example, if we assume that heights of individuals follow a normal law, then heights will form a population which will be a hypothetical population.
6. A population which exists in reality and which can be observed, is called a real population.

16.8.2   Sample

Any subset formed by certain members of a given population will be called a sample.

1. A sample may be finite or infinite. But usually samples of finite sizes from a given population are taken.
2. The number of members in a finite sample will be called as the size of the sample.

3. The sample may be derived from a finite or infinite population.
4. From a given population, we can derive several samples.

16.9   ATTRIBUTES AND VARIABLES

Attribute: If the measurement is made qualitatively, then it is called an attribute. eg, color, sex, etc.

Variable: If the measurement is made quantitatively, then it is called a Variable. eg, height, weight, etc.

Discrete variable: If a variable can take only a specified number of values, then it is called a discrete variable.

Continuous variable: If a variable can take all the possible values between two real numbers, then it is called continuous variable.

16.10   GRAPHS

Graphs are visual presentation of the data which give an immediate visual concept of the trend or comparison.

A graph is a representation of data by a continuous curve on squared paper, while a diagram is any other two-dimensional form of visual representation. Note that a line on a graph is always referred to as a curve—even though it may be straight.

16.10.1   Principles of Graph Construction

Graph construction, like table construction, is in many ways an art. However, like tables again, there are a number of basic principles to be observed if the graph is to be a good one. These are given below:

1. The correct impression must be given.
2. The graph must have a clear and comprehensive title.
3. The independent variable should always be placed on the horizontal axis.
4. The vertical scale should always start at zero.
5. A double vertical scale should be used where appropriate.
6. Axes should be clearly labeled.
7. Curve must be distinct.
8. The graph must not be overcrowded with curves.
9. The source of the actual figures must be given.

16.10.2   Class Interval

The width of a class or a symbol that defines a class is called a class interval (eg, 0–10, 10–20, etc.).

16.10.3   Class Limits

The end values which are included in each class are called the class limits eg, 0–4 then the possible members of this class are 0,1,2,3, and 4. Then the lowest value 0 is called the lower limit of the class interval, and the highest value 4 is called the upper limit of the class interval.

16.10.4   Class Mark

The class mark is the midpoint of the class interval and is obtained by adding the lower and upper class limits and dividing the total by 2.

16.11   SINGLE DIMENSIONAL DIAGRAMS—BAR CHARTS

16.11.1   Simple Bar Charts

In simple bar charts, data are represented by a series of bars: the height (or length) of each bar indicating the size of the figure represented (Fig. 16.1).

Since bar charts are similar to graphs, virtually the same principles of construction apply though, it should be noted that there should never be a "break to zero" in bar charts.

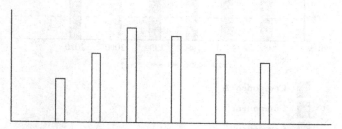

FIG. 16.1   Simple bar chart.

16.11.2   Component Bar Charts

These are like ordinary bar charts, except that the bars are subdivided into component parts. This is constructed when each total figure is built up from two or more components (Fig. 16.2).

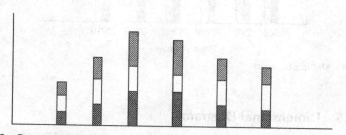

FIG. 16.2   Component bar chart.

16.11.3   Percentage Component Bar Chart

Here the individual component lengths represent the percentage each component forms of the overall total. Note that a series of such bars will all be of the same total height ie, 100% (Fig. 16.3).

16.11.4   Multiple Bar Charts

In this type of chart, the component figures are shown as separate bars adjoining each other. The height of each bar represents the actual value of the component as shown in Fig. 16.4.

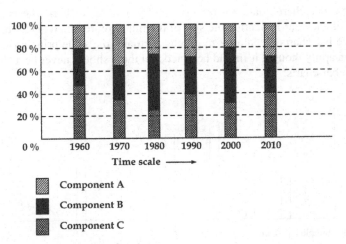

FIG. 16.3   Percentage component bar chart.

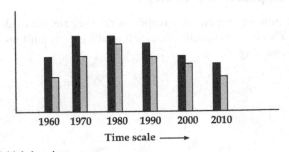

FIG. 16.4   Multiple bar chart.

16.11.5   Dimensional Diagrams

Here the variations are represented by X and Y axes shown in Fig. 16.5

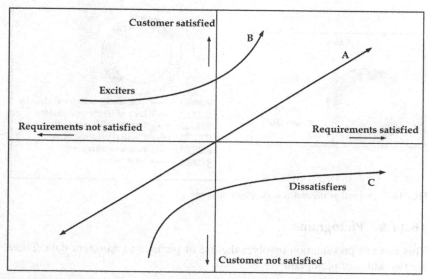

FIG. 16.5   Two-dimensional diagrams—kano model of customer satisfaction.

16.11.6   Pie Diagrams

Circular and pie diagrams eg, circles whose areas are made proportional to given quantities and are of service to show the makeup of the total, its segments representing the ratios of the components parts to the whole as shown in Fig. 16.6.

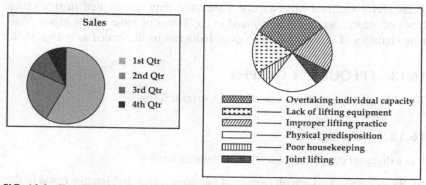

FIG. 16.6   Two illustrations of pie diagrams indicating quarterly sales and another the types of accidents.

16.11.7   Doughnut Diagrams

These are similar to pie diagrams, except the data is indicated between to concentric circles, rather than in a full circle shown in Fig. 16.7.

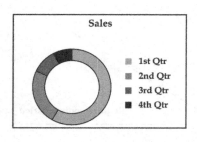

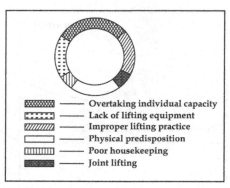

FIG. 16.7   Same data illustrated in doughnut diagrams.

16.11.8   Pictograms

This form of presentation involves the use of pictures to represent data. There are two kinds of pictogram:

(a)   Those in which the same picture always the same size, is shown repeatedly, the value of a figure represented being indicated by the number of pictures shown (see Fig. 16.8a).
(b)   Those in which the pictures change in size, the value of a figure represented being indicated by the size of the picture shown (see Fig. 16.8b).

16.12   INNOVATIVE GRAPHS

These days, you find in newspapers statistics data represented in innovative types of graphs, a recent one noticed in the Times of India on the urban planning statistics of various cities/towns of India can be illustrated as in Fig. 16.9.

16.13   FREQUENCY GRAPHS

These frequency graphs can be drawn in several forms.

16.13.1   Histograms

It is a diagram consisting of a set of rectangles having:

(i)   Bases on the X axis with centers at the class marks and lengths equal to the class interval sizes.
(ii)   Areas proportional to class frequencies.

This gives the frequency per unit length of class interval and is known as the frequency density over that class-interval. In view of the importance given to the histograms as a traditional tool of TQM, this is dealt more in detail in Section 20.3.

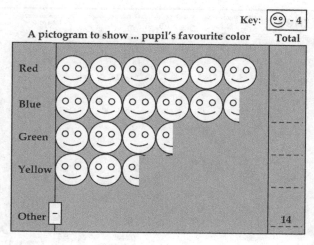

(A)

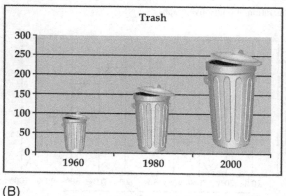

(B)

FIG. 16.8 Diagrammatic representation by pictograms.

16.13.2 Frequency Polygon

This is obtained by joining the midpoints of the top line of each bar in the previously mentioned histograms.

16.13.3 Frequency Curve

This is obtained by smoothing the frequency polygon to obtain a curve shown in Fig. 16.10.

16.14 OGIVE

It is the frequency polygon of the cumulative frequencies. Thus, an ogive is obtained by plotting the cumulative frequencies corresponding to values of the

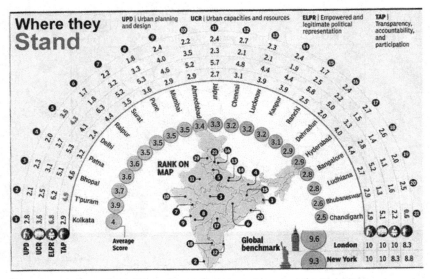

FIG. 16.9   An illustration of innovative graphs.

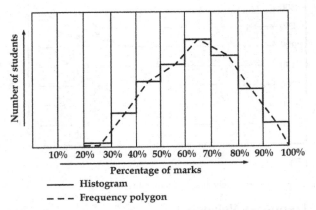

——— Histogram

– – – Frequency polygon

FIG. 16.10   A histogram, superimposed with frequency polygon and frequency curve.

variate, to which they belong. Cumulative frequency polygon when smoothed out will be referred to as an ogive curve is shown in Fig. 16.11.

16.15 "Z" CHART

A "Z" chart (Fig. 16.13) is a combination of three graphs drawn for data over a period of one year incorporating:

(a) Individual monthly figures.
(b) Monthly cumulative figures for the year.
(c) A moving annual total.

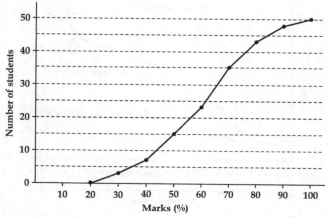

FIG. 16.11   An illustration of ogive.

Construction of Z-charts:

1. The basic curve representing the values of a variable over each of the 12 months of a particular year is drawn in a time scale (curve A).
2. The cumulative values for each month is computed and represented by curve B.
3. The total value of the variable for the 12 months, including that month plus the preceding 11 months, viz February of the previous year to January of this year is computed and indicated against January month. Similarly, the total of the variable from March of the previous year to February of this year is indicated against February. In this manner, the total of the 12 months' values for the particular month plus the preceding 11 months is indicated against each of the months. A line joining these points is called the moving annual total curve (curve C) and since these three lines together look like Z of the English alphabet, it is called z-chart as illustrated in Fig. 16.12.

16.16 LORENZ CURVES

It is a well-known fact that in practically every country, a small proportion of the population owns a large proportion of the total wealth. Industrialists know, too, that a small proportion of all the factories employ a large proportion of the factory workers. This disparity of proportions, which is similar to ABC analysis and Pareto principle, is a common economic phenomenon. American economist Max Lorenz illustrated this in 1905 by a graph called the Lorenz curve, showing the reality of wealth distribution. A straight diagonal line representing perfect equality of wealth distribution is drawn above it for providing contrast. The difference between the straight line and the

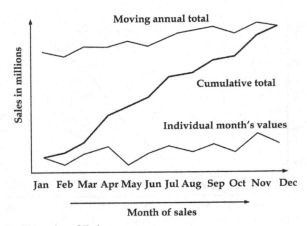

FIG. 16.12   An illustration of Z-chart.

curved line is the amount of inequality of wealth distribution, a figure also called the Gini coefficient. Lorenz curve is shown in Fig. 16.13.

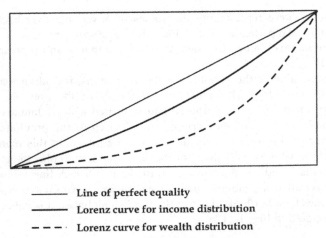

—————— Line of perfect equality

—————— Lorenz curve for income distribution

— — —· Lorenz curve for wealth distribution

FIG. 16.13   Lorenz curve for the wealth and income distribution.

16.16.1   Application of Lorenz Curves

Lorenz curves are applicable in the following situations:

(a) Incomes in the population.
(b) Tax payments of individuals in the population.
(c) Industrial efficiencies.
(d) Industrial outputs.
(e) Examination marks.

(f) Customers and sales.
(g) Production planning.

16.17   FREQUENCY DISTRIBUTION

In earlier paragraphs, we have studied a variety of graphic methods of presentation of data. One of the frequent problems related to statistics is to represent grouped data graphically. Consider the following example (Table 16.1), indicating the daily income distribution among 50 workers, classified into 10 groups between Rs. 80 and Rs. 129.

TABLE 16.1 Frequency Distribution of Daily Income

Class Interval (Daily Income)	Class Midpoint (X)	Class Frequency (No. of Worker) (f)	Cumulative Frequency	f_x
80–84	82	1	1	82
85–89	87	1	2	87
90–94	92	3	5	276
95–99	97	5	10	475
100–104	102	8	18	816
105–109	107	12	30	1284
110–114	112	7	37	784
115–119	117	6	43	702
120–124	122	5	48	610
125–129	127	2	50	254
		50		5370

16.18   CENTRAL TENDENCY

This frequency distribution table is the basic data representation for further computations and statistical analysis. The following few paragraphs explain the subsequent steps in analyzing the central tendency of the figures.

As a second step, this distribution can be represented as frequency polygon illustrated in Fig. 16.14. Where we join the top values of each income group, we get what is called the frequency distribution curve. This is also illustrated in Fig. 16.14, which indicates other parameters of central tendency.

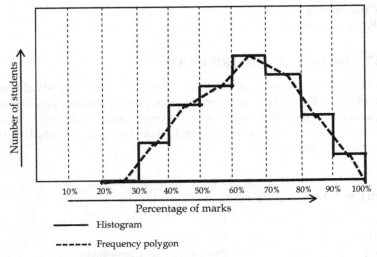

FIG. 16.14   Frequency distribution and frequency polygon.

You can see from the histogram of Fig. 16.14 that while the income group value (or *x* value) increases, the frequency (or *y* value) slowly increases from a low value and attains a high value at somewhere about the center of the value and again decreases towards the end of value. This is called the Central tendency, and is common in most statistical distribution.

16.19   MEASURES OF CENTRAL TENDENCY

The following are some of the measures for the frequency distribution.

1. Mean (or average)
2. Median
3. Mode

Also the other measures for the dispersion of the data (ie, how the individual values are spread out) are

1. Range
2. Mean deviation
3. Standard deviation or variance
4. Kurtosis
5. Skewness

The three measures mean, median, and mode all sound similar, but the frequency curve as illustrated in Fig. 16.15, which has a deviation from the normal curve, explains their differences better.

Range, represented by *d* is the difference between the minimum and maximum values in a series.

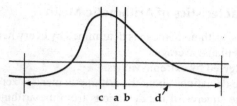

FIG. 16.15 Indication of mean, mode, and median in typical frequency curve.

Mean, represented by *a,* is the average point on either side of which the frequencies are equally distributed, that is around which the areas on either side are equal.

Median, represented by *b,* is the mid-point of the range, that is the point midway between the smallest and largest value of the frequency. The median is the value of the middle item when the items are arranged according to size.

Mode, represented by *c,* is the point corresponding to the largest value in the array, that is the peak of the frequency distribution.

16.20 MEAN OR AN AVERAGE

Mean or an average is a typical value which is intended to sum up or describe the mass of data. It also serves as a basis for measuring or evaluating extreme or unusual values. The average is a measure of the location of the point of central tendency. Mean can be in four types as below:

(a) Arithmetic mean
(b) Geometric mean
(c) Harmonic mean
(d) Quadratic mean

16.21 ARITHMETIC MEAN

Due to ease of computation and long usage, the arithmetic mean is the best known and most commonly used of all the averages. When the word "mean" is used without qualification, the reference is to the arithmetic mean.

Calculation of arithmetic mean: The arithmetic mean of a small group of individual values may be obtained by dividing the sum of the values by the number of items used.

The computation of the arithmetic mean is expressed in formula form as:

$$\bar{X} = \frac{\Sigma(X)}{N}$$

where

$\bar{X}$ = arithmetic mean.
Σ = symbol meaning "sum of."
X = data expressed as individual values.
N = number of items.

16.21.1   Characteristics of Arithmetic Mean

1. The value of the arithmetic mean is determined by every item in the distribution. It is a calculated average.
2. It is greatly affected by extreme values.
3. The sum of the deviations about the arithmetic mean is zero.
4. The sum of the squares of the deviations from the arithmetic mean is less than those computed about any other point.
5. Its standard error is less than that of the median.
6. In every case, it has a determinate value.
7. The sum of the means in case of multiple distributions equals the mean of the sums, whereas the difference between the means equals the mean difference.

$$X_{1+2} = X_1 + X_2$$
$$X_{1-2} = X_1 - X_2$$

16.21.2   Advantages of Arithmetic Mean

1. The arithmetic mean is the most commonly used.
2. Its computation is relatively simple.
3. Only total values and the number of items are necessary for its computation.
4. It may be treated algebraically. For example, where averages for subgroups are available, they in turn may be averaged in order to obtain an average for the whole group.

16.21.3   Disadvantages of Arithmetic Mean

1. The arithmetic mean may be greatly distorted by extreme values, and therefore it may not be a typical value.
2. The arithmetic mean cannot be computed from a distribution containing "Open ended" class intervals, that is, when the items are grouped in "and under" or "and over" class intervals.

16.22   GEOMETRIC MEAN, QUADRATIC MEAN, AND HARMONIC MEAN

The three other means viz., the Geometric mean, Quadratic mean, and Harmonic mean are rarely used in statistics and hence, not covered here. However, for academic interest, the formulae are given below:

If $X_1, X_2, X_3 \ldots$ etc., are the X values, then

Geometric Mean $= \sqrt[n]{X_1 \times X_2 \times X_3 \times X_4}$.

Quadratic Mean $= \sqrt{\dfrac{X^2}{N}}$.

Harmonic Mean is given by $\dfrac{1}{HM} = \dfrac{1}{X_1} + \dfrac{1}{X_2} + \dfrac{1}{X_3} + \dfrac{1}{X_4} \ldots \dfrac{1}{X_n}$.

16.23   MEDIAN

16.23.1   Definition

The median is the value of the middle item when the items are arranged according to size. If there is an even number of items, the median is taken as the arithmetic mean of the values of the two central items.

The median is an average of position while the arithmetic mean is a calculated average of values.

16.23.2   Calculation from Ungrouped Data

The median is computed from ungrouped data as follows:

1. Arrange the items according to magnitude (this arrangement is called an array).
2. Record the size of the middle value. If there is an even number of items in the array, there will be two central values and the arithmetic mean of these two values is taken as the median.

16.23.3   Calculation from Grouped Data

The median is computed from grouped data as follows:

1. Determine the number of the desired middle item by using the Eq. (16.1) where N is the number of items in the distribution. For example, if there are 150 items in an array, the median item is the seventy-fifth item.

$$\frac{N}{2} = \frac{150}{2} = 75 \tag{16.1}$$

2. Find the class interval in which the seventy-fifth item appears by cumulative addition of the frequencies. The value of this class interval is the median.

16.23.4   Characteristics of Median

1. The median is an average of position.
2. The median is affected by the number of items, not by the size of extreme values.
3. The sum of the deviations about the median, signs ignored, will be less than the sum of the deviation about any other point.
4. The median is most typical when used to describe distributions where central values are closely grouped.
5. A value selection at random is just as likely to be located above the median as below. At times, therefore, the median is called the "Probable" value.

16.23.5   Advantages of Median

1. The median is easily calculated.
2. It is not distorted in value by unusual items.
3. It is sometimes more typical of the series than are other averages because of its independence of unusual values.
4. The median may be calculated even when the class intervals of the distribution are "open ended."

16.23.6   Disadvantages of Median

1. The median is not as familiar as the arithmetic mean.
2. The items must be arranged according to size before the median can be computed.
3. It has a larger standard error than the arithmetic mean.
4. The median cannot be manipulated algebraically. The average of the medians of subgroups, for eg, is not the median of the group.

16.24   MODE

16.24.1   Definition

The mode is the most frequent or most common value which occurs in a set of data, provided a large number of observations are available.

The value of the mode will correspond to the value of the maximum point (ordinate) of a frequent distribution, if it is an "ideal," or smooth distribution.

16.24.2   Characteristics of Mode

1. By definition, the mode is the most usual or typical value. Under certain circumstances, it may be considered as the "normal" value.
2. The value of the mode is entirely independent of extreme items.
3. The mode is an average of position.

16.24.3   Advantages of Mode

1. It is in the most typical value, and therefore, the most descriptive average.
2. It is simple to approximate by observation when there are a small number of cases.
3. If there are only a few items, it is not necessary to arrange them in order to determine the mode.

16.24.4   Disadvantages of Mode

1. The mode can be approximated only when a limited amount of data is available.
2. Its significance is limited when a large number of values are not available.
3. If none of the values are repeated, the mode does not exist.

16.25   DISPERSION

The average or typical value has little use unless the degree of variation which occurs about it is given. For if the scatter about the measure of central tendency is very large, the average is not a typical value. It is therefore necessary to develop a quantitative measure of the dispersion (or variation, or scatter) of values about the average.

16.26   RANGE

The range, the simplest of the measure of dispersion, and as defined earlier, is the difference between the minimum and maximum values in a series. It is sometimes given in the form of a statement of the minimum and maximum values themselves.

The difference between the two extreme values indicative of the spread of the series, but quite frequently is misleading, because it gives no information about how the items are dispersed.

16.26.1   Characteristics of Range

1. The range is simple and readily understood.
2. It is easily calculated.
3. Its value is dependent on two items only, the highest and lowest values.
4. It is not necessary to know the distribution of the items between the two extremes in order to obtain the range.
5. Because the range is dependent only upon the two extremes, it is greatly affected by unusual maximum and minimum values.

16.27   MEAN DEVIATION

The range is dependent for its value entirely upon the two extreme values. Obviously, when these end-values are far removed from the remainder of the data, a satisfactory measure of dispersion must be dependent upon the position of every value in the series.

A simple method for determining the scatter of a series of values about a given point is to take the average distance of the items from the given point. The smaller the average distance about this point, the smaller the scatter or dispersion of the values.

In a frequency distribution, the average distances of the items from the measure of control tendency, such as the arithmetic mean, may be used for this purpose. However, since the sum of the deviations about the arithmetic mean is zero, it is necessary to ignore signs in order to obtain the average distance of items from that measure.

16.27.1   Characteristics of Mean Deviation

1. The value of the mean deviation is dependent upon the value of every item in the series.
2. It may be computed about any measure of central tendency.
3. The mean deviation about the median is less than calculated about any other point.

16.27.2   Computation of Mean Deviation

The mean deviation can be computed by the following formula.

$$MD = \frac{\sum d}{N}$$

where

$\sum d$ = sum of the deviations of each value from the arithmetic mean.
N = total number of items.

16.28   STANDARD DEVIATION

The standard deviation represented by the Greek letter sigma, σ, is the most useful value in statistics and in total quality management to understand the deviation of the values from the mean in a distribution. It is computed by taking the quadratic mean of the deviations from the arithmetic mean of those values. A standard deviation close to 0 indicates that all values are close to the mean with a steep curve of high kurtosis. The lower the σ, the wider the variation. While the standard deviation is thus called the root-mean-square, the analysis of the deviation is called analysis of variation or ANOVA in short.

$$\sigma^2 = \frac{\sum \left(X^2 \right)}{N}$$

where,

σ = standard deviation.
X = deviation of individual item from arithmetic mean = $X_n - \bar{X}$.
N = total number of items.

16.28.1   Computation of σ from Ungrouped Data

1. Get the difference between each actual value and the arithmetic mean.
2. Square the values thus obtained. Obtain the average of the squares.
3. Take the square root of the result.

16.28.2   Computation of σ from Grouped Data

Where there are a considerable number of items in the series the calculation of the standard deviation can be more readily performed if the data is first grouped into the form of a frequency distribution.

1. The deviation of the midpoint of each group from the arithmetic mean is used as a measure of the average deviation from the mean of all items in the group.
2. The average deviation of each group is squared to obtain the necessary deviation squared.
3. The average deviation squared is multiplied by the frequency indicated for the group in order to obtain the total of the squared deviations for that group.
4. The totals are then added for the entire distribution.
5. The square root of the sum obtained after dividing by N is the standard deviation.

$$\sigma = \sqrt{\frac{\Sigma f\left(X^2\right)}{N}}$$

16.28.3   Characteristics of Standard Deviation

1. The standard deviation is affected by the value of every item.
2. Greater emphasis is placed on extremes than in the mean deviation; this is because all the values are squared in the computation.
3. In a normal or bell-shaped distribution, the standard deviation shows the following relationship with individual values.
   (a) If a distance equal to one standard deviation is measured off on the X axis on both sides of the arithmetic mean in a normal distribution, 68.26% of the values will be included within the limit indicated.
   (b) If two standard deviations are measured off, 95.46% of the values will be included.
   (c) If three standard deviations are measured off, 99.75% of the values will be included.

16.29   SKEWNESS

Referring to Fig. 16.15, skewness is a term for the degree of distortion from symmetry exhibited by a frequency distribution.

When a distribution is perfectly symmetrical with one mode, the values of the mean, median, and mode coincide. In an asymmetrical (skewed) distribution, their values will be different and it will be farthest from the mode. The mode is not affected at all by unusual values; therefore the greater the degree of skewness the greater the distance between the mean and the mode.

Measure for skewness:

$$S_k = \frac{\left(\text{Mean} - \text{Node}\right)}{\text{Standard Deviation}}$$

16.30   KURTOSIS

This is a measure of the peakness of the distribution as is clear from Fig. 16.16.

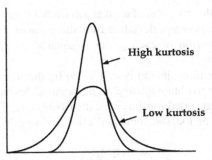

FIG. 16.16 Normal curves with high and low kurtosis.

16.31 CONCLUSION

It can be seen that the understanding of basic statistics and the related terminology is a must for practicing statistical quality control, without which total quality management has no place.

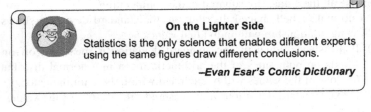

On the Lighter Side

Statistics is the only science that enables different experts using the same figures draw different conclusions.

—Evan Esar's Comic Dictionary

FURTHER READING

[1] http://en.wikipedia.org.
[2] www.investopedia.com/terms.
[3] mathworld.wolfram.com.
[4] everythingmaths.co.za/maths.
[5] https://statistics.laerd.comcentral-tendency.
[6] www.regentsprep.org/regents/math.

Chapter 17

Fundamentals of Statistics—Part II

Chapter Outline

17.1 CORRELATION

We have so far considered series having only one variable or one characteristic, such as the weight or height of a person, or the marks obtained by a student.

But in actual practice, we may have to simultaneously consider more than one variable or characteristic at a time, for example, the height and weight of a person, or quantity of fertilizer applied, and the quantity of yield obtained. Sometimes each item may have three or more variables.

The values of different variables may be interrelated. For example, the weight of a person may depend on the height of a person or their height and weight may depend upon their age. The quantity of yield obtained may depend upon the quantity of fertilizer applied. The relationship between two or more variables is called the correlation and the variables are said to be correlated. Sometimes the relationship is also called co-variation.

Total Quality Management: Key Concepts and Case Studies. http://dx.doi.org/10.1016/B978-0-12-811035-5.00017-9

The following illustration helps us understand this principle better.

A random sample is selected from the population of men between the ages of 40 and 50 in a certain city, who are employed fulltime and the number of years of schooling (X), and the monthly income in thousands of rupees is recorded for each man. Suppose further that the random sample of 12 men yielded data as shown in Table 17.1. The discussions on parameters per Section 17.1.1 help us to understand how (X) schooling years and (Y) income are related linearly.

While Fig. 17.1 tabulates the above data, Fig. 17.2 is drawn to indicate how the income of a person is dependent on the schooling years. The higher the schooling years, the higher would be the income earned, even though it is not so in all cases. This relationship between the two variables is called correlation and helps us know whether one variable is fully dependent on the other, or only partly.

TABLE 17.1 Schooling Years Versus Income

Schooling Years (Thousand of Rupees) (X)	Income (Y)
10	6
7	4
12	7
12	8
9	10
16	17
12	10
18	15
8	5
12	6
14	11
16	13

17.1.1   Scatter Diagram

Below gives a rough idea of how the variables X and Y are related. From the scatter diagram we cannot conclude that there is or is not a linear relationship between the two variables. This is represented by Fig. 17.1.

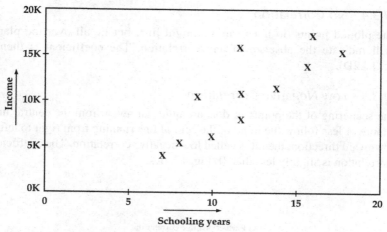

FIG. 17.1   Scatter diagram for data per Table 17.1.

17.1.2   Coefficient of Correlation

We need a measure for knowing how much two variables are related to each other and how much relationship exists between X and Y. This measure is called the coefficient of correlation. Sometimes it is called the Karl Pearson's Coefficient or correlation and is denoted by "r."

17.1.3   Types of Correlation

Fig. 17.2 illustrates the different types of correlation as further explained below. The graph representing these types of correlation is also called a scatter diagram, which is significantly described by the Japanese as one of the seven traditional tools of TQM, as described more in detail in Chapter 20.

17.1.3.1   Perfect Positive Correlation

If all the plotted points or dots form a straight line running from left to right in the upward direction, the correlation is said to be perfect positive, and the coefficient is $r=1$ (Fig. 17.2A).

17.1.3.2   High Positive Correlation

If the points or dots are scattered around a straight line running from left to right in an upward direction, instead of all lying exactly on a straight line, as explained before, the correlation is said to be positive. The coefficient is between 0 and 1 (Fig. 17.2B).

17.1.3.3   Low Positive Correlation

If the scattering of the points or dots are quite far away from the central line, yet more or less follow the trend of the central line, then it is called low positive correlation. The coefficient of correlation is very small, nearer to 0 (Fig. 17.2C).

17.1.3.4   No Correlation

If the plotted points do not form a straight line, but lie all over the plane, it will indicate the absence of any correlation. The coefficient is then 0 (Fig. 17.2D).

17.1.3.5   Low Negative Correlation

If the scattering of the points or dots are quite far away from the central line, yet more or less follow the trend of the central line running from right to left in a downward direction, then it is called low negative correlation. The coefficient of correlation is slightly less than 0 (Fig. 17.2E).

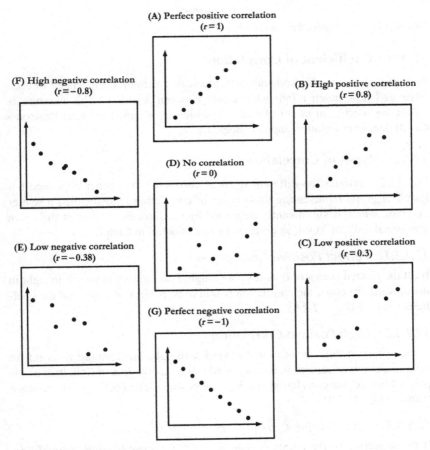

FIG. 17.2   Types of correlation.

17.1.3.6 High Negative Correlation

If the points or dots are scattered around a straight line running from left to right in a downward direction, instead of all lying exactly on a straight line, the correlation is said to be negative. The coefficient is between 0 and −1 (Fig.17.2F).

17.1.3.7 Perfect Negative Correlation

If all the points or dots in a scatter diagram form a straight line running from right to left in a downward direction, the correlation is said to be perfect negative and the coefficient is −1 (Fig. 17.2G).

17.2 REGRESSION

As discussed earlier where a line of best fit is drawn, it is obvious that this is dependent upon the subjective judgment of the person who draws it and the position of the line may slightly differ from person to person. Hence, a line of best fit which is independent of individual judgment will have to be drawn mathematically. Such a line is called a regression line. This is also called the line of best fit.

Difference between Correlation and Regression

Correlation quantifies the degree to which two variables are related. The computed *correlation* coefficient (*r*) that tells you how much one variable tends to change when the other one does. On the other hand, regression line fits through the data points.

The general equation for any straight line graph is

$$y = a + bx \qquad (17.1)$$

where

  x is independent variable and
  y is dependent variable.

The line of best fit should be such that the positive deviations due to dots above the fitted line should be set off against the negative deviations due to dots below the fitted line. That is, an attempt is made to minimize the total divergence of the points from the line. This approach is logically known as the method of least squares, which is based on the principle of minimizing the sum of squares of all the values from the values estimated from the line.

17.3 RELATION BETWEEN CORRELATION AND REGRESSION

1. Both deal with the subject of relationship between two variables.

2. Correlation is a measure of the degree of relationship between two variables, while regression helps to predict the value of one variable of the other is known with the help of correlation.
3. Correlation is numerical measure of relationship while regression establishes a functional relationship which indicates the independent and dependent variables.
4. Correlation does not indicate the cause and effect in relationships while regression does so.
5. Coefficient of correlation is the Geometric mean of the two regression coefficients.

ie, by_x = regression of y on $x = ry_x$
bx_y = regression of x on $y = rx_y$
Therefore $by_x \times bx_y = ry_x \times rx_y = \sqrt{r^2} = r$

17.4   SAMPLING THEORY

17.4.1   Introduction

Sampling theory is a study of relationships between a population and samples drawn from the population. It is an established systematic procedure, by which some members of the population are selected as representatives of the entire population. It is useful in estimating unknown population quantities and parameters, like mean and variance, from a set of corresponding sample quantities, often called the sample parameters. The application of sampling theory is concerned not only with the proper selection of observations from the population, but also involves the use of probability theory, along with prior knowledge about the population parameters, to analyze the data from the random sample and develop conclusions from the analysis.

Some of the terms used here are:

● The *population* is the total set of observations that can be made.
● A *sample* is a set of observations drawn from a population.
● A *parameter* is a measurable characteristic of a population, such as a mean or standard deviation.
● A *statistic* is a measurable characteristic of a sample, such as a mean or standard deviation.
● A *sampling method* is a procedure for selecting sample elements from a population.
● A *random number* is a number determined totally by chance, with no predictable relationship to any other number.

The following paragraphs explain the significance of this systematic procedure and estimating the parent parameters precisely.

17.4.2   Random Number Tables

No matter however much you want to think of some number at random, there will be some unconscious bias in choosing the numbers. Hence, random number tables, where several numbers even up to one million are arranged at random sequence. The randomness can be further increased by choosing them in random order, say the horizontal order, or every fifth number, etc.

17.4.3   The Sampling Process

- Defining the population of concern
- Specifying a sampling frame, a set of items, or events possible to measure
- Specifying a sampling method for selecting items or events from the frame
- Determining the sample size
- Implementing the sampling plan
- Sampling and data collecting
- Data which can be selected

17.4.4   Sampling Methods

1. Deliberate (nonrandom) sampling
2. Simple random sampling
3. Cluster sampling
4. Systematic sampling
5. Stratified cluster sampling
6. Multilevel sampling

Deliberate sampling is also called purposing, or nonprobabilistic sampling. The sample is selected based on the ease of access or based on one's judgment.

Simple random sampling, is one in which the members are drawn independently, using a random number table one after the other, in the order they are printed in the table.

Cluster sampling, where the total population is divided into natural but relatively homogeneous groups. The basic reason for cluster sampling is to reduce the total number of interviews and costs for a desired accuracy and it is often used in marketing research.

Systematic sampling first arranges the population according to some ordering scheme and then selects samples at regular intervals by making a random start and then proceeding with the selection of every kth element from then onwards. As an illustration, to draw a 10% sample from a population of say, 900 cards, first choose a number between 1 and 10 at random, say 6, then select the 6th card, and thereafter, choose every 10th card, like the 16th, 26th, 36th, etc., till 10% of the population are selected as samples.

Stratified sampling is somewhat similar to the cluster sampling, except that each cluster or stratum is treated as a separate population and samples are drawn from each, either by simple random or systematic random scheme. The result for each calculation is given a weightage factor, depending upon the stratum and then the overall arithmetic mean is calculated. It may be noted that in stratified sampling, a random sample is drawn from all the strata, whereas in cluster sampling, only the selected clusters are studied, either in single- or multi-stage. This method is popular in Monte Carlo methods, which is explained in Chapter 19.

Multilevel sampling is a variation of cluster sampling, applicable when the population is too large even to draw representative samples. In this method, the whole population is grouped into a large number of clusters and few clusters are selected at random. Then as a second stage, samples are drawn from each selected cluster and analyzed. Shown below is an illustration of this method.

Australian Bureau of Statistics did a multilevel sampling for a survey of Australian dwelling units by first dividing metropolitan regions into 'collection districts' and selecting some of these collection districts (first level). The selected collection districts are then divided into blocks, and blocks are chosen from within each selected collection district (second level). Next, dwellings are listed within each selected block, and some of these dwellings are selected (third level).

17.4.5   Factors for Selection

Factors commonly influencing the choice between these designs include:

1. Nature and quality of the frame
2. Availability of auxiliary information about units on the frame
3. Accuracy requirements, and the need to measure accuracy
4. Whether detailed analysis of the sample is expected
5. Cost/operational concerns

17.4.6   Frequency of Sampling

The samples must be collected frequently enough to ensure that the process is stable, and to identify the moment the process is going haywire. But at the same time, too frequent sampling only would result in high costs without added value. Hence, the following factors are to be considered before deciding upon the frequency of sampling.

Process stability: When the process is known to exhibit erratic behavior with too many variations in the previously collected samples, then the frequency has to be increased, until the process stability is established.

Frequency of changes in process parameters, like material changes, tool changes, design changes, etc. Sampling must be done immediately after these changes, and frequently thereafter, until no more changes are expected.

*Sampling cost, a*s we said earlier, the more the sampling, the costlier it is. Nevertheless, all the costs must be weighed against each other and a suitable decision should be made. Another cost factor is the duration of sampling, which should be analyzed and decision made.

17.4.7 Estimating the Sample Size

The sample size shall be adequate to give a precise estimation wherever the goal is to make inferences about a population from a sample. Sample size determination is the act of choosing the number of observations or replicates to include in a statistical sample, larger sample sizes would yield increased precision when estimating unknown parameters. At the same time, too large a sample results in wastage of resources such as time, energy, and funds. Hence, it is essential to estimate the size of the sample or the number of sample readings to be taken at the first instant.

Methods of determining the sample sizes are:

- Use of experience as applicable to the situation
- Use of empirical formulae
- Use of statistical tables normally available as a reference
- Use of software such as:
  - nQuery Advisor
  - PASS sample size software
  - SAS power and sample size
- Using a target variance for an estimate to be derived from the sample eventually obtained
- Using a target for the power of a statistical test to be applied once the sample is collected

17.4.8 Factors that Influence the Sample Size Include

1. The amount of confidence required from the results of the samples
2. The variability of the process
3. The cost of sampling and
4. The cost of imprecise estimates

17.5 PROBABILITY

It is defined as a measure of the likeliness that an event will occur. It is used to quantify an attitude of mind towards some proposition of whose truth we are not certain. Many events can't be predicted with total certainty. The best we can say is how likely they are to happen. The word *probability* is derived from the Latin *probabilitasor probity*, which refers to the measure of the authority of a witness in a legal case in Europe during the medieval era.

Instead of saying *"Will a specific event occur?"* the probability concept will say *"How certain are we that the event will occur?"* and this help us in fixing the occurrence in a numeric value between 0 and 1 or as a percentage. It has been our regular experience many times we face a situation of making a decision whose outcome is uncertain. But we have to take a positive or negative decision. In this case, we always ask as above, what would be the probability that the outcome would be as we want. If the probability we think is more than 80%, we will take the decision. If we think it would be around 50%, we just wait for better analysis. If we think that the probability would be 20% or less, we would determine ourselves not to make that decision. Hence, the concept of probability would help us in analyzing and taking a good decision.

The tossing of an individual coin or rolling of a die is a random event, and if repeated many times, the sequence of random events will exhibit certain statistical patterns, which can be studied and predicted. Let us consider an example where all the 52 cards of a deck are well shuffled and laid on the table face down. The probability of picking at random, any card of the hearts suit is 1/4 or 25%, the probability of picking any queen is 1/13 or 7.7%, and the probability of picking a spade king is 1/52 or 1.92%.

Probability theory deals with analysis of random phenomena. It was first conceived by Gerolamo Cardano in the sixteenth century when he was attempting to analyze chances in the betting games. He also wrote a book on *Games of Chances* in 1564. Later Pierre de Fermat and Blaise Pascal of the seventeenth century worked further on probability of theory. Others who are credited with working further and developing this theory were Christiaan Huygens (1657), Jakob Bernoulli's (1713), Abraham de Moivre (1718), Ian Hacking, James Franklin, and Roger Cotes (1722).

17.6 LAWS OF PROBABILITY

17.6.1 The Law of Addition

This law states that in case of two mutually exclusive events, A and B, the probability of occurrence of either A or B is the sum of the individual probabilities of A and B,

That is

$$P(A \text{ or } B) = P(A) + P(B)$$

If three cards A, B, and C, are to be drawn, the probability of drawing any of the three cards is given by

$$P(A \text{ or } B \text{ or } C) = P(A) + P(B) + P(C)$$

17.6.2 Mutually Exclusive Versus Mutually NonExclusive

Mutually exclusive means one event has no effect on the other event. For example, if we have to draw either a spade ace or a diamond king, either has no effect on drawing of the other. Once we draw a spade ace, it can never be drawn in the second draw. Hence, they are mutually exclusive.

However, if we have to draw either a king or a hearts card, and if by first draw we draw a king, then it can either be a hearts or of other suits. If it is a hearts king, then it will affect the probability of drawing a hearts card next. Thus drawing cards from different groups is not mutually exclusive.

The law of addition in case of mutually nonexclusive events states that:

$$P(A \text{ or } B) = P(A) + P(B) - P(A \text{ and } B)$$

In case of three cards, it is given by

$$P(A \text{ or } B \text{ or } C) = P(A) + P(B) + P(C) - P(AB) - P(AC) - P(BC) + P(ABC)$$

17.6.3   Law of Multiplication

The probability of drawing both two independent and mutually exclusive events is given by the product of the individual probabilities.
or

$$P(A \text{ and } B) = P(A) \times P(B)$$

This is similar to any number of events.

17.6.4   Law of Conditional Probability

This law is a variation of the Law of multiplication for dependent events, that is when B can occur only when A is known to have occurred already as symbolized by (B/A) or the probability of B given that A had occurred.

$$P(B/A) = P(AB)/P(A)$$

For three events,

$$P(ABC) = P(A) \times P(B/A) \times P(C/AB)$$

While the above paragraphs give a basic understanding of the probability concept, it is felt that further explanation and advanced topics of the theory are beyond the scope of this book.

17.7   CONCLUSION

Correlation and regression indicate the statistical relationship between two random variables or two sets of data. Though many events can't be predicted with total certainty, probability helps us in predicting how likely an event would happen. These two concepts are expected to develop a basis for the reader for predicting the events to a certain degree of accuracy.

On the Lighter Side

In English, the word 'fat' implies thick or large. One would assume then that the phrase 'a fat chance' refers to a very high probability of something occurring. But the English language has fooled us. 'A fat chance' refers just to the opposite and means that there no chance of occurring or with a low probability of occurrence.

—adapted from the Times of India of April 17, 2015.

FURTHER READING

[1] http://en.wikipedia.org/wiki/Correlation_and_dependence.

[2] www.investopedia.com/terms/c/correlation.asp.

[3] http://en.wikipedia.org/wiki/Sample_size_determination.

[4] www.surveysystem.com/sscalc.htm.

[5] http://en.wikipedia.org/wiki/Probability_theory.

[6] www.britannica.com.

[7] www.mathgoodies.com.

[8] http://ads.harvard.edu/books/.

Chapter 18

Process Capability

Chapter Outline

18.1 STATISTICAL PROCESS CONTROL

Statistical process control (SPC) is a statistical method of quality control for monitoring and controlling a process to ensure that it operates at its full potential. It determines the stability and predictability of a process. It can be applied to any process where the output of the product conforming to specifications can be measured. Control charts, continuous improvement, and the design of experiments are some of the key tools, which are further explained in Chapters 20, 22, and 31, respectively. Of these, control charts are most significant to SPC. The superiority of SPC over other TQM tools such as inspection, is that it emphasizes early detection and prevention of problems, rather than the correction of problems after they have occurred.

The Awarding Committee of Deming Application Prizes defined Statistical Quality Control (SQC) as "the integrated activity of designing, manufacturing and supplying the manufactured goods and services at a quality demanded by the customer at an economic cost." The committee also added that "the customer-oriented principle is the basis, in addition to paying keen attention to public welfare. The company's aim should be to succeed through the repetition of planning, execution, evaluation, and corrective action by applying the statistical concepts of activities of survey, research, design, procurement, manufacture, inspection, sales, etc., both inside and outside the company."

18.2 WHY CONTROL CHARTS?

SPC must be practiced in three phases, for which the control charts play a significant role in all the three phases, as explained in the next paragraph.

1. Understanding the process and the specification limits.
2. Eliminating assignable (special) sources of variation, so that the process is stable.
3. Monitoring the ongoing production process, assisted by the use of control charts, to detect significant changes of mean or variation.

During the 1920s, Dr. Walter A. Shewhart professed that every problem has assignable-cause and chance-cause variation, and introduced the control chart as a tool for distinguishing between the two. Subsequently, the control chart has become one of the most technically sophisticated tools of SPC, which is the process wing of SQC for the purpose of improving the economic effectiveness of the process. While Fig. 18.1 illustrates a control chart, it is explained more in detail in Chapter 20, as one of the seven traditional tools of TQM.

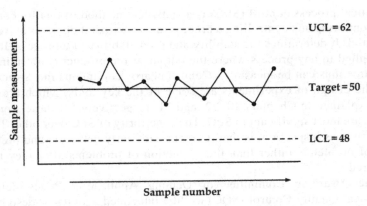

FIG. 18.1   Control chart.

18.3 REASONS FOR VARIATIONS

1. Lot to lot variation
2. Stream to stream variation
3. Time to time variation
4. Piece positional variation
5. Measurement variation
   - Instrumental errors
   - Human errors

18.4   PROCESS CAPABILITY

Process capability is a measurable property of a process which generally determines if the process is meeting its desired specification or not. The two main functions of process capability are:

- to measure the variability of the output of a process, and
- to compare that variability with a proposed specification, or product tolerance.

The output of this measurement is represented by a histogram and predicts how many parts will be produced out of specification (OOS) and is expressed as a process capability index.

18.5   PROCESS CAPABILITY INDEX

Process capability index measures the extent of variation a process experiences relative to its specification limits. It also helps us to compare different processes with respect to the optimal situation or if they come up to our expectations. The most common process capability index is given by C_p, which is an estimation of what the process is capable of producing if the process mean were to be centered between the specification limits, assuming that the process output is approximately normally distributed. We can say it indicates how many times the tolerance zone is larger than the 6σ value of the variation.

That is

$$C_p = \frac{(\text{UCL} - \text{LCL})}{6\sigma}$$

18.6   ONE-SIDED AND TWO-SIDED SPECIFICATIONS

In most cases, both the upper and the lower limits would be specified like the nominal diameter of a bar to be turned. However, in some cases the process mean may not be centered between the specification limits and C_p alone overestimates process capability. For example, for emission levels, only the upper limits would be specified while in cases like the strength, only the lower limit would be specified. These are called one-sided specifications. In such cases, the C_p for the former instances with upper limit is given as:

$$C_{p,\text{upper}} = \frac{(\text{UCL} - \mu)}{3\sigma}$$

while the C_p for the former instances with lower limit is given as

$$C_{p,\text{lower}} = \frac{(\mu - \text{LCL})}{3\sigma}$$

However, in such cases, the index is given by C_{pk}, which is given as the lesser value of the above two.

C_p is the capability the process could achieve if the process was perfectly centered between the specification limits. On the other hand, C_{pk} is the capability the process is achieving whether or not the mean is centered between the specification limits.

18.7   TAGUCHI CAPABILITY INDEX

Another index C_{pkm}, also known as the Taguchi capability index, estimates process capability around a target, T, and assumes process output is approximately normally distributed. It is given by

$$C_{pkm} = \frac{C_{pk}}{\sqrt{1+\left(\dfrac{\mu-T}{\sigma}\right)^2}}$$

where

C_{pkm} = Taguchi capability index,
C_{pk} = Process capability index
μ = Process average
T = Expected target and
σ = Standard deviation of the process

The distinction between C_{pkm} and C_{pk} lies with target and the specification limits. As explained in Chapter 31, Taguchi emphasized more on the target value than the specification limits. Thus, while C_{pk} relates to the specification limits, C_{pkm} relates to the target value, which is narrower than the former.

18.8   RECOMMENDED MINIMUM VALUES OF C_{PK}

If $C_{pk} = 1$, 99.73% of the process output will be expected to be within specification.
Montgomery and Douglas in their book, *Introduction to Statistical Quality Control*, recommend the following minimum values for C_{pk} (Source: Wikipedia).

Sl. No.	Situation	Minimum C_{pk} for Two-Sided Specifications	Minimum C_{pk} for One-Sided Specifications
1	Existing process	1.33	1.25
2	New process	1.50	1.45
3	Safety or critical parameter for existing process	1.50	1.45
4	Safety or critical parameter for new process	1.67	1.60
5	Six Sigma quality process	2.00	2.00

18.9 CONCLUSION

The process capability index is significant in the quality control to such an extent that these days, several corporate buyers include this in the terms of contract for the purchase of components.

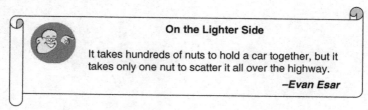

On the Lighter Side

It takes hundreds of nuts to hold a car together, but it takes only one nut to scatter it all over the highway.

—Evan Esar

FURTHER READING

[1] en.wikipedia.org/wiki/Process_capability_index.
[2] http://blog.gembaacademy.com/.
[3] www.isixsigma.com/%85/process-capability.
[4] blog.minitab.com/%85/process-capability.
[5] www.sixsigmastudyguide.com/process-capability.
[6] www.six-sigma-material.
[7] www.qualitygurus.com.

Chapter 19

Inward Inspection

Chapter Outline

In Chapter 2, we traced how the quality control function has undergone a metamorphosis from inspection tasks to the philosophy of TQM. The latest trend is the supplier partnership, wherein the supplier is given the confidence to take full responsibility of supplying quality goods straight to the shop floor without the inward inspection. Though such practices are in vogue in quite a number of Indian industries, the function of inward inspection as a basic function of inspection at several stages of manufacture continue to exist in a majority of industries. A basic knowledge about the inspection and other parameters involved is necessary to understand the processes and practices of TQM in a company. In the following paragraphs we shall discuss the stages of inspection performed in a production facility.

19.1 DEFINITIONS OF INSPECTION

Inspection is the process of measuring the quality of a product or service in terms of established standards.

W.R. Spreigal

Inspection is the art of comparing materials, products, or performance with established standards.

Kimball

Total Quality Management: Key Concepts and Case Studies. http://dx.doi.org/10.1016/B978-0-12-811035-5.00019-2

Inspection is the art of applying tests preferably by the aid of measuring instruments to observe whether a given item or product is within the specified limits of variability.

Alford and Beatly

19.2   OBJECTIVES OF INSPECTION

Inspection is used to aid managerial decisions whether to accept or reject a process to sustain quality.

- To sort out errors in a manufacturing process and to report to the concerned department/manager.
- To stop production of defective goods over and above the accepted norms.
- To ensure that substandard quality products do not reach the customer, thereby affecting further marketing prospects.
- To help the firm in increasing its reputation and brand quality by maintaining good quality standards.
- To aid the design and development department by compiling information on the performance of the product.
- To reduce the overall cost of production.
- To collect information for later decisions and analysis.
- To satisfy regulatory or procedural requirement.
- In case of existence of a quality problem, it helps in:
  - Identifying the problem
  - Preventing its occurrence
  - Eliminating the problem

19.3   STEPS INVOLVED IN INSPECTION

- Interpretation of quality requirements
- Sampling
- Examination
- Decision
- Action

The following inspection standards should be established in a company prior to undertaking inspection:

1. Inspection standards for new materials
2. Inspection standards for work in progress
3. Finished products
4. Working inspection standards
5. Inspection standards for finished products
6. Inspection standards for complete mechanism

19.4   CLASSIFICATIONS OF INSPECTION METHODS

The inspection processes can be classified into categories:

A. Based on the method adapted
B. Based on the quantum of inspection
C. Based on the purpose of inspection
D. Based on the stage at which the inspection is performed

A. Based on method adapted
   • Dimensional measurements, to determine if the numerical value of the parameter is as per the drawing or not.
   • Go, no-go checking is as a short cut to the above with the use of fixed limit gages, which can determine if the dimension especially the diameter is within the allowed tolerance limits.
   • Functional checking by testing the component/product performs the desired function per the norms, by simulating the usage.
   • Visual inspection, involving the human judgment by visually checking the parameters like color and surface finish. Visual standards like color charts can be used for comparison.
B. Based on the quantum of inspection
   • 100% inspection, when each and every item is checked for one or more parameters.
   • *On-spot inspection* when only few of a lot are checked, may be or may not be at random.
   • *Random inspection* when few pieces are selected from a lot at random and all required parameters are checked.
   • *Statistical sampling inspection*, where the statistical principles like considering the statistically estimated sample size, use of random tables and application of probability principles in the analysis etc., are adapted.
C. Based on purpose of inspection
   • Process control to prevent defectives from being produced unchecked. It involves taking samples from the process periodically and taking corrective action to reduce further occurrences.
   • Acceptance inspection to classify a lot as acceptable or rejected, especially for the purpose of informing the suppliers.
   • Process development, because in several cases, the inspection report provides a basis for introspection and revision of the process and manufacturing standards.
D. Based on the stage at which the inspection is performed
   • Source inspection
   • Inward inspection
   • In process inspection
   • Trial-run inspection
   • First piece inspection

- Patrol inspection
- Bench inspection
- Key point inspection
- Operator inspection
- Last piece inspection
- Final inspection
- Pre-shipping inspection

19.5   SOURCE INSPECTION

Source inspection is a technique used to prevent product defects by controlling the conditions that influence quality at their source. It is the performance of the supplier's facilities to increase customer confidence with the supplier's product quality. These days, with development of supplier partnership, as described in detail in Chapter 9, the vendor himself conducts the inspection and the buyer accepts these goods without further inspection at his place, unless unacceptable defects are discovered at the assembly or other stages.

A striking illustration is the system adapted by Chennai's TVS group companies during the 1990s itself by inspecting the goods at the suppliers' premises before shipment. It may be noted that this system has now evolved into supplier partnership as detailed in Chapter 12, wherein even the source inspection is no longer done by mutual trust.

The following elements are essential parts of source inspection.

- The quality history of suppliers.
- Any possible effects that occur during purchasing, based on the performance, safety, and reliability of the final product.
- Product complexity.
- The ability to measure the product quality from buyer data.
- The availability of special measuring equipment either at the buyer's or supplier's premises to perform the required inspection.
- The product's nature and its quality.

It is important to have either external or internal company inspectors to assure adequate product control. A sources inspection is performed to insure that the decision-making is correct and unbiased. Furthermore, source inspection can be divided into two categories as follows;

1. Vertical source inspection inspects the process flow to identify and control external conditions that affect quality.
2. Horizontal source inspection inspects an operation to identify and control interval conditions that affect quality.

19.6   INWARD INSPECTION

Before the buyer authorizes the payment to the vendor, the incoming goods are physically inspected to check their specifications. Either 100% or sampling inspection is carried out, depending upon the size and quality of the incoming items, as illustrated in Table 19.1. If the inspection is carried out at the buyer's place after the goods are received, it is called incoming inspection or inward inspection. Sometimes the inspection is carried out by the buyer at the vendor's place before they leave when it is called source inspection.

Objectives of Inward Inspection

- To separate those items that do not meet required specification before they enter the factory.
- Evaluate the vender's quality and ability to supply goods per required specification.
- Reduce the cost of wasteful transport by inspecting large-sized items at the vendor's premises.

TABLE 19.1 Inward Inspection Norms

Type	Procedure	Where Appropriate
100% Inspection	Each item of the lot is inspected for all or some of the design features	Critical items where the cost of inspection is justified by the cost of risk of defects. Also used to establish the quality level of new vendors
Sampling inspection	A sample of each lot is evaluated by a predetermined sampling plan and a decision made to accept or reject the lot	Important items where the vendor has established an adequate quality record history
Identity inspection	The product is examined to assure that the vendor sent the correct product. No inspection characteristics is made	Items of lesser importance where the reliability of the vendor has been established in addition to the quality level of the product

19.7   SINGLE AND DOUBLE SAMPLING INSPECTION

Single and double sampling in inward inspection.

If the inward material is large-sized but a handful in number, 100% inspection is carried out. But in most cases, the components like fasteners are received in large numbers of 10,000 or more. In that case, sampling insertion would be done.

This sampling can either be single sampling or double sampling as detailed below.

Let us assume that the lot size is 1000 and the computed sample size is 100 with an acceptance limit of 3%, then the acceptance/rejection decision is illustrate in Fig. 19.1A. If out of the 100 random samples, only 3 pieces are defective, accept the whole lot and if the number of defectives is 4 or more, reject the lot.

In double sampling, a two-stage sampling is done, especially in case of specialty items and you want make doubly sure before rejecting the whole lot. Here you fix a range of 3–5% defectives for the rejection. In the first sample, if the number of defectives is three or less, accept the whole lot and if the number of defectives is six or more, reject the whole lot forthwith. However, if the number of defectives is either four or five, then draw another sample of 100 and follow the decision as in the first sample, as illustrated in Fig. 19.1A.

Multiple sampling is an extension of double sampling with more than two sampling stages used as per the above illustration.

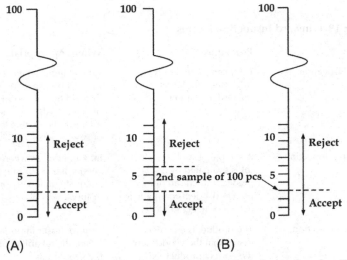

FIG. 19.1 Acceptance sampling plans for inward material: (A) single sampling plan, (B) double sampling plan.

19.8 IN PROCESS INSPECTION AND FINAL INSPECTION

In Process inspection is the inspection performed by Quality Assurance on production or process materials or assemblies, all along the progress of the manufacture. This is a vital stage after incoming Inspection and before final inspection.

The inspection of a part during production is to detect errors. Errors that are detected early may allow the part to be reworked or prevented from continuing through the manufacturing process. Hence, it is a measure aimed at checking, measuring, or testing of one or more product characteristics and to relate the results to the requirements to confirm compliance.

Traditionally, this task is usually performed by specialized personnel and does not fall within the responsibility of production workers. However, as explained in Chapter 13 on TPM, a major portion of this task such as the routine checking of the dimensions etc., are now performed by the operator himself.

Trial-run inspection: The tools, jigs, fixtures, gauges, etc., are inspected for accuracy prior to the commencement of production. The trial-run products are tested meticulously and set-up is being adjusted until an acceptable product is produced. This is generally significant in cases of automatic machines. Once the trial-run is successfully carried out, the quality products continue to come out for a longer period than the manual or semi-automatic machines.

First piece inspection is done more meticulously on the initial production or the first items that come out of machine working on a particular operation. This is done for the purpose of setting up and making adjustments on the machine. The aim is to detect defects of a non-random type or repetitive nature and eliminate them early. A majority of these defects occur due to faulty set-up of the machine, jigs, fixtures, tools measuring instruments or from misinterpretation of the drawings or the instructions given. This is mostly done by the operator or his supervisor. The quality department is generally not involved in this unless called for or defects identified during the patrol inspection which is explained in the next paragraph.

Patrol inspection involves regular or periodic sampling inspection performed during the progress of an operation. As per Taylor's principle of Scientific Management, a team of inspectors from the quality control department keep moving among the machines, continuously performing inspection of the components under manufacture. Hence this is called the patrol inspection. Again in view of the principle of TPM, wherein the operator does most in-process inspections, this patrol inspection is becoming less significant, though this practice is continuing in several factories world over, especially those involving specialized skills and instruments.

Bench inspection is similar in principle to patrol inspection, but if the checking instruments are heavy, the components under manufacture are carried at random to the inspectors' table and checked. Even under TPM, also when the testing instruments are heavy or table-mounted, the operator carries them to the bench to do the inspection himself.

Key point inspection: In the process sequence, certain complicated and expensive operations are selected as key points, when prior to that operation, all rejects are segregated so that defective pieces are not unnecessarily sent for this expensive operation.

Operations in the process that are generally skipped for inspection: While literally process inspection has to be carried out after every operation, sometimes only certain operations in the process are selected for inspection. For example, if a reaming operation is to be followed after drilling on a separate machine, the inspection is done after reaming only.

Last piece inspection is carried out in the final pieces of a batch of components being manufactured. This allows any modifications necessary to be made in the set-up to rectify the faults in the equipment before commencing the next batch.

Final inspection is done after the final operation on the component. The purpose is to prevent faulty components being sent to assembly or getting shipped to the customer. Visual and functional testing are performed during the final inspection.

Endurance inspection: This is done mostly on assembled components to test how long an assembly will withstand its use and to determine its weaknesses for correction.

Pre-shipping inspection: Several companies give wide publicity to the systems they adapt in final product testing and the buyers receive the QC report to create confidence among the buyers. The objectives of final inspection can be summarized as:

- Ensure product safety prior to shipping
- Minimize the quantum of defective merchandise
- Reduce customer complaints due to inferior purchased components
- Detect merchandise having non-standard or non-compliant components
- Eliminate late shipments

19.9   TOOLS OF INSPECTION

The following are some tools and instruments and methods that are commonly used in inspection.

Measuring instruments, where the exact dimensions are determined by measuring with micrometer, Vernier calipers, torque indicating wrenches, etc.

Limit gauges, which can be used even by the unskilled operator to know whether the dimensions are within tolerance limits.

Multiple gauges, like electrical current operated instruments which can simultaneously measure more than one parameter.

Air gauges, useful in the inspection of mass production work.

Optical sensors for on-line testing of the dimensional accuracy, as illustrated Fig. 19.2 can be used to check the run-out even at high speeds.

NDT gadgets: Non destructive testing (NDT) is basically an off-line testing for material defects, such as cracks or air bubbles inside the cast or welded components. There are basically four categories of NDT inspection, the dye penetrant testing, Magnetic particle testing, ultrasonic testing, and radiographic testing, besides thermal graphing, etc. More details of these NDT methods can be had by referring to Chapter 6 in the book on Maintenance Engineering and Management by this author.

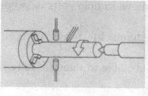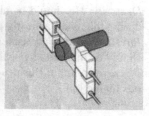

FIG. 19.2   Optical sensors.

19.10   NORMAL JOBS OF A QUALITY CONTROL INSPECTOR

1. Read blueprints and specifications
2. Monitor operations to ensure that they meet production standards
3. Recommend adjustments to the assembly or production process
4. Inspect, test, or measure materials or products being produced
5. Measure products with rulers, calipers, gauges, or micrometers
6. Accept or reject finished items
7. Remove all products and materials that fail to meet specifications
8. Discuss inspection results with those responsible for products
9. Report inspection and test data

19.11   REQUIREMENTS OF AN INSPECTOR

1. *Knowledge and skill*: He should have full working knowledge of quality standards and possess skill in the inspection process is essential. He should know his job thoroughly.
2. *General knowledge*: He should be intelligent and have good grasping power.
3. *Statistical processes*: He should have the basic understanding of statistics and statistical quality control.
4. *Mechanical skills*: He must be able to use specialized tools and machinery when testing products.
5. *Technical skills*: He must understand blueprints, technical documents, and manuals, ensuring that products and parts meet quality standards.
6. *Mathematical skills*: He should have basic mathematical and computer skills, because measuring, calibrating, and calculating specifications are major parts of quality control testing.
7. *Responsibility*: He should understand his responsibility fully and work with patience. Any negligence on his part would cause severe losses.
8. *Cost-consciousness*: He should be cost-conscious and should not allow unnecessarily strict and tight control limits.

9. *Dexterity*: He should be able to quickly remove sample parts or products during the manufacturing process.
10. *Physical strength*: Because workers sometimes lift heavy objects, he should be in good physical condition.
11. *Physical stamina*: Besides strength, he must be able to stand for long periods on the job.

19.12   CONCLUSION

As explained in Chapter 12, the corporate buyers and suppliers should forge a partnership with all commitments and accountability from both sides. This would enable direct flow of components from the suppliers directly to the assembly shop, eliminating the inward inspection. This was actually witnessed by this author in one of the TVS groups companies, when a supplier's truck went straight to the assembly shop and unloaded the components bypassing the inward inspection.

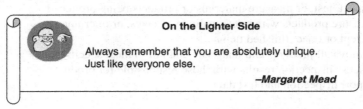

On the Lighter Side

Always remember that you are absolutely unique.
Just like everyone else.

–Margaret Mead

FURTHER READING

[1] Kiran DR. Maintenance engineering and management. Hyderabad: BS Publications; 2014.
[2] Inward inspection. https://scn.sap.com/.
[3] Kiran DR. Inward inspection at the bicycle factory, a report dated, 1978.
[4] National Productivity Council. A report on Inspection process of 1989.
[5] Material inspection. http://www.academia.edu/.
[6] Inspetion fundamentals. https://www.faa.gov/regulations_policies/.

Chapter 20

Seven Traditional Tools of TQM

Chapter Outline

20.1 INTRODUCTION

Kaoru Ishikawa, the father of cause and effect or Fishbone diagram, and other TQM exponents proposed seven methods of analyzing TQM problems graphically, in line with *the* Japanese folklore which says that the Samurai or the warrior always went to a battle armed with seven tools. These have come to be known as the seven tools of Total Quality Management. Because subsequently seven more tools for TQM have been propounded as indicated in the next chapter, Ishikawa's tools have come to be known as the Seven Traditional Tools while the latter are known as the Seven Modern TQM tools, which are discussed in a later chapter. The Seven Traditional Tools are:

1. Cause and Effect, Fishbone, or Ishikawa diagram
2. Pareto principle
3. Scatter diagram
4. Control chart
5. Flow chart
6. Histogram or bar graph and
7. Check sheet

Total Quality Management: Key Concepts and Case Studies. http://dx.doi.org/10.1016/B978-0-12-811035-5.00020-9

We can group these into:

A. *Statistical Tools*
  1. Check sheet, also called stratification by some authors
  2. Histogram
  3. Scatter diagram
  4. Control chart
B. *Analytical Tools*
  5. Pareto analysis
  6. Flow chart
  7. Cause and Effect analysis

It can be seen that the statistical tools are very much interrelated and form the sequence to be followed for any statistical analysis. An illustration is the Teacher Assessment Form used in engineering colleges, as explained in the later paragraphs. On the other hand, the tools listed above as analytical tools can be independently used for any analysis.

20.2 CHECK SHEETS AND CHECKLISTS

A Check Sheet or a Tally Sheet is a data recording form that has been designed to readily interpret results from the form itself. It needs to be designed for the specific data it is to gather. Being adaptable to different data gathering situations, it is easy and quick to use and requires only minimal interpretation of results. It is free from various forms of bias—exclusion, interaction, perception, operational, non-response, estimation.

This check sheet may not be confused with a checklist which contains items that are important or relevant to a specific issue or situation. Checklists are used under operational conditions to ensure that all important steps or actions have been taken. Their primary purpose is for guiding operations, not for collecting data.

As one of Ishikawa's basic quality tools, check sheets are an effective means of gathering data in a helpful, meaningful way. They are easy to use and allow the user to collect data in a systematic and organized manner. Many types of check sheets are available. The most common are the defective item, defective location, defective cause, and checkup confirmation check sheets.

Sheets, apart from several types as illustrated below.

When the recording of a particular parameter is done in several sheets, the tally sheet helps in summarizing all the parameters from all the pages into a single sheet without missing a single parameter.

Objectives of check sheets

- Clearly identify of what is being observed.
- Keep the data collection process as easy as possible.
- Group the data. Collected data should be grouped in a way that makes the data valuable and reliable. Similar problems must be in similar groups.
- Create a format that will give the most information with the least amount of effort.

Types of check sheets

1. *Distribution check sheets*, used to collect data in order to determine how a variable is dispersed within an area of possible occurrences.
2. *Location check sheets*, used to highlight the physical location of a problem/ defect in order to improve quality.
3. *Cause check sheets*, used to keep track of how often a problem happens or records the cause of a certain problem.
4. *Classification check sheets*, Used to keep track of the frequency of major classifications involving the delivery of products or services.

Basic steps to construct a check sheet

1. Clearly define the objective of the data collection.
2. Determine other information about the source of the data that should be recorded, such as shift, date, or working point.
3. Determine and define all categories of data to be collected.
4. Determine the time period for data collection and who will collect the data.
5. Determine how instructions will be given to those involved in data collection.
6. Design a check sheet by listing categories to be counted.
7. Pilot the check sheet to determine ease of use and reliability of results.
8. Modify the check sheet based on results of the pilot.

SI. no	Class Interval	Tally	Frequency
1	40 to 44	/	1
2	45 to 49	//	2
3	50 to 54	//// /	6
4	55 to 59	//// ////	10
5	60 to 64	//// //// //	12
6	65 to 69	//// //// ////	15
7	70 to 74	//// /	6
8	75 to 79	////	4
9	80 to 84	///	3
10	85 to 89	/	1

FIG. 20.1 Tally sheet for marks obtained.

Illustration No. 1

Step 1: Let us consider the marks obtained by each of the 60 students in a class in the order of their roll number are:

57, 56, 54, 73, 83, 48, 60, 50, 67, 62, 74, 56, 84, 78, 60, 67, 79, 53, 50, 76, 66, 65, 65, 72, 52, 87, 42, 75, 66, 68, 68, 69, 70, 51, 61, 72, 64, 63, 54, 74, 66, 58, 57, 68, 81, 74, 67, 68, 64, 47, 64, 63, 63, 65, 67, 62, 67, 55, 58 and 59.

Step 2: Arrange a tally sheet with 10 class intervals, each containing 5 marks like 40 to 44, 45 to 49… up to 85 to 90, as per Fig. 20.1.

This is an illustration of simple tally sheet for a single variable. This is further explained in Section 20.3 on how the histogram can be drawn based on this tally sheet.

Illustration No. 2

Another typical illustration for the tally sheet is the *"Subject Evaluation by Students"* exercise being done at engineering colleges, as illustrated below.

The principle of the check sheet or tally sheet is illustrated by Figs. 20.2 and 20.3 of the Teacher Assessment Form used in engineering colleges as per the following procedure.

1. In a class of 60, each student assesses the capacity of the teacher by giving 5 gradings of Excellent, Very good, Fair, Poor, and Very Poor for 4 or 5 parameters, each student filling up one sheet. Here we have considered only 5 parameters, whereas there can be as many as 15 parameters that can be assessed. Fig. 20.2 illustrates a blank format in which the student records his assessment.
2. The class representative then prepares the tally sheet as per Fig. 20.3, transferring the data from the 60 sheets into a single sheet, each grading/parameter entered by a tick.
3. He then counts the number of ticks in each box and enters it in a corresponding box per Fig. 20.3, which will give the frequency of that grade.
4. By assigning a value for each grade and a weight for each parameter, the total points scored by the teacher can be calculated and recorded.
   1. The format is as shown in Fig. 20.2.

.........ENGINEERING COLLEGE
Subject evaluation by Students

Department   Class   Class strength

Name of the subject

Name of the staff

Sl. No	Parameter	Excellent	Very Good	Fair	Satisfactory	Poor	Remarks
1	Proficiency of subject						
2	Systematic presentation						
3	Clarity & Audibility						
4	Encouraging interaction						
5	Preparedness						

Name and signature of class representative   **Head of the Department**

FIG. 20.2 Format for subject evaluation by students.

2. The tally sheet is done on the same format with larger sized boxes for clear tallying, as illustrated in Fig. 20.3.

..........ENGINEERING COLLEGE

Subject evaluation by Students

Department
Class
Class strength
Name of the subject
Name of the staff

SI. No	Parameter	Excellent	Very Good	Fair	Satisfactory	Poor	Remarks
1	Proficiency of subject	++++ ++++ ++++ ++++ ++++ ++++ **30**	++++ ++++ // **12**	/// **3**			
2	Systematic presentation	++++ ++++ ++++ ++++ ++++ //// **29**	++++ ++++ /// **13**	/// **3**			
3	Clarity and Audibility	++++ ++++ ++++ ++++ ++++ /// **28**	++++ ++++ //// **14**	/// **3**			
4	Encouraging interaction	++++ ++++ ++++ ++++ ++++ // **27**	++++ ++++ // **12**	//// **4**	// **2**		
5	Preparedness	++++ ++++ ++++ ++++ ++++ ++++ // **32**	++++ //// **9**	// **2**	/ **1**		

Name and signature of class representative   Head of the Department

FIG. 20.3   Tally sheet for subject evaluation by students.

20.3   HISTOGRAM OR BAR GRAPH

Histogram, derived from the Greek word *histos* meaning anything set upright, was first introduced by Karl Pearson in 1891. It is a graphical representation of the distribution of data in a set of rectangles having:

(i) The X axis with centers at the class marks and lengths equal to the consecutive, non-overlapping intervals of a variable class.

(ii) The height of a rectangle is also equal to the frequency density of the interval.

This gives the frequency per unit length of class interval and is known as the frequency density over that class-interval.

Histograms are used to plot the density or frequency of occurrence of a particular event or parameter, estimating the probability density function of the underlying variable or parameter.

Histograms are the basic diagrams, the smoothing of which give us normal curves, the most significant curve in statistics. Smoothing is done by joining the midpoints of the peaks of each of the bar by a smooth curve. Thus, just like normal curves, the histograms, too, can be smoothened (Figs. 20.4 and 20.5):

1. Symmetrical or
2. Skewed right or
3. Skewed left or
4. Flat or
5. Peaked or
6. Bimodal.

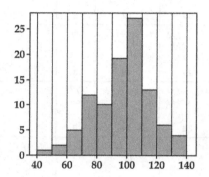

FIG. 20.4   Histogram.

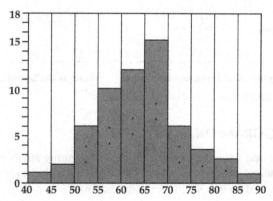

FIG. 20.5   Histogram for marks.

How to build a histogram

1. Arrange all the values or readings in the order they appear.
2. Determine the range which is the difference between the highest and the lowest value.
3. Distribute this range into a reasonable number of groups, say 10. These are also called the arrays and the magnitude of each group is called the class interval.

4. Tabulate the frequency of each value by placing it alongside of the class interval in tally form.
5. Total the tallies of each class and indicate this against each class as frequency.
6. Construct the graph with the class intervals in the X-axis and the frequency value in the Y-axis., erecting bars for each class interval, the height of each bar being equal to the frequency.

Principles of histogram construction

1. The correct impression must be given by giving a clear and comprehensive title.
2. A reasonable number of class intervals should be chosen to provide a meaningful, diagram that is not overcrowded.
3. The histogram must have clearly labeled axes.
4. The independent variable should always be placed on the horizontal axis.
5. The vertical scale should always start at zero.
6. A double vertical scale should be used where appropriate.
7. The bars should of the same width, meaning the class interval should be same.
8. The source of the actual figures must be given.

20.4   SCATTER DIAGRAM

As scatter diagram, scatter lot, or scatter graph is a type of statistical diagram using Cartesian coordinates to display values for two variables for a set of data and shows how much one variable is affected by another. It is effectively a line graph with no line—ie, the point intersections between the two data sets are plotted, but no attempt is made to physically draw a line. It is used when a variable exists at random which is beyond our control. The Y-axis is conventionally used for the characteristic whose behavior we would like to predict. A scatter plot (Fig. 20.6) can suggest various kinds of correlations between two variables' weight and height with a certain confidence level. It is the basic tool to explain the correlation between two variables as explained in more detail in Section 17.1.4, of Chapter 17.

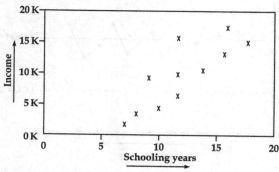

FIG. 20.6   Scatter diagram.

This relationship may be correlated, uncorrelated, positively related (rising), or negatively related (falling). If the points are close to making a straight line in the scatter plot, the two variables have a high correlation, but if they are equally distributed in the scatter plot, the correlation is low, or zero. Several types of scatter graphs shown in Fig. 17.2 of Chapter 17 indicate how a certain variable represented by the Y-axis changes with respect another variable represented by the X-axis. More noticeable is Fig. 17.2(iv) which indicates absolute no correlation while 17.2(i) and 17.2(vii) indicate perfect positive and negative correlation.

20.5   CONTROL CHART

Walter Shewhart (pronounced as *shoe-heart*), in his book, *Statistical Method from the Viewpoint of Quality Control* (1939), put a question, "*What can statistical practice, and science in general, learn from the experience of industrial quality control?*" This question and his explanation formed a major turning point in the concept of quality control. Walter Shewhart visited India in 1947–48 under the sponsorship of P. C. Mahalanobis of the Indian Statistical Institute. Shewhart toured the country, held conferences, and stimulated interest in statistical quality control among Indian industrialists.

Shewhart framed the problem in terms of assignable cause and chance-cause variation and introduced the control chart as a tool for distinguishing between the two. Shewhart stressed that bringing a production process into a state of statistical control, where there is only chance-cause variation, and keeping it in control, is necessary to predict future output and to manage a process economically. He concluded that while every process displays variation, some processes display controlled variation that is natural to the process, while others display uncontrolled variation that is not present in the process causal system at all times. That is how the control charts came to be treated as the most significant tool of quality control analyses. A typical control chart is illustrated in Fig. 20.7.

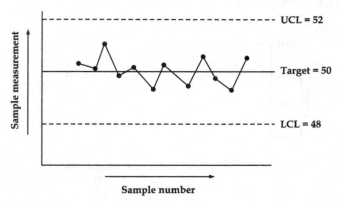

FIG. 20.7   Control chart.

A control chart consists of:

- Points representing values of a quality characteristic, such as mean, range, proportion, in samples taken from the process at different times.
- The mean of this parameter for all the samples is calculated.
- On the graph, a Center line is drawn at the value of the mean of the statistic. All other points representing the respective values are plotted on the graph.
- Upper and lower control limits, sometimes called natural process limits, beyond which the values are not expected to go, or the threshold at which the process output is considered acceptable, are calculated.
- The upper value and the lower value are designated as upper control limit and lower control limit.
- The space between the control limits is divided into zones depending upon the requirement.
- As long as the values of the points are within these limits, the process is assumed as going as expected, but if the points are seen beyond these limits, there we can suspect that either the value measurement is wrong or that there is something wrong in the process that needs our attention.

20.6 PARETO PRINCIPLE

The Pareto principle (also known as (i) the 80–20 *rule*, (ii) the *law of the vital few*, and (iii) the *principle of factor sparsity* states that, for many events, roughly 80% of the effect comes from 20% of the causes (machines, raw materials, operators, etc.). In other words, in any population, 20% of the people contribute to 80% of a parameter, say the GDP. This is similar to the Principle of ABC analysis which states that in an engineering industry, 10% of the production items contribute to 70% of the total annual consumption. The origin of Pareto Diagram is indicated in Section 3.1 (Fig. 20.8).

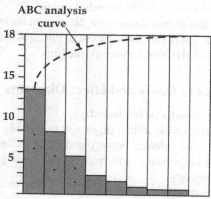

FIG 20.8   Pareto diagram.

When applied to the TQM situation, we can say that as per Pareto principle, 20% of the effects cause 80% of the total machine downtime. Effort aimed at the right 20% can solve 80% of the problems.

This principle would help in identifying which of the products or processes that fail more often, so as to prioritize and concentrate our efforts more. Effort aimed at the problematic 20% can solve 80% of the problems, just as we concentrate our inventory control effort on the 10% items to control 70% of the annual value. A typical Pareto chart indicating the significance of the shrink effect in casting defects is illustrated in a figure from Wikipedia. Double (back to back) Pareto charts can be used to compare "before and after" situations.

20.7 CAUSE AND EFFECT DIAGRAM

When you have a serious problem, it's important to explore holistically all of the things that could cause it, before you start to think about a solution, rather than just addressing part of it and having the problem run over and again. Ishikawa diagrams were proposed in the 1960s, by Kaoru Ishikawa, one of the founding fathers of modern management, who pioneered quality management processes in the Kawasaki shipyards. This forms one of the seven basic tools of TQM along with the histogram, Pareto chart, check sheet, control chart, flowchart, and scatter diagram.

Because of its principle of showing the *causes* of a certain *event* for further analysis, it is called the *Cause and Effect diagram* and also *Ishikawa diagram*, after its pioneer. Since the diagram looks like a fish skeleton with its bones spreading out from the vertebral bone (Fig. 20.9), it has become more popular as the *fishbone diagram*.

This method can be used on any type of problem, and can be tailored by the user to fit the circumstances. Use of this tool has several benefits to process improvement teams:

- Straightforward and easy to learn visual tool.
- Involves the workforce in problem resolution—preparation of the fishbone diagram provides an education to the whole team.
- Organizes discussion to stay focused on the current issues.
- Promotes "System Thinking" through visual linkages.
- Prioritizes further analysis and corrective actions.

20.7.1 Categories of Cause and Effect Diagrams

A. *The 6 Ms (used in manufacturing industry)*

Toyota Production System had originally used 6 Ms to which two more Ms, viz Management (money power) and maintenance have later been added to make it 8 Ms, as stated which, of course, is not globally recognized.
- Machine (technology)
- Method (process)
- Material (includes raw material, consumables, and information.)

- Manpower (physical work)/mind power (brain work): Kaizens, Suggestions
- Measurement (inspection)
- Milieu/mother nature (environment)

The other 2 Ms added later on as above are:
- Management/money power
- Maintenance

B. *The 7 Ps (used in the marketing industry)*
- Product/service
- Price
- Place
- Promotion
- People/personnel
- Positioning
- Packaging

C. *The 4 Ps (used in service industry)*
- Policies
- Procedures
- Process technology
- People

20.7.2   Basic Illustrations of Cause and Effect Diagrams

The six parameters of every event called 6 Ms, as stated above, are analyzed in manufacturing industries. These are Machines, Methods, Materials, Measurements, Manpower, and Mother Nature (Environment) (Fig. 20.9).

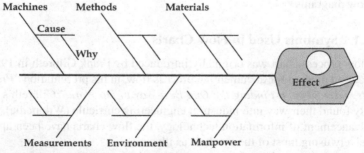

FIG. 20.9   Fish bone diagram for manufacturing industries.

In service industries there can be four parameters called *4 Ps* as stated above. These are the company policies, the procedures adapted, the person working on it, and the Process Technology adapted where the event has taken place. Fig. 20.10 illustrates this situation.

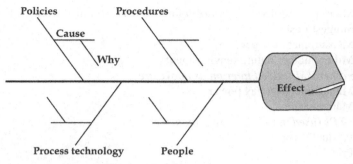

FIG. 20.10   Fish bone diagram for service industries.

20.8   FLOW CHARTS

A flowchart is a type of diagrammatic representation to illustrate a solution to a given problem. Basically, it is a line diagram representing the flow of a particular product along different operation sequences without reference to the geographical location of these work stations. This will give a bird's eye view of the several operations that the product would undergo. It indicates the steps involved in a process in the form of boxes of different kinds, and their order by connecting them with arrows. The boxes show the process operations, the various steps and actions, while the arrows show the order of the steps. Fig. 20.11 illustrates some of the boxes used in flow charts.

There are basically 3 categories of these charts:

- Process Flow Charts, as explained above
- Operation process charts, as explained in the next paragraph
- Flow diagrams

20.8.1   Symbols Used in Flow Charts

The flow process chart was originally introduced by Frank Gilbreth in 1921 as a structured method for documenting process flow, in his presentation *"Process Charts, First Steps in Finding the One Best Way to Do Work."* Gilbreth's tools quickly found their way into industrial engineering curricula (Wikipedia). With the advancement of information technology, the flow charts have been applied extensively using most of the symbols as illustrated.

Flow chart is also known as process flow diagram or chart, or system flow diagram or simply flow chart, a term used by industrial engineers, depending upon the application basically a manufacturing system. It is used primarily in process engineering and the chemical industry where the complex relationship between major components and how the material flows through various stages and components is depicted to provide an easy comprehension of the user. This generally is a combination of outline process chart and the flow diagram, where each operation is represented by the appropriate shape of the equipment as illustrated in Fig. 20.11. This gives a visual picture of the equipment, as well as the operation sequence. Sometimes the equipment is represented by functional blocks. Even in computer programming, flow charts are used to represent a series of decisions and the corresponding actions taken. A typical such flow diagram for a nuclear power plant is illustrated in Fig. 20.12.

Sl. no	Symbol	Explanation	Sl. no	Symbol	Explanation
1		Process/operation	16		Card
2		Alternative process	17		Punched tape
3		Decision	18		Summing junction
4		Data	19		Or
5		Predefined process	20		Collate
6		Internal storage	21		Sort
7		Document	22		Extract
8		Multi document	23		Merge
9		Terminator	24		Stored data
10		Preparation	25		Delay
11		Manual input	26		Sequential access
12		Manual operation	27		Magnetic disc
13		Connector	28		Direct access
14		Off-page connector	29		Display
15		Transfer of materials	30		Direction of flow

FIG. 20.11   Symbols used in flow charts.

It may be noted that in industrial engineering and work study, nomenclatures like flow diagrams, operation flow charts, etc. are used with slightly different applications. These charts are explained in the appendix so that we can distinguish clearly between the nomenclature used in TQM books and those used in Industrial Engineering books.

20.8.2   The Benefits for Process Flowchart

The process flow chart provides a visual representation of the steps in a process. Constructing a flow chart is often one of the first activities of a process improvement effort, because of the following benefits:

- Helps in making process flow charts
- Gives everyone a clear understanding of the process
- Helps to identify non-value-added operations
- Facilitates teamwork and communication
- Keeps all information on the same page.

It may be noted that the terms Flow chart or flow diagram are used to represent the flow of material along the several equipment especially in a process industry. In TQM, several books use the words *charts* and *diagrams* interchangeably almost as synonyms. However, in industrial engineering where these tools have been extensively used since the days of W. Taylor, charts signify the operations, movements, inspection, and delays that occur during the flow of materials from the first operation to the final operation in a group of operations performed on the material between work stations. On the other hand, the diagrams are geographic representations which may or may not be to a scale depicting the positioning of the equipment and the flow of material. It is hence felt to be appropriate to detail the various charts and diagrams used in industrial engineering projects to have a clear appreciation of the terminology used in TQM, as explained in the annexure at the end of this chapter.

20.8.3   Operation Process Chart

It is a simplified form using only the five parameters of operations, movements, inspections, delays, and storage. These charts are used for method study analysis in industrial engineering applications to identify and eliminate the wasted elements and operations, while the original form of a flow process chart or the

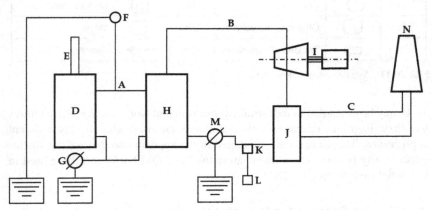

FIG. 20.12   Flow chart of a nuclear power plant. **A, B & C**: Primary, secondary and cooling water circuits, **D**: Core reactor, **E**: Control Rods, **F**: PORV, **G**: Primary booster pump (EIW), **H**: Steam generator (boiler), **I**: Turbine—Generator, **J**: Heat Exchanger, **K**: Demineralizer, **L**: Condensate tank, **M**: Secondary water booster pump, **N**: Cooling Tower.

flow charts are more used in analyzing quality control issues. Since method study is very significant in an industrial engineer's day-to-day performance, the operation process chart and flow diagram are explained more in detail in the annexure at the end of this chapter.

20.8.4   Flow Diagram

It is a plan, substantially to scale, of the factory or shop with the location of the machine, workplace, etc., indicated. On this, the movement of each product or component can be graphically represented, as illustrated in Fig. A5.

20.9   CONCLUSION

Undoubtedly a proper diagrammatic representation of the flow of materials and the operations performed on them gives us a better understanding of the physical flow, more than the narrative. It is essential that these diagrams and charts should be drawn to an international standard so that they convey the same concept to everyone. The appendix below illustrates, in addition to those specified in the 7 traditional tools, some other charts and diagrams normally used by engineers, so that the confusion in the terminology is avoided.

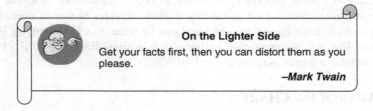

On the Lighter Side
Get your facts first, then you can distort them as you please.
–Mark Twain

APPENDIX

Apart from the flow Process chart that has been cited as one of the 7 traditional tools of TQM, industrial engineers have been using different types of process charts from the days of W. Taylor. These are illustrated and explained in this annexure.

PROCESS CHART

This has been used by Industrial engineers from the days of W. Taylor. Basically a flow chart is a line diagram representing the flow of a particular product along different operation sequences without reference to the geographical location of these work stations. This will give an idea of the several operations that the product would undergo. On the other hand, a flow diagram depicts the geographical position of the various workstations and may or may not be to scale. This would give a bird's eye view of the general location of the activities carried out as well as the route followed by the product, the operators, the materials and the equipment. There can be several types of flow charts or diagrams as follows.

1. Outline process charts
2. Flow process charts
   - Man type
   - Material type
   - Equipment type
3. Two-handed process charts
4. Flow charts
5. Process charts

OUTLINE PROCESS CHART

The outline process chart (also referred as operation process chart) is a chart that gives a bird's eye view of the various operations, inspections, and storage done in sequence for all the components that go into a particular product or assembly.

B.S. 3138.

It may be noted here that most European authors, including ILO, refer to this chart as outline process chart where as American authors including HB Maynard, in his *Industrial Engineering Handbook*, refer to this as operations process chart. Both names are valid and can be treated as synonyms as illustrated in Fig. A1. As mentioned earlier, this is a simplified form of the Flow chart and the symbols used are as per Fig. A2.

FLOW PROCESS CHART

Flow process can be defined as follows:

Flow process chart sets out the sequence of the flow of a product or an equipment or a man, by recording all the events under review using appropriate process symbols. These flow process charts can either be material type or equipment type or man type, depending whether the subject being charted is the material, equipment or man.

As the definition indicates, the flow process charts can be of three types.

(a) Material types, wherein the flow of the materials, the products, or the components are charted as illustrated in Figs. A3 and A4. A majority of the flow process charts are of this type.

(b) Equipment type, wherein the movement of certain equipment like welding equipment, portable drills, cranes, fork lift trucks, and air compressors, that are taken from workplace to workspace on a regular basis. This chart would be useful if these equipment cause bottlenecks or excessive waiting time for the work place that needs it.

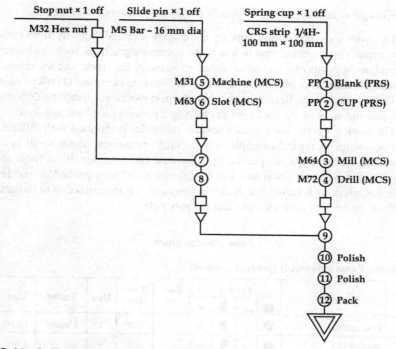

FIG. A1 Outline process chart.

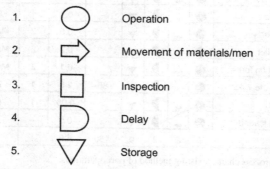

FIG. A2 Symbols for operation process chart.

(c) Man type, wherein the activities of a single worker or a gang moves from place to place as a part of their work, like the maintenance gang or the road laying gang, this type of chart would be used.

Fig. A3 and A4 illustrate the material type flow process chart for pallet handling in a port.

Flow process charts are specifically useful the following 3 situations

- The work sequence or elements are not exactly identical, but vary from cycle to cycle. This variation may be due to the operator's practices, such as cleaning equipment or may be due to the inherent nature of the work, like the maintenance operations or like the powdering of the hardened fertilizer in bulk storage.
- The work is not cyclic, but unique like the port workforce handling different types and sizes of the packages in clearing a general cargo storage area.
- The work is cyclic, but includes several subcycles performed with different frequencies. A typical example is in packing operations, where small automobile components are packed in individual cartons, then 10 of them are packed in a larger carton and 4 or 6 of the larger cartons packed in wooden boxes, all these forming one cycle. In this case, it is essential that to indicate the frequency of each element and the subcycle.

Flow process chart

Subject: Pallet movement (present method)

No.	Description	Symbol				Dist. (m)	Time (m)	Men	Eqpmt	load
1	Load pallet	●	→	D	▼		1.0	2	Manual	1.5 MT
2	Wait for FLT	●	→	D	▼		0.5	–	–	–
3	Insert forks in pallet	●	→	D	▼		0.05	1	FLT	–
4	Carry to wagons	●	→	D	▼	80	2.5	1	FLT	1.5 MT
5	Incline forks	●	→	D	▼	–	0.1	1	FLT	–
6	Push out bags	●	→	D	▼		0.5	2	Manual	1.5 MT
7	Pull out pallet	●	→	D	▼		0.2	2	Manual	20 kg
8	Load pallet on forks	●	→	D	▼		0.1	2	Manual	20 kg
9	Move to chute wagon	●	→	D	▼	80	2.5	1	FLT	20 kg
10	Drop pallet	●	→	D	▼		0.05	1	FLT	20 kg
11	Wait being loaded	●	→	D	▼		0.5	–	–	–
	Totals	7	2	2	0	160	8.0			

FIG. A3 Flow process chart—existing method of port operations.

FLOW DIAGRAM

A flow diagram is a plan, substantially to scale, of the factory or shop with the location of the machine, workplace, etc., indicated. On this, the movement of each product or component can be graphically represented.

Flow process chart

Subject: Pallet movement (proposed method)

No.	Description	Symbol				Dist. (m)	Time (m)	Men	Eqpmt	Load
1	Load pallet	●	→	D	▼		1.0	2	Manual	2.0 MT
2	Wait for FLT	●	→	D	▼		0.5	–	–	–
3	Insert forks in pallet	●	→	D	▼		0.05	1	FLT	–
4	Carry to wagons	●	→	D	▼	80	2.5	1	FLT	20 MT
5	Incline pallet	●	→	D	▼	–	0.05	1	FLT	–
6	Push out bags	●	→	D	▼		0.5	2	Manual	2.0 MT
7	Move to chute wagon	●	→	D	▼		0.2	2	Manual	20 kg
8	Land pallet	●	→	D	▼	80	0.5	1	FLT	20 kg
		●	→	D	▼					
		●	→	D	▼					
		●	→	D	▼					
Totals		5	2	1	0	160	5.30			

Description	Present	Proposed	Diff
Operations	7	5	2
Movements	2	2	–
Delays	2	1	1
Storage	–	–	–
Total Dist. in meters	160	160	
Total time in MTs	8.0	5.30	2.70

FIG. A4 Flow process chart—for the existing and proposed methods of Port operations.

As the definition implies, the flow diagram is a supplement to the flow process chart as illustrated in Fig. A5. This can also be used for the movement of men and tools. The general steps involved in drawing a flow diagram are:

(a) Draw to scale the plan of the work area,
(b) Mark the relative positions of the machines and all equipment like benches, booths, racks, etc.
(c) From the different observations and recordings made, draw the path and direction of the movement of the material or men on the diagram. The paths of different components can be marked in different colors.

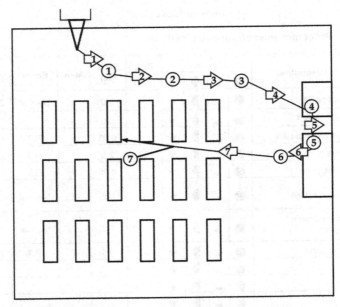

FIG. A5   Flow diagram for a stores operation.

FURTHER READING

For Cause and Effect Diagrams

[1] en.wikipedia.org/wiki/Ishikawa_diagram.
[2] www.isixsigma.com/tools-templates/cause-effect.
[3] http://www.mindtools.com/pages/article/newTMC_03.htm#sthash.G3x2za0S.dpuf.
[4] Tague NR. The Quality Toolbox. Milwaukee: ASQ Quality Press; 2004.

For other Charts and Diagrams

[1] ILO—An introduction to Work Study.
[2] Maynard et al.—Industrial Engineering Handbook.
[3] D.R. Kiran—Lecture notes on Work Study.

Chapter 21

The Seven Modern Tools of TQM

Chapter Outline

Total Quality Management: Key Concepts and Case Studies. http://dx.doi.org/10.1016/B978-0-12-811035-5.00021-0

21.1 THE SEVEN TRADITIONAL TOOLS OF TQM

The 7 traditional TQM tools, also called the Old Seven, the First Seven, or the Basic Seven Tools, which were developed earlier by Kaoru Ishikawa are discussed in Chapter 20. They are a set of graphical techniques identified as being most helpful in troubleshooting issues related to quality. They are called *basic* because they are suitable for people with little formal training in statistics, and because they can be used to solve the vast majority of quality-related issues.

The seven basic tools are:

1. Cause-and-effect diagram
2. Check sheet
3. Control chart
4. Histogram
5. Pareto chart
6. Scatter diagram
7. Stratification (alternately, flow chart, or run chart).

21.2 THE SEVEN MODERN TQM TOOLS

In 1976, the Union of Japanese Scientists and Engineers (JUSE) developed management and planning (MP) tools, or simply the seven management tools, to promote innovation, communicate information, and successfully plan major projects. Though the traditional and the modern tools are of a similar nature, the Japanese thinkers, to conform to the folklore rule of 7 tools as stated in Section 20.1, preferred to group them as two sets of 7 tools each, rather than a single set of 14 tools.

The Seven New TQM tools are:

1. Affinity diagram (KJ Method)
2. Interrelationship diagraph (ID)
3. Tree diagram
4. Prioritization matrix
5. Process decision program chart (PDPC)
6. Activity Network Diagram
7. Single Minute Exchange of Die (SMED).

21.3 AFFINITY DIAGRAM (KJ METHOD)

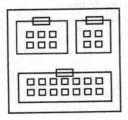

Affinity diagram, basically a project management tool, is a brainstorming technique that organizes large amounts of disorganized data on a large number of ideas, process variables, concepts, opinions, and other information into groupings based on natural relationships. It is called KJ diagram after Japanese anthropologist Jiro Kawakita, who developed it in the 1960s.

Affinity diagram provides visual presentation of a grouping of a large number of related items or data to help in organizing action plans. They come in very handy when you are confronted with many facts or ideas in apparent chaos, and when issues seem too large and complex to grasp.

21.3.1 Guidelines

- Ensure ideas are described with phrases or sentences. Minimize the discussion while sorting—discuss while developing the header cards.
- Aim for 5–10 groups. If one group is much larger than others, consider splitting it.

21.3.2 How to Conduct an Affinity Sort

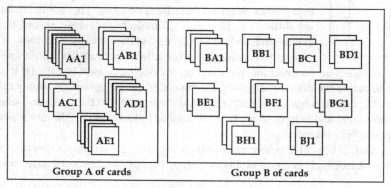

FIG. 21.1 Illustration of card arrangement for affinity diagram.

Conduct a brainstorming session on the topic under investigation. Group the ideas that seem to belong together.

1. Identify and list the ideas. Record each of them on small cards or Post-It notes.
2. Lay out cards on a table, flipchart, wall, etc.
3. Sort the cards into "similar" groups you think best. Discuss with your team members and keep shuffling them until consensus is reached. This is illustrated in the Fig. 21.1.
4. When most of the ideas have been sorted, give a title and a header card for each group in a concise 3–5 word description which should reflect the unifying concept for the group. Place header card at the top of each group.
5. Discuss the groupings and try to understand and record how the groups relate to each other.

21.3.3 Checklist

1. Inquire whether any more ideas can come from the participants until consensus is reached.
2. Check if all ideas are adequately clarified.

3. Use only 3–5 words in the phrase on the header card to describe the group.
4. If possible, have groupings reviewed by non-team personnel.
5. While sorting, physically get up and gather around the area where the cards are placed.
6. Sorting should not start until all team members are convinced that all points are covered.
7. If an idea fits in more than one category or group, and consensus about placement cannot be reached, make a second card and place it in both groups.

21.4   INTERRELATIONSHIP DIAGRAM

 In practical life, we come across situations where several events occur, each having an interrelationship with each and every other event in some form or other. These relationships are difficult to be perceived initially unless we record these events and their relationships systematically and look at it critically, considering what effect one can have on another. This is called an interrelationship diagram (or diagraph, as it is called). It can also be called the network diagram, the matrix relation diagram, or simply relations diagram. IDs basically generate visual presentation of the cause-and-effect relationships, and help us to analyze the natural links between different aspects of a complex situation.

A look at Fig. 21.2 below makes this point clear. Some causes directly create a problem such as cause numbers 2, 7, 9, 10, etc., while some causes result

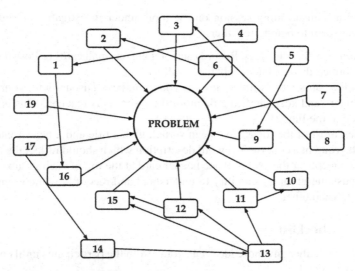

FIG. 21.2   Interrelationship diagraph drawn between several events.

in some other problem that would result in the problem in questions such as 5 through 9, 1 through 16, etc.

Interrelationship diagrams more or less looks like a critical path diagram used in a critical path method (CPM), or the Program Evaluation and Review Technique (PERT), except that the main problem in question is drawn at the center. In a nutshell, ID is problem-oriented while CPM is sequence-oriented.

21.4.1   Objectives of the Interrelationship Diagraph

- To encourage team members to think in multiple directions rather than in linear and unilateral thinking.
- To explore the cause and effect relationships among all the issues.
- To allow key issues to emerge naturally, rather than to be forced by a dominant or powerful team member, referred to as *The Lion* in Section 11.4.5 of Chapter 11.
- To systematically bring out the basic assumptions and reasons for disagreements among team members.
- To identify root cause(s), even when credible data does not exist.

21.4.2   Procedure for Constructing an Interrelationship Diagraph

1. Place the problem statement or desired outcome in the middle of a flipchart or a large piece of paper.
2. Arrange the major items (if the data as collected from an affinity diagram, use the title cards) in a circle around the problem statement. Place the cards which have ideas most closely related to the problem nearer the problem, if this can be determined.
3. Draw lines between ideas that are related. Put an arrowhead on the end of the line that shows the direction of the cause and effect relationship. Use only one-way arrows. The arrow should originate from the cause and point toward the effect. Each of the cards should have an arrow leading toward the problem, statement or to another cause statement that would ultimately lead to the problem.
4. Count the number of arrows leading into and out of each idea card. A value can be given for each cause by indicating the number of arrows going out and those coming in. For example, cause number 11 can be represented as −2/1, meaning 2 arrows are going out and one arrow in coming in.
5. The card with the most arrows going out like Box 13, is the key cause factor. Place a double box around it.
6. Give each team member a copy of the results and discuss it with them. Attempt to reach consensus on what are the key drivers to the problem.
7. The key idea categories from the ID may be used in the systematic tree diagram as discussed in the next paragraph.

21.5 TREE DIAGRAM

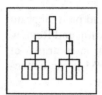

The tree diagram systematically links ideas, targets, objectives, goals, or activities in greater detail. It maps out the full range of paths and tasks that need to be taken to achieve the primary goal and related sub-goals and presents them in a visual form.

The tree diagram is used in strategic decision-making, valuation, or probability calculations. The diagram starts at a single node, with branches emanating to additional nodes, which represent mutually exclusive decisions or events. It is called a tree diagram as it looks like a tree, with a trunk and multiple branches. It is also called as tree analysis, analytical tree, or hierarchy diagram.

This TQM tool is used to break down broad categories into finer and finer levels of detail. It can map levels of details of tasks that are required to accomplish a goal, solution, or task. Developing the tree diagram helps one to broaden their thinking from generalities to specifics. It can map specific tasks to primary and secondary goals and also indicate the methods required to achieve corporate goals. The tree diagram shows the key goals, their sub-goals, and key tasks. It can help identify the sequence of tasks or functions required to accomplish an objective. The tree diagram can help translate customer desires into product characteristics. It can also be used like an Ishikawa diagram to uncover the causes of a particular problem.

Applications of tree diagrams

- To show the relationship between the subject and its component elements
- To show means and procedures for achieving a goal
- To identify potential root causes of a problem

Typical subjects for tree diagrams

1. Finding ways and means of implementing TQM
2. How to make and market a product
3. Reduction in power consumption
4. Improve health standards in the village
5. Requirements of a telephone answering machine.

The tree diagram analysis can also be likened to the fault tree analysis (FTA), which was originally developed in 1962 at Bell Laboratories by H.A. Watson, and is used mostly in logical fault location in electronic circuits, as illustrated by Fig. 6.17 in Chapter 6.

There can be two modes of this analysis. Event tree analysis and FTA, both of which are explained below.

21.5.1   Event Tree Analysis

Here the engineer presumes a faulty system, starts from an initial faulty event and traces in the forward direction to identity the possible effects of the failure. This is similar to the FTA, except that the tracing is in the reverse direction. This is very useful in hazardous situations. This also can be called a mathematical version of the scenario analysis. The Fig. 21.3A below is an illustration of the event tree analysis.

21.5.2   Fault Tree Analysis

Here the engineer starts from a definite system failure or an undesirable event and goes backwards to trace possible causes of the fault. This method is very useful in emergency situations. This is similar to the troubleshooting charts as a part of the operator's instruction manual given by all automobile manufacturers and most domestic appliance manufacturers, especially those for electronic goods. Fig. 21.3B is an illustration of the FTA. The symbols in Fig. 21.3 can be explained as per Table 21.1.

Sometimes this FTA is called decision-tree as applied in logical fault location of electronic components, as illustrated in detail more in Chapter 7, Section 7.17. This analysis hence comes very much useful in maintenance management.

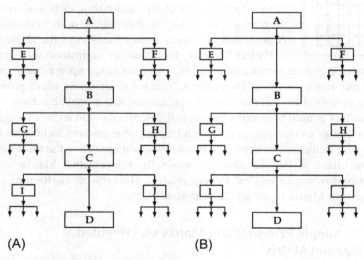

(A)   (B)

FIG. 21.3   (A) Event and (B) fault tree analysis.

TABLE 21.1 Fault Tree Analysis, per Fig. 21.3

Event Tree Analysis (Fig. 21.3A)	Fault Tree Analysis (Fig. 21.3B)
Problem: Effects of heavy pressing of brake pedal	*Problem*: Causes for brake failure
Event sequence	Fault sequence
A: Brake pedal pressed too heavily and suddenly	*A*: Brakes fail to hold
B: The brake tube gets punctured	*B*: The brake fluid leaks
C: The brake fluid leaks	*C*: The brake tube gets punctured
D: Brakes fail to hold	*D*: Brake pedal pressed too heavily and suddenly
E to J: Other possible effects of the respective stages	*E to J*: Other possible faults that contribute to the fault in the consecutive stage

21.6 PRIORITIZATION MATRIX

A matrix shows the relationship between items. At each intersection, a relationship is either absent or present. It then gives information about the relationship, such as its strength, the roles played by various individuals, or measurements.

A prioritization matrix is an extension of the above used in Project Planning. It emphasizes brainstorming to get the opinion of everyone to evaluate a number of issues based upon certain criteria and to arrive at a prioritized list of items. This helps determine which problems need to be solved first, in order to meet project/organizational objectives.

This tool is used to prioritize items and describe them in terms of weighted criteria. It uses a combination of tree and matrix diagramming techniques to do a pair-wise evaluation of items and to narrow down options to the most desired or most effective. Popular applications for the Prioritization Matrix include Return-on-Investment or Cost-Benefit analysis (Investment vs. Return), Time management Matrix (Urgency vs. Importance), etc.

21.6.1 Simple Prioritization Matrix vs. Weighted Prioritization Matrix

A simple prioritized matrix is one in which each factor has an equal weightage while a weighted prioritized matrix is one in which some factors may be more important than others, and may therefore be given a higher weightage. A weighted prioritization matrix is more commonly used, as it has more relevant widespread applications (because some evaluation criteria are usually more important than others). It is specifically useful in the Improvement Phase of the DMAIC Process.

21.6.2   When to Use a Prioritization Matrix

- Use it to prioritize complex or unclear issues, where there are multiple criteria for deciding importance.
- Use it when there is data available to help score criteria and issues.
- Use it to help select items to be in action from a larger list of possible items.
- When used with a group, it will help to gain agreement on priorities and key issues.
- Use it, rather than simple voting, when the extra effort that is required to find a more confident selection is considered to be worthwhile.

21.6.3   The 4 Basic Steps Involved in Creating a Prioritization Matrix

(a) *Brainstorming*
- Identify the problems which need to be addressed. Create a list of problems to be discussed.
- Identify the key criteria on the basis of which of these problems are to be evaluated. Also, determine the relative weightage given to each criterion.

(b) *Draw the prioritization matrix chart* (as indicated in the sample given in Fig. 21.4.)
- In the first column, list the problems which have been identified in the brainstorming session and code them.
- In the horizontal axis from the 2nd column onwards, list the criteria for evaluation along with their weightage (if any).

(c) *Rank the problems*
- Make all the participants rank each of the problems on a pre-determined scale against each of the evaluation criteria.
- Repeat this process if required to obtain multiple votes from each participant.

(d) *Total the results*
- Compute the total ranking for each problem.
- This helps obtain a prioritized list of problems to work upon.

21.6.4   Symbols Used in the Prioritization Matrix

FIG. 21.4   Prioritization matrix.

⊚ Strong relationship :: ◯ Medium relationship

△ Weak relationship :: (Blank) No relationship

21.6.5 WSA's 6-Step Detailed Procedure to Create a Prioritization Matrix

Work Systems Affiliates International Inc. of the United States has suggested the following 6-Step procedure to create a Prioritization Matrix

Step 1: *Create a matrix* (Fig. 21.5).
- Along the vertical axis, list the set of objectives, tasks, activities and code them A1, A2, A3, etc.
- Along the horizontal axis, record the criteria that the factors identified in Step 1 will impact, such as listing the organization's Critical Success Factors.

		Critical success factors					
ID	Key activity	Customer care	Systems design	New business opportunities	Excellent supply channels	Count	Performance index
A1	Customer feedback						
A2	Prompt response to customers						
A3	Training						
A4	Sales force reporting						
A5	Competitor profiling						
A6	Recruitment						
A7	Supplier profiling						
A8	Accurate customer information						
A9	Capacity planning						

FIG. 21.5 Step 1 critical success factors.

Step 2: *Analyze the impact of the achievement* (Fig. 21.6).

Analyze the impact the factor would have on each element listed on the horizontal axis. If the impact is positive, then shade the box with a color or put a tick in it. For example, since Customer Feedback would have positive impact on Customer Care, put a tick in the respective box, whereas it may have less impact on System design, supply channels, etc. You may also use the symbols as indicated in Fig. 21.4.

		Critical success factors					
ID	Key activity	Customer care	Systems design	New business opportunities	Excellent supply channels	Count	Performance index
A1	Customer feedback	X					
A2	Prompt response to customers	X		X			
A3	Training	X	X	X	X		
A4	Sales force reporting	X		X			
A5	Competitor profiling	X		X	X		
A6	Recruitment	X	X	X	X		
A7	Supplier profiling	X		X	X		
A8	Accurate customer information	X		X			
A9	Capacity planning	X	X				

FIG. 21.6 Step 2 impact analysis.

Step 3: *Count the positive impacts and rate the Performance Level* (Fig. 21.6).

Count the number of positive impacts and enter the same in the last but one column.

Step 4: *Rate the current Performance Level* (Fig 21.7).

Set up a scale of the Performance Level based on how the company is progressing with the objectives, such as:

A for Not occurring currently,

B for Started, but without noticeable effect

		Critical success factors					
ID	Key activity	Customer care	Systems design	New business opportunities	Excellent supply channels	Count	Performance index
A1	Customer feedback	X				1	B
A2	Prompt response to customers	X		X		2	E
A3	Training	X	X	X	X	4	B
A4	Sales force reporting	X		X		2	C
A5	Competitor profiling	X		X	X	3	E
A6	Recruitment	X	X	X	X	4	C
A7	Supplier profiling	X		X	X	3	D
A8	Accurate customer information	X		X		2	D
A9	Capacity planning	X	X			2	F

FIG. 21.7 Steps 3 and 4 performance index.

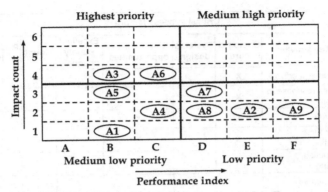

FIG. 21.8   Step 5 impact—performance snapshot matrix.

C for Poor,
D for Fail,
E for Progressing well to the expectation,
F for Excellent.

Step 5: *Impact—performance snapshot matrix* (Fig. 21.8)

Step 6: *Assign priorities to the activities*

From the figure, it can be seen that priority shall be allotted to the following activities in this order, A3, A6, A5, A4, A1, A7, A8, A2, and A9.

21.7   PROCESS DECISION PROGRAM CHART

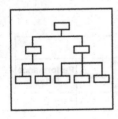

Process Decision Program Chart (PDPC) helps to identify what might go wrong in a plan under development and the consequential impact of failure on activity plans. It lays down the various steps that are needed to be taken to achieve a goal or objective in the form of a flow chart and helps in creating appropriate contingency plans to limit risks, in the following three stages.

1. Identifying what can go wrong (failure mode or risks)
2. Consequences of that failure (effect or consequence)
3. Possible countermeasures (risk mitigation action plan)

A useful way of planning is to break down tasks into a hierarchy, using a tree diagram. The PDPC extends the tree diagram a couple of levels to identify risks and the possible abnormal happening as soon as to prepare countermeasures for the bottom level tasks. Different shaped boxes are used to highlight risks and identify possible countermeasures. Fig. 21.9 illustrates PDPC.

In a way, PDPC can be likened to the failure mode and effects analysis (FMEA) in that both identify risks and consequences of failure, and suggest

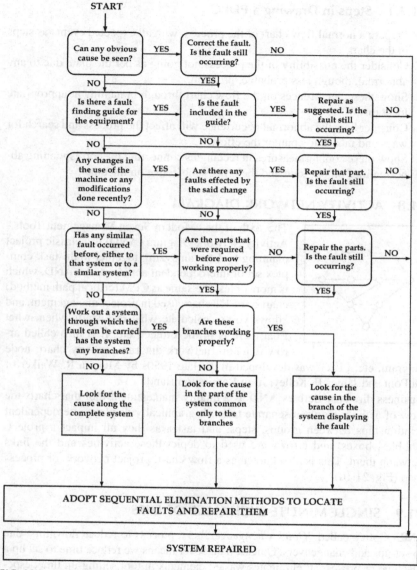

FIG. 21.9   An illustration of PDPC for fault location.

contingency actions. However, while FMEA adds prioritized risk levels through rating the relative risk for each potential failure point, PDPC identifies the risk factors in the appropriate sequence or schedule for a set of tasks and related subtasks in the critical path. In this context of emphasizing the critical path, PDPC can also be likened to PERT or the CPM.

21.7.1 Steps in Drawing a PDPC

1. Prepare a normal flow chart of the process, with all expected events as steps in the chart.
2. Consider the possibility of the process not going as per the plan, due to any abnormal, though less probable, problem.
3. Show these occurrences on the flow chart through branching at appropriate locations.
4. Consider how the abnormal occurrence will affect the process and search for ways and means to counter the effect.
5. Show these countermeasures in rectangles connecting the corresponding abnormalities on one side and the process objective on the other.

21.8 ACTIVITY NETWORK DIAGRAM

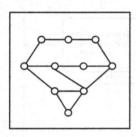
The sixth of the "Modern Seven Management Tools," Activity Network Diagram (AND) is a classic project planning tool for analyzing together multi-task complex set of interdependent activities. An AND, which is more or less the same as CPM (critical path method) chart or PERT chart, used in project management, and allows you to calculate what task starts when, what float there is in the project, etc. It is also called arrow diagram, network diagram, activity chart, node diagram, etc. PERT was developed in the late 1950s by Morgan R. Walker of DuPont and James E. Kelley, Jr. of Remington Rand.

Business directory defines AND as a quality management tool that charts the flow of activity between separate tasks. It graphically displays interdependent relationships between groups, steps, and tasks, as they all impact a project. Bubbles, boxes, and arrows are used to depict these activities and the links between them. This is also known as a flow chart, project network, or process map (Fig. 21.10).

21.9 SINGLE MINUTE EXCHANGE OF DIES

SMED, often called "*Quick Changeover*," is a process to reduce downtime due to set-ups and changeovers. Quick Changeover means we reduce time to set up a machine or process. It eliminates wasted elements in tool setting-up time, especially while changing a die of a press machine. This rapid changeover is key to reducing production lot sizes and thereby, improving flow. In fact, this is the principle behind the Japanese term *Mura*, which is explained further in Chapter 25.

Shigeo Shingo in his book, *A revolution in Manufacturing: The SMED System*, explains how in 1950, while analyzing the die changing operations during an efficiency improvement program in Toyo Kogya, the Mazda three-wheeler manufacturer, he conceived the idea of reducing the die-changing time

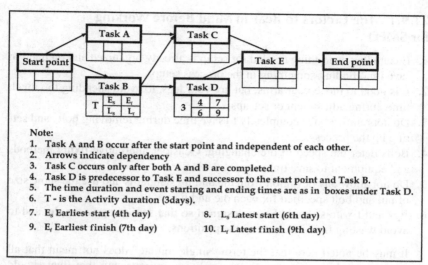

FIG. 21.10 Illustration of activity network diagram.

by industrial engineering methods, which he later called SMED. He observed that the dies on the large transfer-stamping machines that produce auto vehicle bodies are the most difficult tools to change and the dies—which must be changed for each model—weigh many tons, and must be assembled in the stamping machines with tolerances of less than a millimeter. He redesigned factory fixtures and vehicle components to maximize their common parts and standardize assembly tools and steps. He also classified all the set-up operational elements into two categories.

- IED—Internal set-up elements that can be done off line only, that is, cannot be done when the press work is being performed, such as the basic set-up times like fixing the die into the press base, and
- OED—External set-up elements that can also be done on line, that is when the press work is being performed These elements can include
- Keeping all the components ready, such as fixing screws and the new die ready by the side during the last stages of the previous operation,
- Preheating of the die, and
- Keeping the die transfer equipment ready that can locate the die in the press body in a few seconds.

This concept resulted in development of common tooling that reduced change-over time. This resulted in reducing the original die changing time from 12 h to only 10 min. The success of this program contributed directly to just-in-time manufacture, which is part of the Toyota Production System. SMED makes load balancing much more achievable, and also reduces economic lot size and thereby the stock level.

21.9.1   The Factors to Bear in Mind Before Working for SMED

1. Ensure that everything needed for setup is already organized and on hand to save time finding something in the process setup.
2. It is good to move your arms, but not your legs, to avoid spending too much time during adjustment or set-ups.
3. Do not remove bolts completely to save time during removing bolts and setting up the process.
4. Bolts deter the speed of die changing. Consider alternative fixing methods and equipment to save time.
5. Do not allow any deviation from die and jig standards and use the same size of nut and bolt specified for each die and jig.
6. Jigs and fixtures should be so designed so that setting up is simple, and to avoid wasting time in adjusting the positions.

It may be noted here that the term "single minute" does not mean that all changeovers and startups should take only *one* minute, but that they should as low as possible. Even if it takes less than 10 min, it is big achievement. Sometimes this is also called "One-Touch Exchange of Die" (OTED).

21.9.2   Internal and External Activities

1. Separate the internal and external activities before planning for SMED. As explained before, the internal activities are those that can only be performed when the process is stopped, while external activities can be done while the last batch is being produced, or once the next batch has started. For example, go and get the required tools for the next job before the automated machine stops.
2. Where possible, convert internal activities into external ones, such as preheating tools required for the next operation or batch.
3. Streamline or simplify the internal and external activities.
4. Document the new procedure, and actions that are yet to be completed.
5. Do it all again with improvement in mind. For each repetition, keep a goal of at least 25% improvement in set-up times.

21.9.3   FACTORS STRESSED UPON BY SHIGEO SHINGO, THE ORIGINATOR OF SMED

1. Separate internal from external setup operations
2. Convert internal to external setup
3. Standardize function, not shape
4. Use functional clamps or eliminate fasteners altogether
5. Use intermediate jigs
6. Adopt parallel operations

7. Eliminate unnecessary adjustments
8. Adapt mechanization where possible.

21.9.4   Benefits of SMED per Shigeo Shingo

1. Stockless production which drives capital turnover rates,
2. Reduction in footprint of processes with reduced inventory, freeing floor space,
3. Productivity increases with reduced production time,
4. Increased machine working time results from reduced setup times, even if number of changeovers increases,
5. Elimination of setup errors and elimination of trial runs reduces defect rates,
6. Improved quality from fully-regulated operating conditions in advance,
7. Increased safety from simpler setups,
8. Simplified housekeeping from fewer tools and better organization,
9. Lower expense of setups,
10. Preferred by operator because it is easier to achieve,
11. Lower skill is required because changes are now designed into the process, rather than a matter of skilled judgment,
12. Elimination of unusable stock from model changeovers, thereby reducing estimate errors.

21.10   FORCE FIELD ANALYSIS

Though force field analysis is not listed as one of the 7 modern tools, this author feels it is equally significant as an effective quality tool and hence is cited and briefly explained here.

The force field diagram (Fig. 21.11) is an analysis tool for identifying the obstacles in the path of reaching the goal and analyzing the possible causes and solutions to the problems. It is a systematic procedure of weighing pros and cons, and looks at all the forces for and against a decision. Since creativity is the main key for this tool, brainstorming is used during the procedure.

Force Field Diagram: Procedure

1. Prepare the team, indicating the time limit.
2. Identify the present situation, using the brainstorming procedure in both cases.
3. Identify some possible goals by brainstorming and select the appropriate one among them. Come to an agreement on this goal. There should be only one goal.
4. Create a flip chart so that everyone in the team can see what is being written and add their ideas.

5. Draw a chart as indicated below.
6. Determine the restraining forces (things that keep you from reaching the goal by brainstorming—see Section 22.11 of the next chapter on Kaizen)
7. This flip chart can later be used as a basic document in a larger meeting involving the deciding authorities.

For example, if you are deciding whether to install new manufacturing equipment in your factory, you might draw up a force field analysis like Fig. 21.11:

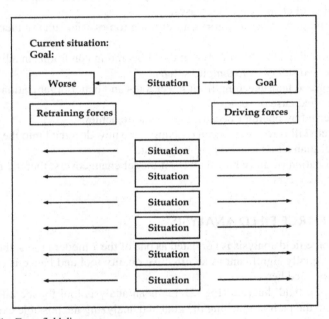

FIG. 21.11   Force field diagram.

By adapting this procedure you could:

- Reduce the strength of the forces opposing a project,
- Increase the forces pushing a project, or
- Change the forces from say, 15 (against): 10 (for) to say, 5 (against):20 (for).

21.11   CRITERIA RATING FORM

Like the Force Field diagram, Criteria Rating Form, though not listed among the 7 modern quality tools, it is an effective interpretation tool for objectively selecting ideas and solutions among several alternatives, as explained below. It is more or less similar to the plant location analysis as illustrated below.

TABLE 21.2 Criteria Rating Form

Degree Rank	Definition of the Grading	Grading Value	Points for a Weightage of 60
1 (Best)	Located at the heart of the raw material source	3	60
2	Located within easy reach of the raw material source	2	40
3	Located far away from the reach of raw material source,	1	20
4	Nowhere near the raw material source, with practically no transport	0	0

Procedure for the Criteria Rating Form:

1. Prepare for the session with a chart as per Table 21.2.
2. List out all the criteria by the brainstorming method. For example, for plant location, the criteria can be nearness to raw materials. Nearness to markets, transport facilities, etc.
3. Discuss and decide upon the weightage for each criterion. For example, for locating a sugar factory, nearness to raw materials has to be given the highest weightage, say 50. The other criteria can be assigned weightage as per their importance to the particular situation, such as 30, 20, 10, 5 and for the least important criterion, it may even be 1 or 2.
4. Discuss and decide upon the grading of situations (degree) for each criterion. For example, for the criterion of nearness to raw materials sources are as indicated in Fig. 21.2.
5. Assigning the full weightage as points for the most favorable grading, assign points to each grading.
6. The definitions, weightage, etc., for each criterion may depend upon the situation, but once defined, would be same for all alternatives considered.
7. Indicate such points for each alternative in the same table.
8. Add up the points for each alternative.
9. Select the one with the maximum total points as the best one.

21.12 MODELS THAT CAN BE USED TO REPRESENT A PROBLEM

In fact, most of the tools cited in Chapter 20 and this chapter are based on the following basic models that can be further developed and analyzed.

1. *Verbal models*, like explaining a problem in a discussion form.

2. *Schematic models*, like the graphical presentation in the form of charts, tables, or graphs.
3. *Physical models*, which are scale models, either in 2 dimensions or 3 dimensions.
4. *Analogue models*, which are physical systems having characteristics similar to the Actual problem. For example, the flow of water through a pipe under varying conditions of pressures can be represented as an analogy for the flow of electrical current in a wire.
5. *Mathematical models*, having the advantage of the precision of mathematics and are in the form of mathematical equations. These can either be deterministic or probabilistic.
6. *Deterministic models*, where the variables and their relationships are stated, presuming static and ideal conditions. For example, the economic order quantity for a purchased item can be given by

$$Q = \sqrt{\frac{2Pr}{i}}$$

where Q = economic order quantity, P = annual consumption quantity, r = rate of consumption, $i = 1$.

Probabilistic models, which take into consideration the uncertainty of the variables, and other dynamically changing situations by introducing the probability factor.

21.13 OTHER ANALYTICAL TESTING METHODS FOR SAFETY

Apart from the above modern tools, the following tools can be cited as being useful in TQM projects.

One of the earliest methods used for testing a product with respect to its strength, was the destructive methods such as UTS testing or impact testing, or even the crash test performed by the automobile manufacturers. However, these cannot be done for the majority of cases, especially from the safety point of view. The analytical methods of testing detailed below are useful tools in this regard. These have been discussed in detail earlier in this chapter, but again cited here because these tools are useful in designing for safety, in addition to design for quality.

- Scenario analysis
- Failure mode and effect diagram
- Fault tree analysis
- Event tree analysis
- Risk benefit analysis

21.14 CONCLUSION

CPM is the fundamental tool of project engineers to analyze the occurrence of events in multi-task problems, consisting of several interrelated events and identifying the critical events that contribute to overall delay in the completion of the project. Later, this is refined into PERT, which considers the possible variations in the expected event durations, as well as the time of their occurrence. Quality proponents have adapted this PERT to analyze TQM problems as AND.

On the Lighter Side

Kaoru Ishikawa had seven analytical methodologies in his mind, which he called "Seven Tools of TQM," the figure seven having no significance except for Japanese folklore. Later when others wanted to add more methodologies, they termed them as "Seven Modern Tools" just to make them sound in line with Ishikawa's tools. Perhaps the next methodologies that would be developed would be limited to seven and called "Seven Ultra-Modern Tools" and later development would be called "Seven Super-Ultramodern Tools"!

–Margaret Mead

FURTHER READING

[1] mot.vuse.vanderbilt.edu/mt322/Affinity.
[2] www.skymark.com/resources/tools/affinity_diagram.
[3] www.asq.org/learn-about-quality/idea-creation-tools/overview/affinity.html.
[4] www.qimacros.com/quality-tools/tree.
[5] www.wsa-intl.com/250-prioritization-matrix.

Chapter 22

Kaizen and Continuous Improvement

Chapter Outline

22.1 WHAT IS KAIZEN?

In Japanese,

	Kai	means	Change
and	Zen	means	Good
Thus	Kaizen	means	Change for the good

Total Quality Management: Key Concepts and Case Studies. http://dx.doi.org/10.1016/B978-0-12-811035-5.00022-2

Kaizen emphasizes continuous improvement as compared to innovation, which is a one-time improvement. Fig. 22.1 below indicates how kaizen attains high performance levels at no or marginal costs, as opposed to innovations that need heavy investments. Its philosophy implies that whatever we do can be improved continuously, whether it is the workplace or at home.

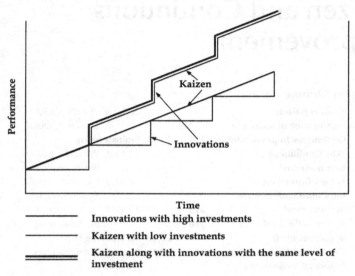

FIG. 22.1   Kaizen vs. innovation.

22.2   SIGNIFICANCE OF KAIZEN IN CONTINUOUS IMPROVEMENT

- Kaizen is similar to methods improvement studies undertaken by industrial engineers.
- It is a continuous improvement tool and not a one-time approach.
- Its motto is *"Do 100 things 1% better, than doing 1 thing 100% better."*
- It does not involve substantial investment.
- It aims at excellence at the shop floor level.
- It involves everyone in the factory.
- It involves identification of wastes involved in the production processes and operations.

22.3   WHY CONTINUOUS IMPROVEMENT?

Richard Y. Chang and Matthew E. Neidzwieckin their book, *Continuous Improvement Tools*, observed that

> *The single most destructive force in the move to improving the quality of American organizations today is the lack of commitment and understanding of how to make quality happen on the job.*

Thus, continuous improvement is the hallmark of TQM. It has been in vogue ever since man started living as a social animal and constantly strived to live better, and provide better living conditions for his dependents. Einstein, Edison, et al. are known for their inventions and innovations, but even they had failed on their first attempt and finally succeeded because of their quest for continuous improvement and betterment.

Juran observed in his famous Quality Trilogy, that among the three basic elements of quality management, viz, quality planning, quality control, and quality improvement, enough attention is being given to quality control, but not much attention to quality planning and quality improvement. He emphasized that the importance given to continuous improvement in Japanese industries had ultimately resulted in Japanese economic upsurge.

Juran recommended the following eight steps or road map for the methodology for this breakthrough, which word he preferred to use in place of improvement.

Step Number	Juran's Methodology	What It Means
1	Proof of the need	Selection of a problem
2	Organization of diagnosis	Task force or quality circle
3	Diagnosis	Finding out the root cause
4	Breakthrough in the knowledge	Finding out solutions
5	Remedial action in the findings	Implementing solutions
6	Breakthrough in cultural resistance	Getting acceptance
7	Control at new level	Monitoring
8	Holding on to the gains	Defining the standard procedure

Masaki observed in his book on Kaizen that

The starting point for improvement is to recognize the need. This comes from recognition of a problem. If no problem is recognized, there is no recognition of the need for improvement. Complacency is the archenemy of KAIZEN.

Readers' attention is drawn to Section 28.3 of the chapter on Business Process Reengineering, citing the statement of Thomas Devonport, one of the proponents of BPR that, *"Today firms must seek not fractional, but multiplicative, levels of improvement—10× (ten times) rather than 10%."* This observation may sound contradictory to Kaizen, but should be treated as complementary. Nevertheless, it may be pointed out here that the concept of BPR in recent times could not sustain its initial enthusiasm among the industry managers because of the factor of drastic changes. This fact even more highlights the superiority of the continuous improvement professed by Kaizen over drastic innovations. Table 28.1 supports this view point.

Changes have taken place all along history, mostly for the better, especially the in the post-World War era. The latter half of the 20th century had seen tremendous changes, not only in manufacturing processes and procedures, but also in management thinking.

22.4 SOME ILLUSTRATIONS OF THE CONTINUOUS PROCESS IMPROVEMENTS

1. By improving efficiency
2. By combining ideas from other fields
3. By trying to get things right the first time
4. By eliminating rejections and rework by using new materials and technologies
5. By benchmarking with competitive process or products
6. By adopting JIT (Just-in-Time) systems of inventory
7. By constant customer feedback and review
8. By using Quality Improvement Teams to solve problems
9. By using Six Sigma concepts
10. By adopting SPC and TPM techniques by using quality circles
11. By introducing five S
12. By introducing Kaizen (small improvements)
13. By maintaining people to do their best by foolproof designing using Poka-Yoke
14. By training all employees as per their needs.

22.5 KAIZEN IS THE UMBRELLA

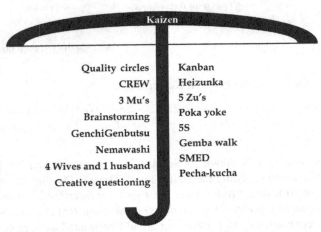

FIG. 22.2 The umbrella of Kaizen.

Kaizen can be called the umbrella enveloping most of the Japanese practices involving creativity, as illustrated by the Fig. 22.2.

22.6   REQUIREMENTS FOR CONTINUOUS IMPROVEMENT

1. Accepting that the problem lies in the inadequacy of the present level in performance of the product or service
2. Right attitude to solve the problem
3. Proper organization to solve the problem
4. Adequate knowledge and practice in using problem-solving tools and techniques
5. Structured method of problem-solving
6. Problem definition and analysis to be based in hard facts
7. Solutions to the cause, more than for the symptoms of the problem
8. Implementing and continuous monitoring until consistent result is obtained
9. Overcoming resistance to change
10. Control system for reversible changes.

22.7   INDUSTRIAL ENGINEERING PRINCIPLES vs. KAIZEN PRINCIPLES

In fact, it was during the 19th century that Frederick Taylor propounded the Scientific Management principles from which industrial engineering had developed into a science of studying a process and improving it on a continuous basis. It may be noted that most of these practices were in use in the Western world also, but the emphasis in Japan was the importance given to the core worker, which was absent in the Western world. It may therefore be said that the use of Japanese terms in place of the English terms created interest and indirectly helped young managers to understand and appreciate these practices better. Thus, we can say that Kaizen is the Japanese adaptation of industrial engineering, where creativity is the fundamental requirement. Hence, we will talk about the principles of creativity in the following sections.

22.8   IMPORTANCE OF CREATIVITY

- Shakespeare said *"Creative imagination is the gift that makes the man the paragon of animals."*
- Einstein went further by saying *"Imagination is more important than knowledge."*
- Alex Osborne put the same thing in a different way, *"Knowledge may be power, but can be more powerful if creatively applied."*
- Robert T. Ross, in his contribution on *Suggestions Schemes in Industrial Engineering Handbook* edited by Maynard, termed it as the *Idea Power.*

Have you ever wondered why the engineers, after their Master of Engineering get a Ph.D. (Doctor of Philosophy) degree and not a Doctor of Engineering? This is true all over the world, but the concept is based on Hindu philosophy.

Esa sarveshu bhuteshu gudho atma na prakasate,
Drsyate tvagryaya buddhya sukshmathi suksmam darshibi.

Atma is in all beings, but hidden and therefore, is not manifest. It can be realized, however, by the concentrated reasoning of those who have trained themselves in perceiving subtle and even more subtle truths. This is the basic principle behind all the research analyses and it is this philosophy that is highlighted in awarding Ph.D. degrees. This is more so in the industrial engineering techniques, whether method study or reengineering, or any analytical process. We call it creativity, the basic principle of kaizen.

Subtle truth can also be attributed to mother pearl found only in the deep seas. Nothing will come to you on its own. You have to go very deep and investigate thoroughly to achieve the pearl. This is the truth.

Kaizen ideas are obviously the key to the solutions for all kinds of problems, whether of production, materials handling, advertising, selling, human relations, or more significantly, the housewife's planning the monthly household expenditure or managing the kitchen.

Let us consider the various powers of the human brain which generates all these ideas. They are

Absorptive power	–	the ability to observe and apply attention
Retentive power	–	the ability to memorize and recall
Reasoning power	–	the ability to analyze and judge
Creative power	–	the ability to visualize, foresee, and generate ideas

By the first two, we learn and by the latter two, we think. A computer may be able to perform the first three better than the human, but the creative power is the specialty of the human brain. Again, it is an established fact that the creative processes are the right-brain activities and rational thinking processes are left-brain activities, as illustrated in Fig. 22.3.

Whatever the right brain imagines, the left brain applies logic and erases the illogical imaginations, even before they are formed. That means the left brain prevents us from getting ideas. So we should forget the presence of the left brain, think of ideas from the right brain, even though they appear illogical, and then apply logic with the help of the left brain. This process is explained further in Section 22.10.3, which advises us to let our mind loose.

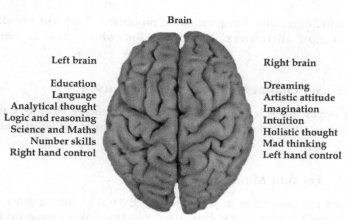

Brain

Left brain

Education
Language
Analytical thought
Logic and reasoning
Science and Maths
Number skills
Right hand control

Right brain

Dreaming
Artistic attitude
Imagination
Intuition
Holistic thought
Mad thinking
Left hand control

FIG. 22.3   The human brain.

22.9   CREATIVE METHODOLOGY

In method study, the basic analytical wing of industrial engineering, the well-established procedure is called SREDDIM. Creative application methodology, too, forms almost the same steps, which we may rename as below:

1.	Select	Orientation	–	Locating and selecting the problem
2.	Record	Preparation	–	Data collection
3.	Examine	Analysis	–	Breaking down the problem into sub-problems
4.	Develop	Ideation	–	Thinking of all the possible ideas and alternatives
5.	Define	Synthesis	–	Selecting and defining the optimal alternatives
6.	Install	Evaluation	–	Testing and modifying the results and solutions
7.	Maintain	Follow up	–	Ensuring that the new method is sustained

We numbered them one, two, three, etc., but the sequence as far as creativity is concerned is not rigid. We may analyze them, even during the evaluation. We may still be digging for facts even after ideation. We may start guessing even while preparing.

22.10   THE PRINCIPLES OF CREATIVITY

22.10.1   Divide and Conquer

Even though it sounds similar to the negative principle of *Divide and Rule*, the way East India Company has widened its roots in India by splitting the Indian rulers, this is a positive principle in creativity. After selection, split each and every problem into sub-problems and deal with each sub-problem individually, especially for questioning procedure. This will broaden the scope for more alternatives. The idea is somewhat similar to elemental analysis.

22.10.2 Set Quotas and Deadlines for Yourself

The practice of setting up a quota of a minimum number of ideas to be thought over before a definite time always helps in getting larger and larger number of ideas.

22.10.3 Let Your Mind Loose

Do not put any constraints to your thinking. Even if an idea appears silly, it doesn't matter, list it first, think about it, and then judge. When it comes to thinking, try to act as if you have two different personalities, first a blind thinker and then a judge, only one at a time. The point is, even the silliest idea if analyzed further and tailored down, may lead to an ideal solution. This is emphasized in Section 22.8.

22.10.4 Blue Sky Thinking

A new idiom that has found a place in English vocabulary is *Blue Sky Thinking*, which means to get creative ideas out of the box that are not limited to current thinking or beliefs. This expression stems from *opening one's mind as wide as the blue sky*, or letting loose your mind as the previous paragraph emphasizes. London's *Guardian* newspaper lists the following 10 blue sky ideas that changed the course of history:

1. Plato's philosophy,
2. Freud's theory of unconsciousness,
3. Theory of universe by Copernicus and Galileo,
4. Newton's universal gravitation theory,
5. Einstein's theory of relativity,
6. Descartes's theory of "I think, therefore I am."
7. Marx's analysis of capitalism,
8. Adam Smith's Laissez-Faire concept, cited in Chapter 4,
9. Women's Liberation, and
10. World Wide Web.

22.10.5 Two Heads are Better Than One

This is the popular expression that gave birth to brainstorming, which is a kind of group activity that is used to generate a lot of ideas. Participants are encouraged at the beginning to think of and list ideas, even if they sound silly, or far-fetched, as explained in the previous paragraph.

Kaizen and Continuous Improvement Chapter | 22   321

22.10.6   Question Each and Every Detail

This is the cream of creativity. Rudyard Kipling said in his famous poem

I had six stalwart serving men,
They taught me all I know,
Their names were What and Where and When
and Why and How and Who.

This is similar to the Japanese metaphor *one husband and four wives* where the four Wives (W) are **What, Where, Why and When** and the one Husband (H) is **How**. This is explained further in Chapter 24 on Japanese Management Terminology.

Question each and every detail. Why? Why? Why?... It should be emphasized without hesitation that a majority of the activities in our day-to-day life are based on this principle. It may also be stressed that while creativity is an art that depends upon the IQ, it can always be developed by the preceding guidelines, and also by constant practice.

Industrial engineers, as a result of their years of experience in continuous improvement, never speak of the best method, but only of the best available method, or the best method now desired. Even the Japanese word Kaizen speaks of continuous improvements as compared to innovation, which is a one-time improvement. In short, wherever manual work is involved, there is a continuing opportunity to improve the method.

If this concept is accepted, the method is examined and critical questions are put forth with an open mind, all the resistance to changes cited earlier such as, "*It won't work,*" or "We *tried it before,*" and "*It cannot be done,*" do not form obstacles to the goal of methods improvement.

The man who constantly asks questions and takes nothing for granted would certainly generate new ideas for doing the work if done in a systematic procedure. The questioning technique is the means by which the critical examination is conducted, each activity being subjected, in turn, to a systematic and progressive series of questions, while the questioning attitude is a state of mind which takes nothing for granted during the investigation of a procedure or an operation.

In a nutshell, this systematic questioning results in determination of solutions based on facts and guards against the influence of emotions, opinions, habits, or prejudices. These are illustrated in Figs. 22.4 and 22.5.

This author nostalgically remembers how during his post-graduation of 1967, his professor illustrated some of the above with everlasting memory.

22.11   BRAINSTORMING

Wikipedia defines brainstorming as *a group or individual creativity technique by which efforts are made to find a conclusion for a specific problem by gathering*

322   Total Quality Management: Key Concepts and Case Studies

a list of ideas spontaneously contributed by its members. In principle, you pose brainstorms to a group of thinkers. List out each and every idea put forward by them. It would be surprising to note that the number of ideas generated by a group would be more than the sum of all the ideas that each would be able to think of if left alone. Hence, the expression *two heads are better than one.*

Osborn, who calls this ideative efficacy, says that the two principles that contribute to this are (a) to defer judgment, and (b) reach for quantity. This has given rise to the four factors of brainstorming which are,

1. Focus on quantity
2. Withhold criticism
3. Welcome unusual ideas
4. Combine and improve ideas

22.11.1   When to use Brainstorming

1. Brainstorming helps your team generate a large number of ideas and to determine possible causes and/or solutions to problems.
2. Planning out the steps of a project. Although not for primary use, it can be used to identify the different steps in implementing a project.
3. When deciding which problem to work on, you can use brainstorming in any situation where many ideas need to be generated in a relatively short period of time.
4. You want to include all opinions. Round-robin brainstorming, as explained below, helps to ensure equal participation in an idea generating session.

22.11.2   Freewheeling vs. Round Robin

- *Freewheeling*, is when anyone who has ideas is allowed to say them aloud. They are listed as they are said.
- *Round Robin*, is where everyone takes a turn to offer ideas.

22.11.3   Techniques of Brainstorming

Wikipedia suggests the following methods of brainstorming:

1. *Nominal group technique*: Participants are asked to write their ideas anonymously. Then the facilitator collects the ideas and the group votes on each idea.
2. *Individual brainstorming*: Here each person writes his ideas using free writing with free word association and at the end all the papers are collected and studied.
3. *Group passing technique*: Each person in a circular group writes down one idea, and then passes the piece of paper to the next person, who adds some more related thoughts and improvements and then passes on to a third, till each participant writes one idea in each of the paper passed around.
4. *Team idea mapping method*: Each participant brainstorms individually, then all the ideas are merged onto one large idea map.

5. *Breaking the rules technique*: Participants first list the formal or informal rules that govern a particular process, and then try to develop alternative methods to bypass or counter these established protocols.
6. *Directed brainstorming*: Each participant is asked to produce one response only and stop, and then all of the papers (or forms) are randomly swapped among the participants, so the second person creates a new idea that improves on the first idea. This is the passed on to a third, and the process repeats. This is quite similar to the group passing, except that the passing is at random.
7. *Guided brainstorming*: Same as the breaking the rules method, but with the facilitator explaining some constraints in the mind set, and also in time allowed.
8. *Question brainstorming*: This involves brainstorming the *questions*, rather than trying to come up with immediate answers and short-term solutions. Then the answers are written and consolidated.

22.12   SIX THINKING HATS

Edward de Bono put forth a thinking process called *Six Thinking Hats*, to separate thinking into six clear functions and roles. Each thinking role is identified with a colored symbolic *thinking hat*. By mentally wearing and switching hats, you can easily focus or redirect thoughts.

1. The White Hat calls for information known or needed.
2. The Yellow Hat symbolizes brightness and optimism. Under this hat, you explore the positives and probe for value and benefit. This may be likened to the right half of the brain or "*Let loose your mind*," as per third point in Section 22.10.4.
3. The Black Hat is judgment—the devil's advocate or why something may not work. Spot the difficulties and dangers; where things might go wrong. It can be likened to the left half of the brain, and probably the most powerful and useful of the Hats.
4. The Red Hat signifies feelings, hunches, and intuition. When using this hat, you can express emotions and feelings and share fears, likes, dislikes, loves, and hates.
5. The Green Hat focuses on creativity; the possibilities, alternatives, and new ideas. It's an opportunity to express new concepts and new perceptions.
6. The Blue Hat is used to manage the thinking process. It's the control mechanism that ensures the Six Thinking Hats guidelines are observed.

22.13   PRIMARY AND SECONDARY QUESTIONS

Primary questions cover the first stage of the questioning technique, when the very purpose, the place, the sequence, the person, and the means of doing the

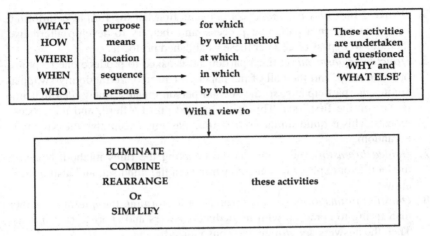

FIG. 22.4 Primary and secondary questions.

operation are systematically questioned. Secondary questions cover the second stage of the questioning technique, during which the answers to the primary questions are subjected to further query to determine whether possible alternatives of place, sequence, persons, and/or means are possible and preferable as a means of improvement over the existing method. Alluding to the previous paragraph, we can say that what, when, where, who, and how are the primary questions, whereas why and what else, are the secondary questions. This is where the mind is let loose, and ideas created before we apply reasoning commenting upon the feasibility of each alternative. Fig. 22.4 illustrates a systematically structured methodology for the critical examination, integrating the primary and secondary questions.

22.14   DEVELOP

Though Develop is listed as a separate step of method study, it is inseparable with the step "Examine" and both are generally done together. This is where each of the generated alternatives is considered with logic and weighed and evaluated for the economic, technical feasibility, and other factors. A typical critical examination chart is illustrated in Fig. 22.5, wherein the first and second columns correspond to primary questions, the third corresponds to the secondary questions, the fourth to the logic, while the fifth column corresponds to Develop. This step determines the following as per the last column of the format:

- What is to be done?
- How and by what means it should be done?
- Where should it be done?
- When should it be done? and
- Who should do it?

DESCRIPTION OF THE OPERATION (Original/Proposed method)			Op No. Date	Chart ref. Charted by Sheet No.	REMARKS
WHAT Explain the operation in one sentence	**WHY** Give reasons for doing this	**WHAT ELSE** List all possible alternatives	**COMMENT** Comment on each alternative	**WHAT SHOULD** Confirm whether the operation should be dine it not	
HOW a) Specify the material b) Specify the equipment c) Explain the present method in detail d) Specify they extra safety precautions	**WHY THAT WAY** Give reasons for each	**HOW ELSE** List all possible alternatives	**COMMENT** Comment on each alternative	**HOW SHOULD** Suggest one or two procedures for each of a, b, c etc.	
WHEN a) After what operation b) Before what operation c) Frequency d) How long	**WHY THEN** Give reasons for each	**WHEN ELSE** List all possible alternatives	**COMMENT** Comment on each alternative	**WHEN SHOULD** Specify when it should be dine	
WHERE a) Exact spot b) Generallocation c) Size etc.	**WHY THERE** Give reasons for each	**WHERE ELSE** List all possible alternatives	**COMMENT** Comment on each alternative	**WHERE SHOULD** Specify where it should be dine	
WHO a) No of hands used b) Skilled/unskilled c) Men/women d) Day/night shift e) Other details	**THY THEY** Give reasons for each	**WHO ELSE** List all possible alternatives	**COMMENT** Comment on each alternative	**WHO SHOULD** Specify who should do it	

FIG. 22.5   An illustration of critical examination chart.

At this stage, we have initiated the development of alternative methods, may be one, or two or three, and our proposals should be defended by a report that is to be submitted. After all, our suggestions have to be approved by someone else, maybe the top management, or the production manager, or even the supervisors, before we can install them.

The report should satisfy the following:

Give the details of the existing and proposed methods and should give full justification for the proposal,

(a) Precede by a summary of the recommendations and cost savings,

(b) Reflect the systematic procedure or compare the relative costs of materials, labor, overhead, etc., of the present and proposed methods, and the savings affected,

(c) Show the estimated cost of installation of the new or modified equipment, layout, etc.

(d) Indicate the road map of action, that is "who should do what, and when" in the process of implementation and

(e) Remember to discuss the report with the persons concerned, especially the production manager, before the final submission.

22.15 DEFINE

All the recording charts, such as OPC, FPC, and flow diagrams, whichever were drawn for, the existing method should now be drawn and charted for the proposed method also. Each and every operation should be detailed with reference to the procedure, tools, jigs, inspection gages, etc., so that the shop personnel can perform the operation exactly the way you set it, without any wrong interpretation. This is drawn in the form of Written Standard Practice (WSP), also called Operative Instruction Sheet (OIS), or Standard Operation Procedure (SOP). The Main purpose of this is to

(a) record the improved method for future reference

(b) explain the new method to the operatives, foremen, and management without any difficulties in interpretation

(c) act as an aid for training

(d) form a basis for time studies and standards setting

(e) help in preparing material warrants and planning of inspection gages, special equipment, etc.

22.16 INSTALL

Once the process is defined and well understood, the next step is to implement the new method or install it by putting into practice as per the written standard practice. This is the basic purpose of the whole method study assignment from the management, as well as the work study engineer's point of view. By leading the implementation of the proposals, the work study engineer will have an opportunity to test and modify the details based on the feedback from the initial stages of implementation. It may be pointed out here that method study is one wing of work study, the major function of an industrial engineer. His job is also to train, guide, and advise the operatives and the supervisors until such time the new method is adapted without any problems. Fig. 22.5 is a typical learning curve, which indicates that the worker often requires long practice to obtain the high and consistent speed of working.

We can summarize that the step installation consists of 4 stages.

1. Gaining acceptance of the changes by the departmental personnel, right from the production manager to the supervisor.
2. Getting approval of the top management.
3. Getting acceptance by the concerned worker and/or their representative, giving due attention to the fact of resistance to the change from them. Reference can be made to Section 11.6 of the chapter on *Total employee involvement.*
4. Training and retraining the worker to operate in the new method. Even though the worker is experienced and normally gives over 100% output in the existing method, he cannot immediately attain the full standard performance in the very first trial of the new job (Fig. 22.6).

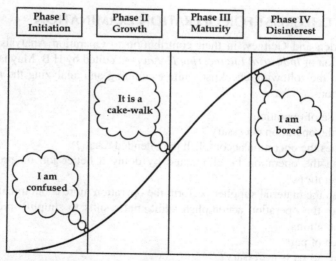

FIG. 22.6   A typical learning curve.

In carrying out the first three stages, the importance of preliminary education and training of all personnel concerned with the change, including the workers, supervisors and, if necessary, the management, in work study and its techniques, is significant. Their reception to the idea of change is more proactive when they know what and how the change is made, than when they are merely presented with the result. Sometimes it may be necessary to decide if the savings by the new method justifies the training and retraining the old workers, who find it hard to come out of their old practices. In that case, it is better to concentrate on training the new workers on the new methods, letting the old workers continue to work in the way they know.

22.17 MAINTAIN

The new method, after being installed successfully, should not be allowed to slip back into the old method during the course of time or to introduce elements not allowed for. It is absolutely wrong to presume that the work study man's job is finished the moment the proposal is approved and installed. There is a general tendency to drift away from the installed method. He should maintain the proposed new method for fairly a long period until the new method gets set. It is imperative he keeps visiting the workplace and inquiring into any difficulties in the developed method, and sincerely solving any of the difficulties they have been facing. The supervisor must be encouraged to understand the new method and to assist the workers in case of problems, but the work study man must keep visiting to solve the difficulties. ILO has specifically recommended maintaining close contact with the job until the progress is satisfactory.

22.18 CHECKLIST FOR OPERATION EXAMINATION

Stegemerten and Geitgey, in their contribution to Operation Analysis in the third edition of *Industrial Engineering Handbook*, edited by H.B, Maynard, has suggested the following checklist while examining and analyzing the method of operation.

1. Purpose of operation
   - Is the operation necessary?
   - Does the operation accomplish the intended result?
   - Can the operation be eliminated by doing a better job on preceding operations?
   - Can the material supplier perform the operation more economically?
   - Can the operation accomplish additional results to simplify preceding operations?
2. Design of part
   - Are all parts necessary?
   - Could standard parts be substituted?
   - Does design permit least costly processing and assembly?
   - Will design allow eventual automation?
3. Process analysis
   - Can the operation being analyzed be eliminated?
   - Or combined with another?

- Or be performed during the period of another?
- Is the sequence of operation the best possible?
- Should the operation be done in another department to save cost or tooling?

4. Inspection requirements
   - Are tolerances, allowances, finish and other requirements necessary, or too costly or suitable for the purpose?
   - Should statistical quality control be used?
   - Is the inspection procedure effective and efficient?

5. Materials
   - Consider alternative size, suitability, straightness, and condition.
   - Can cheaper materials be used?
   - Will tool modifications permit using lighter materials and thinner sections?
   - Would a more expensive material lower the machining and processing costs?
   - Is packaging suitable?

6. Materials handling
   - Can incoming materials be delivered directly to the work station?
   - Can signals, such as lights and bells be used to notify the material movers that the material is ready to be moved?
   - Should cranes, gravity conveyors, tote pans, or special trucks be used?
   - Consider the layout with reference to the distance moved.
   - Are containers correctly sized?

7. Work Place layout, setup, and tool equipment
   - Is the arrangement of work area, location of tools materials optimal?
   - How are drawings and tools secured?
   - Can setup be improved?
   - Are the machine adjustments and trial runs optimally executed?

8. Methods
   - Are hand motions symmetrical?
   - Are parts transferred between hands?
   - Is more detailed motion study needed?
   - Has safety been considered?
   - What is the working posture?
   - Does the method follow the laws of motion economy?
   - Is the lowest class of movement used?
   - How does it compare with other operator on the same or identical job?
   - Can foot-operated mechanisms be used?

9. Working conditions
   Consider the following for improvement
   - Light
   - Heat
   - Ventilation
   - Drinking fountains
   - Washrooms, etc.

22.19   OTHER CONTINUOUS IMPROVEMENT TECHNIQUES

Continuous improvement can also be effected by other concepts, such as CREW, 3M's, DFSS, and SMED and several others as cited in Section 22.5 and are explained more in detail in later chapters.

22.20   CASE STUDIES ON KAIZEN APPLICATIONS

(a) *Unbalanced workload on an assembly line*: A ceiling fan manufacturing company had a conveyor belt assembly line for its stator assembly, where the workloads among the different operatives on their line was unbalanced with one operative being occupied for 20% of the total cycle time, waiting for the work to come from the previous operator. During a study, the industrial engineer broke up all the operations into transferable elements of work and presented it before the operatives, who studied them and redistributed the elements among themselves so that each operative was more or less equally occupied during the cycle time, and the output of the line increased from 15 to 40 fan units per hour. This can be seen as an application of Heizunka, as listed in Fig. 22.2 under the umbrella of Kaizen, although this author was not aware of this Japanese term in 1973 when the above project was implemented.

(b) *Fitness equipment at a peoples' park*: Look at the following photographs. The gym equipment used in most gymnastic clubs is so expensive that the

(A)   (B)

(C)

FIG. 22.7   Kaizen applied to fitness equipment.

patronage is poor due to high cost of usage. However, a Chennai park purchased and installed simpler gym equipment that cost far less and can be used by the park users free of cost. Photograph 1 shows the costly gym equipment at a gym club with poor patronage, compared to the park equipment shown in photograph 2. Photograph 3 shows a simple turning wheel that is good exercise to the shoulders and back muscles (Fig. 22.7).

(c) *Problem of unfilled soap boxes in assembly line*: A soap manufacturing factory received complaints of unfilled soap boxes being passed on to the market. The factory management considered several alternatives to identify the unfilled cartons on their way out of the soap-filling machine by suggesting sophisticated and expensive equipment, such as X-ray scanners with high resolution monitors. However, a shop level operative suggested mounting an air circulator pointing it to the conveyor belt, so that the empty boxes would fly off.

(d) *Automobile rearview mirrors*: Kaizen can also be affected by continual improvement of minor accidental innovative thinking. There was no concept of the rearview mirror when the first cars were introduced. One lady in the passenger seat was making up her face through her vanity mirror when her husband was driving the car. Suddenly, she saw in the mirror something at the back of the car and drew his attention. Then she developed the practice of constantly looking into the mirror and informing him. Thus the need arose, for a permanently fixed rearview mirror, which today is the most used part of the car.

(e) *Pen that works well in zero gravity*: When the first US astronaut was launched into space, NASA discovered that the pens did not work there due to zero gravity preventing the free flow of ink. After spending years on research and US$ 12 million, they developed a pen that worked well in zero gravity, under water and in any temperature varying from 0°C and 200°C. *How did the Russians solve this issue? Just by using a pencil!*

22.21   SOME QUOTATIONS ON CHANGE

1. Change alones is eternal, perpetual and everlasting.
   –Arthur Schopenhauer
2. Change begets change. Nothing propagates so fast.
   –Charles Dickens
3. If you want things to stay as they are, things will have to change.
   –Giuseppe di Lampedusa
4. Be the change you wish to see in the world.
   –Mahatma Gandhi
5. Every organization has to prepare for the abandonment of everything it does.
   –Peter Drucker
6. Without deviation, progress is not possible.
   –Frank Zappa

7. Great discoveries and improvements invariably involve the cooperation of many minds.
–Alexander Graham Bell

22.22 CONCLUSION

Right from Taylor's days, SREDDIM worked as the mantra of continuous improvement, enabling the global industrial production to attain a high level of efficiency. Today KAIZEN, the same mantra in different words, has enabled the global industrial production to attain a high level of efficiency.

On the Lighter Side

In the park cited in Section 22.20b, a notice board at the gym says "only for persons over 15 years," while a board at the neighbouring children playpen says "only for persons below 14 years." What about children aged 14 or 15 years? This reminds us of Rabindranath Tagore's "Boy of Fourteen" wherein he humorously describes the travails of a 14 years old, who is neither given the benefit of being a child nor given the respect of an adult.

FURTHER READING

[1] Chang RY, Neidzwick ME. Continuous improvement tools. New Delhi: Wheeler Publishers; 1998.

[2] Imai M. Kaizen: the key to Japan's competitive success. New York: McGraw-Hill; 1986.

[3] Kiran DR. How to be More Creative. House Journal of Ralli Group, 1975;6–7.

[4] Kiran DR. Method study, a necessary tool for productivity improvement—a case study. Ind Eng J 1982;29–34.

[5] Kiran DR. Material layout planning. Ind Eng J 1980;11–6.

[6] Kiran DR. Work study in Transport Sector, Proceedings of Work Study Seminar, Tanzania, 1982.

Chapter 23

5S

Chapter Outline

23.1 INTRODUCTION

5S is more a practice than a technique used to establish and maintain the quality environment in an organization. It is basically a method for organizing the workplace on a shop floor, or even in an office. It advocates *what* to keep, *where* to keep it, and *how* to keep (maintaining, cleaning, etc.) the needed items in the workplace. This is based on the principle that a good environment motivates the worker to produce quality products or services with little or no waste and with high productivity. A bad and dirty environment or workplace distracts the attention of the worker, making him produce more defects. This concept is used today not only to improve the work environment, but also to improve the awareness, thinking, and philosophy of working.

That is why a majority of quality oriented organizations today emphasize this and give specialized training to their workers. Even though every worker

Total Quality Management: Key Concepts and Case Studies. http://dx.doi.org/10.1016/B978-0-12-811035-5.00023-4

is conscious of these as good habits, an emphasis on these in the training program would align them to practice the habits in their day-to-day working. 5S instills a sense of ownership among the workers to be more accountable for their workplace. It may sound too simple and nothing but common sense, but the commitment and meticulous practice of these in any workplace go a long way in achieving higher quality levels in industry.

By formalizing this technique, the Japanese established the framework which enabled them to successfully convey the message across the organization and achieve total employee involvement, and successfully implement the practice. Surprisingly, though these principles have been taught at the elementary education level globally as civic habits, only after realizing how the practice of 5S helped the Japanese to successfully maintain quality output, the Western world and later the rest of the world, adopted this as a management technique.

The name 5S stands for the five Japanese terms *Seiri, Seiton, Seiso, Seiketsu,* and *Shitsuke,* which are further explained in Table 23.1, comparing them with some equivalent English terms. In short, we can say that 5S is a systematic and rational approach for a clutter-free and safe workplace with the objective of reducing waste and preparing ground for further improvement.

23.2   EXPLANATION OF THE 5Ss

23.2.1   Seiri (Structuring—Distinguish Between the Necessary and the Unnecessary—Adopt Red Tagging)

It is the action of separating all the tools, materials, etc., which are necessary for the job from those that are not, and keeping only essential items. Everything else is stored elsewhere or discarded.

A few examples of unwanted items are:

1. Machines to be scrapped,
2. Rejected materials,
3. Expired goods,
4. Broken tools, pallets, bins, and trolleys,
5. Old notices,
6. Scrap heaps outside.

How to distinguish between wanted and unwanted:

At times, you would not be able to decide what and when to discard. In this case, make a list of all the items around you in respect of their usage.

(a) Identify things constantly in use, required hourly, or even once daily, and keep them close to the machine.
(b) Things that are used quite frequently, say once daily or once weekly—keep them on racks near the machine.

TABLE 23.1 Meaning of the 5S's

Japanese Word	In Japanese Script	Equivalent English Words Generally Used	English Words Preferred to be Used by This Author	Meaning
Seiri	整理	Sort	Structurizing	Distinguish between wanted and unwanted items. Remove unwanted items. Adopt red-tagging
Seiton	整頓	Straighten	Set in order or Systemize	Organize the workplace (**PEEP**-A place for everything and everything in its place). Keep items in their correct place for easy and immediate retrieval
Seiso	清掃	Sweep or sanitize	Shine	Be committed to keeping the workplace clean and look for ways to keep the workplace clean
Seiketsu	清潔	Standardize	Standardize	Maintain workplace or storage as per the established standards. Exhibit all the procedures posters for everyone to see, understand, and follow
Shitsuke	躾	Sustain the discipline	Self-discipline	Follow the rules of the company and honor established procedures. Do things in a way they are supposed to be done

(c) Things that are rarely used, say once in a month or so, but still needed—keep them in the stores.
(d) Things that are never used (large ones for the past year and small ones for the past 4 or 6 months) and which you think would never be used—discard them.

> **Remember the golden rule of Seiri to classify the unwanted items:**
> • **Wanted but not here**
> • **Wanted but not in this quantity**
> • **Wanted but not now**
> • **Not Wanted at all**

FIG. 23.1   Golden rule of Seiri.

23.2.2   Seiton: Systemize: (Or PEEP, A Place for Everything and Everything in its Place)

Once you have decided which items should be kept near the workplace, the next step is to arrange the items in a manner to facilitate work flow. For manual jobs like assembly, the components and tools should be kept in the order they are required, in the sequence of the elements of the operation. It focuses on work efficiency. An example is to hang the pneumatic screwdrivers, etc., in front of your assembly table, so when required, you just have to pick it up. After screwing, the pneumatic screwdriver gets retracted and goes back to its hanging position, as illustrated by Fig. 23.1. Also illustrated in Fig. 23.2 is an arrangement of all required components like screws, nuts, and washers that should be kept in pre-located positions, so that they can be picked up without spending any time searching for them. The same would be true in other situations, like operating a machine (Fig. 23.3).

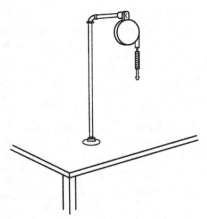

FIG. 23.2   Retractable screwdriver for simple assembly operations. *Section 2 of Industrial Engineering Handbook edited by H.B. Maynard, 1963.*

This Seiton principle, coupled with simultaneous motion of the two hands, has been a classic example of simple nut-bolt assembly, as demonstrated to students of workstudy during the 1960s. Fig. 23.4 from ILO's 1969 book on *Work Study* illustrates how the assembly workplace layout should be organized

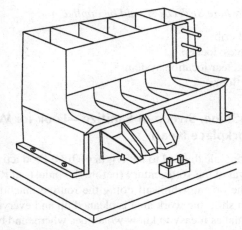

FIG. 23.3 Bin to provide preplanned locations for assembly components. *Section 2 of Industrial Engineering Handbook edited by H.B. Maynard, 1963.*

for simple and logical movement of the hands. This figure also emphasizes the fact that the Seiton Principle was practiced in the pre-world war era itself, but the Japanese gave it a good impetus by putting it in simple language to be understood clearly and practiced wholeheartedly by the operatives.

FIG. 23.4 Seiton principle applied to simple assembly operations.

In case of machine operation the location of the needed tools and equipment near the workplace, should be well demarcated and the locations suitably labeled.

Few examples where SEITON should be applied are:

1. Unlabeled tool crib
2. Clustered shelves, lockers
3. Stores with no clear location system
4. Things on the floor.

23.2.3 Seiso (Shine, Sweep or Sanitize—Look for Ways to Keep the Workplace Neat)

This is the very principle that led to the origin of Total Productive Maintenance. In fact, it is a revival of the 16th century (before the Industrial Revolution) practice of the machine operator himself doing the routine machine maintenance. At the end of each shift, the work area is cleaned up and everything is restored to its place. This makes it easy to know what goes where, and be confident that everything is where it should be. The operator feels responsible, not only for the output and quality of his output, but also for proper working of his machine. This motivates him to take part wholeheartedly in the quality circle meetings and put forward his suggestions for the improvement of machine performance. The fact highlighted here is that maintaining cleanliness should be part of the daily work—not an occasional activity initiated when things get too messy.

Few examples of dirty workplaces that need cleaning are:

1. Dirty machines
2. Dust on products, parts, and raw materials
3. Dirty jigs, fixtures molds, and bobbins
4. Dusty walls, roofs, littered floor, etc.
5. Untidiness outside the factory

Fig. 23.5 is a cartoon from the *New York Times* Magazine of 1964, indicating that this **Seiso** is so much common sense that even in those days neatness in the office or factory was being emphasized.

THIS ..   not THAT!

FIG. 23.5   Good workplace vs. bad workplace.

23.2.4   Seiketsu (Standardize—Keep the Workplace as per the Established Standards)

This principle encompasses the duties of the supervisor or the engineer in order to assist the operatives to adhere to and effectively follow 5S. Everything from work areas to storage locations, equipment, tools, etc., should be clearly marked and identified with clearly drawn labels and signs. They should create a standardized and consistent 5S workflow and assign tasks and clear schedules so that everyone knows their responsibilities.

Other visual controls include work instructions, hazard warnings, cautions, and reminders and indicators of where things are kept, equipment, and tool designations. Posters depicting what happens when, provide effective communication. Locate them in such places and manner that anyone can see and use them easily and conveniently.

For everyone to understand a standardized system, they should be trained on it, and perhaps regularly tested to ensure adequate understanding. The design of the system should ease learning.

Visual controls include work instructions, hazard warnings, indicators of where things are kept, equipment and tool designations, cautions and reminders, and plans and indicators of what happens when.

A particular technique of Seiketsu is *visual management*. Visual management or graphical display leverages all of these so that when we are looking for something, it stands out indicating the needed information like the location, distance, shape, brightness, color, and contrast.

Shown alongside is a safety poster at a beach, communicating what you should do and what you should not do. Also shown on the next page are some of the posters kept at strategic locations of the factory to draw the attention of the operatives (Fig. 23.6).

FIG. 23.6   Seiketsu principle applied to a poster at beachside.

Why visual control display?

1. To help people avoid making operating errors
2. Alert dangers
3. Indicate where things should be put
4. Preventive maintenance displays
5. Other instructions

Examples of visual controls

1. Warning lights
2. Safety posters
3. Transparent windows
4. Color coding
5. Labels
6. Position marks
7. Okay marks
8. Visualize conditions
9. "What is where" charts
10. Who is where charts
11. Inspection labels, etc. (Fig. 23.7).

FIG. 23.7 Seiketsu principle applied to posters in factories.

23.2.5 Shitsuke (Sustain the 5S Practices by Work Discipline—Follow the Rules)

Shitsuke, meaning *discipline*, denotes commitment to maintain orderliness and to practice the first 4 Ss as a way of life. The emphasis of shitsuke is elimination of bad habits and constant practice of good ones. The responsibility for shitsuke is shared between management and the workforce. Management must take responsibility for continuing to communicate the 5S message, and for regular inspections to enforce the standards. Employees should be held accountable for doing the work and creating the results.

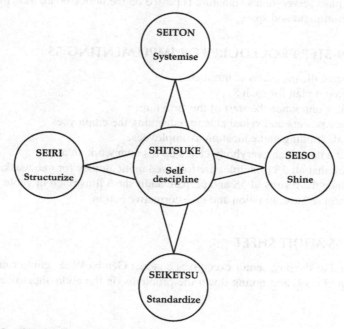

FIG. 23.8   Shituke is the foundation for 5S.

The following are some of the points to be followed by the operatives to the core.

1. Wearing appropriate safety gear at all times is essential
2. Name plates (badges) should be worn.
3. If all members of an officer or workshop are not present when they are supposed to be, it is poor 5S, punctuality is the backbone of 5S.
4. Timeliness also applies to preparation of reports, filling out the charts, etc.
5. Supervisors must correct wrong practices everyday on the spot.
6. Training the employees is a must. This must be done through classroom, actual practice, and also by attractively illustrated instructions and displays as explained in Seiketsu.

23.2.6 Significance of Shitsuke in 5S

As can be clearly visualized, shitsuke forms the most important foundation for 5S. Unless you maintain the discipline, you cannot be committed to 5S. Unless you are committed to 5S, you cannot achieve 5S. This significance is illustrated in Fig. 23.8.

In many Japanese companies, the employees take a pledge at the beginning of each campaign for 5S that they will adhere to the standards of putting things in order, proper arrangement, cleaning, and maintaining the neatness of the workplace. Everyone's signature is posted on the notice board as symbol of their commitment and spirit.

23.3   9-STEP PROCEDURE FOR IMPLEMENTING 5S

1. Organize the programs committee.
2. Develop a plan for each S.
3. Publicly announce the start of this program.
4. Prepare posters and visual aids for educating the employees.
5. Provide training and education to employees.
6. Select a day when everybody clears up his own work area.
7. Ensure that all 5S principles are followed meticulously for one week.
8. Evaluate the results of 5S and prepare audit sheet illustrated in Table 23.2.
9. Perform self-examination and take corrective action.

23.4   5S AUDIT SHEET

At the end of the day, senior executives conduct Gemba Walk, going round the factory premises, and noting down the problems on the audit sheet illustrated below.

TABLE 23.2 5S Check Sheet

Date	5S Step	Problem Found	Remedial Action Suggested	Rating
	SEIRI	Unneeded tools and equipment are present		
		Items are present on aisle ways, stairways, etc		
		Unneeded inventories, supplies, etc., are present		
		Safety hazards exist		
		Unneeded and irrelevant notices/slips are present on the bulletin boards		

TABLE 23.2 5S Check Sheet — cont'd

Date	5S Step	Problem Found	Remedial Action Suggested	Rating
	SEITON	Items are not in correct places. Correct places are not indicated or understood		
		In three minutes of questioning, the operator could not locate 10 regularly used items		
		Aisle ways and equipment locations are not identified		
		Items are not put away immediately after use		
		Height and quantity limits are not clear		
	SEISO	Machine is dirty and greasy		
		Chips are not cleared from workplace		
		Floor, walls, and surroundings are dirty		
		Cleaning materials are not accessible		
		Bathrooms are stinking		
	SEIKETSU	Signboards are missing or broken		
		Labels, signboards are dirty and not clearly visible		
		Necessary information is not provided		
		Checklists do not exist		
	SHITSUKE	... operatives out of ... in the section had no 5S training		
		5S exercise was not done for the past ... days		
		Average late coming by the operatives is ... minutes		

23.5 AN EASY WAY OF REMEMBERING THE 5S TERMS

Look back at the list of 5S—so many sei's! Seiri, Seiton, Seiso—or is it Seiso before Seiton? Ooph, it is hard to remember these Japanese terms especially the order in which these appear.

Now let us see if the following illustration gives us is an easy way of remembering them!

A rolling Operation (abbreviated as Ri) is being done on a rolling machine by an operator called Krishna. After completing the operation, he has to approach the switch box to switch off the machine. Let us call this switch box "K."

That is, after completing the operation called "Ri," Krishna has to go towards the switchbox "K." (Fig. 23.9).

FIG. 23.9   Rolling operation (Note the operator's hand near the switch).

In other words,
The operation "Ri" is done. So Krishna goes to Switch "K."
Let us put these words this way –
Ri—done—so—Krishna—(goes to) Switch K.
In Tamil script "T" and "D"; are the same and generally the nickname for Krishna is Kitchu. Again *"goes to"* is not important here, but is written within brackets only to complete the sentence.
Hence, we can rewrite this as,
Ri—Ton,—so—Kitchu—Switch K.
By slightly modifying the spelling, this becomes

ri—ton—so—ketsu—Shitsuke.
Now add Sei to the first four:
Seiri, Seiton, Seiso, Seiketsu, and Shitsuke (Table 23.3).
Summary

TABLE 23.3 Summary of Easy Way of Remembering 5S Terms

Word as Above	Modified Word	5S Word
Ri	ri	*Seiri*
Done	ton	*Seiton*
So	so	*Seiso*
Kitchu	ketsu	*Seiketsu*
(goes to) Switch K	Shitsuke	*Shitsuke*

You can practice reciting these words fast as you would do in the tongue twister games. Initially, remember the above illustration to check your order, and gradually you would master the terms and the order. As one of the professors during the author's college days used to say,

Even if you are awoken in the middle of the night, you should be reciting these words before you open your eyes.

23.6 CONCLUSION

The concept of machine maintenance during the earlier part of post-Industrial Revolution was that the machine operator was fully responsible for its maintenance. By the principle of specialization, Taylor transformed this concept with maintenance gangs, even for machine oiling. The Japanese precept of 5S during post-World War II era had brought a complete U-turn to this practice, with the operator again being made responsible for the basic machine maintenance. This gave birth to Total Productive Maintenance.

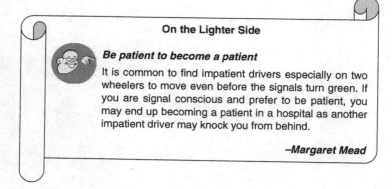

On the Lighter Side

Be patient to become a patient

It is common to find impatient drivers especially on two wheelers to move even before the signals turn green. If you are signal conscious and prefer to be patient, you may end up becoming a patient in a hospital as another impatient driver may knock you from behind.

—Margaret Mead

FURTHER READING

[1] Hirano H. 5 Pillars of the visual workplace. Cambridge, MA: Productivity Press; 1995.
[2] Osada T. The 5S's: five keys to a total quality environment. USA: APO; 1995.
[3] Weber A. Lean workstations: organized for productivity. Assembly, 2005.
[4] Kiran DR. An easy of remembering the 5S words and their order. NIQR J 2014; XI, p. 33.

Chapter 24

Six Sigma

Chapter Outline

24.1 INTRODUCTION

Six sigma is a business management strategy that allows companies to drastically improve their bottom line by designing and monitoring everyday business activities in ways that minimize waste and resources, while increasing customer satisfaction. The usual quality control programs focus on detecting and correcting commercial, industrial, and design defects. Six sigma encompasses something more. It provides specific methods to recreate the process so that defects and errors never arise in the first place. It should be noted that six sigma does not bring an overnight change. It is a long-term forward-thinking initiative. It is a philosophy.

This approach seeks to improve the quality of process outputs by identifying and removing the causes of defects (errors) as soon as they occur and minimizing variability in manufacturing and business processes. Once it gets established and consolidated, it results in almost zero defects by incorporating defect prevention right from the design stage.

Six sigma was heavily inspired by six preceding decades of quality improvement methodologies, such as quality control, TQM, and zero defects, based on the work of pioneers and quality gurus, as detailed in Chapter 3.

It may be noted that in recent times, the concept of Zero-Effect-Zero-Defect has become popular, as cited in Section 2.4, emphasizing industrial production without any effect on environment apart from defect-free production.

24.2   DEFINITIONS OF SIX SIGMA

Six sigma can be defined as:

1. A statistical measure of the performance of a process or a product that establishes a measurable status on yield, as defects per million opportunities (ppmo).
2. A goal that reaches near perfection for performance improvement and reduces the variation to achieve a small standard deviation, so that almost all products and services meet or exceed customer expectations.
3. A management system to achieve lasting business leadership and world-class performance, with a management philosophy focused on elimination of mistakes, rework, scrap, and other wastes.
4. Six sigma is a business management strategy that seeks to improve the quality of process outputs by identifying and removing the causes of defects (errors) and minimizing variability in manufacturing and business processes.
   –Wikipedia
5. Six sigma is a rigorous and disciplined methodology that uses data and statistical analysis to measure and improve a company's operational performance by identifying and eliminating "defects" in manufacturing and service-related processes.
   –iSix Sigma Orgn
6. Six sigma is a data-driven method for achieving near perfect quality. Six sigma analysis can focus on any element of production or service, and has a strong emphasis on statistical analysis in design, manufacturing, and customer-oriented activities.
   –UK Department for Trade and Industry

24.3   HISTORY OF SIX SIGMA

- Since the 1920s, the word "sigma" has been used by mathematicians and engineers as a symbol for a unit of measurement in product quality variation.
- The latter half of the 20th century has seen significant advances in quality improvement methodologies, such as quality control, TQM, and zero defects, based on the work of pioneers such as Shewhart, Deming, Juran, Ishikawa, Taguchi, and others.
- Inspired by these works, the engineers of Motorola developed the principle of Six Sigma in 1981. They used "Six Sigma" as an informal name for an in-house initiative for reducing defects in production processes.

- By the late-1980s, following the success of the above initiative, Six sigma was extended to other critical business processes. Six sigma became a formalized branded name for a performance improvement methodology in any field of business process.
- Motorola started the system of grading their six-sigma experts by colored belts like *black belt*, in line with the popular practice of the karate experts being graded by the color of their belts, such as blue belt, green belt, and black belt. In 1991, the company certified its first "Black Belt" six sigma experts, which indicates the beginnings of the formalization of the accredited training of six sigma methods.
- According to Wikipedia, Motorola held the trademark for six sigma and reported to have saved over US$ 17 billion until 2006.
- In 1995, General Electric implemented Six Sigma and by 1998, it reported savings of over US$750 million.
- Almost simultaneously, the Japanese industries like Toyota have successfully implemented this technique and some of them have achieved the six sigma level.
- During the late 1990s, Indian industries, too, began taking this seriously and some of them succeeded in achieving six sigma level by the middle of the last decade.
- Nowadays, some organizations have integrated six sigma with the concept of lean manufacturing and called the methodology *Lean Six Sigma*.

24.4 REQUIRED SKILLS FOR BLACK BELTED EXPERTS IN SIX SIGMA

A black belt expert is the team leader and is expected to be most skillful in planning and implementing six sigma projects. He is responsible for:

- identifying and understanding these processes in detail, and also
- understanding the levels of quality (especially tolerance of variation) that customers (internal and external) expect, and then
- measuring the effectiveness and efficiency of each process performance—notably the "sigma" performance—ie, is the number of defects per million operations (ppm).

24.5 THE CONCEPT OF SIX SIGMA IN THE CONTEXT OF TQM

Since the 1920s, the word "sigma or σ" has been used by mathematicians and engineers as a symbol for standard deviation, which is a unit of measurement in product quality variation that is how far a given process deviates from perfection. The higher the sigma is, the higher the variation.

Let us see the normal curve used in statistics. For a definite value of σ, the shape of the curve or the Kurtosis is the same. When we say 6σ here, we mean spread within six times the σ value arithmetically. The % number of items included between μ+6σ and μ−6σ is 99.99966% as illustrated in Fig. 24.1 and detailed in Table 24.1. Prior to the evolution of the term *six sigma*, people used the term *Five Nines* or 99.999% which almost represents today's six sigma level.

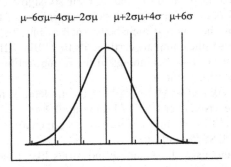

FIG. 24.1   Statistical concept of σ.

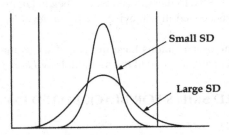

FIG. 24.2   TQM concept of sigma.

However, in quality control, the concept of the term six sigma is different from the above. Here the upper and lower control limits for a process are fixed and are very significant. Thus, when we say three sigma level quality, we mean that the 99.73% of the items equal to those that are included within μ+3σ and μ−3σ values in a normal curve lie within the control limits, with a lower kurtosis curve. When we say six sigma quality, we mean that 99.9999966% of the items, equal to those that are included within μ+6σ and μ−6σ values in a normal curve, lie within the control limits, so that the curve is steeper with a higher kurtosis value.

To make it clear, we can say that *in TQM, the six sigma is referred to a situation when the % number of items that are contained within the specified control limits are equal to those contained within the μ+6σ and μ−6σ limits in a normal curve of statistics.* Figs. 24.1 and 24.2 clearly explain the difference between the two concepts.

In TQM, the central idea behind Six Sigma is that if you can measure the number of defects in a process, you can eliminate them and get as close to "zero defects" as possible. To achieve Six Sigma Quality, a process must produce no more than 3.4 defects per million opportunities. An "opportunity" is defined as a chance for nonconformance, or not meeting the required specifications. This means we need to be nearly flawless in executing our key processes. Six Sigma basically involves the following factors.

1 Customer perception of quality
2 Defects that prevent delivery of what the customer wants
3 Process capability
4 Variation as seen and felt by the customer
5 Consistent and predictable processes for improvement
6 Designing to meet customer needs and process capability.

24.6   ORIGIN OF THIS CONFUSION BETWEEN STATISTICAL 6σ AND TQM SIX SIGMA

The illustration below shows how the use of the word 6σ in the concept of six sigma has created a general confusion.

> In a guest lecture on six-sigma in one of the technical meetings of a professional association, the speaker gave a very good case study from an industry. He illustrated how the quality, thereby the profitability, has tremendously improved by increasing the sigma level from the 4-sigma to 6-sigma. Really, it was a good and interesting case study. However, to illustrate the higher sigma levels, he was progressively increasing the horizontal space between his hands. One of the listeners pointed out that the wider difference in higher sigma level should only indicate higher variation, in which case, how could it mean higher quality. The speaker had no answer.

The concept with which we use the term σ (sigma) in statistics is different from that it is used in TQM, as can be clearly understood by Figs. 24.1 and 24.2. The figure on the left explains the statistical meaning of **6σ** without reference to the upper and lower control limits. It indicates the percentage of the values that lie within the certain σ values. On the other hand, from the TQM point of view, the upper and lower control limits are very important and it is the kurtosis of the control curve that decides the percentage of the values that lie within the two limits. The higher the kurtosis value, the higher the sigma level per TQM concept as explained above. Hence, it is very important to distinguish between σ of statistics and **sigma** of TQM. In statistics σ is represented by "**value**," while in TQM, sigma is represented by "**level**."

The engineers of Motorola in the United States used the term "Six Sigma" for the first time in 1981 as an informal name for an in-house initiative for reducing defects in production processes. Whereas in statistics **σ** is a measure of the deviation from the normal, Motorola engineers meant to imply the values contained within that sigma limits for the same normal curve, without perhaps realizing the confusion that would be created by the word "sigma."

Nevertheless, this concept was extensively applied first in Toyota Motor Corporation to achieve higher quality levels, and has come to stay as the most important tool world over to achieve higher quality standards. This term *Six Sigma* has now become a hallmark that many organizations have used, and continue to use, to improve quality, and to provide quality and performance improvement services and training. Several Indian industries have achieved this level.

Motorola itself puts it,

We think about Six Sigma at three different levels:

- *As a metric*
- *As a methodology*
- *As a management system*

Essentially, Six Sigma is all the three at the same time.

In a nutshell, Six Sigma is a vertical (top-down) method, initiated at CEO-level for executing business strategy by using and optimizing these process elements:

- Aligning critical improvement efforts to business strategy.
- Mobilizing teams to attack high-impact projects.
- Accelerating the improvement of business results.
- Governing efforts of the teams to achieve and sustain improvements.

It is widely believed that introducing Six Sigma is likely to produce far greater returns in organizations that need to achieve these things, compared to organizations that are already doing them.

Remember while using sigma in TQM:

1. Call it **six sigma** (at the most 6-sigma) and do not use the symbol 6σ.
2. To refer to the extent of its application, use the term "level" and not "value." That is to say "I achieved six sigma level quality," and not "I achieved a quality of six sigma value."

24.7   SIX SIGMA ACCORDING TO GENERAL ELECTRIC

According to General Electric, "Six Sigma is a highly disciplined process that helps us focus on developing and delivering near-perfect products and services. The word sigma is a statistical term that measures how far a given process deviates from perfection. The central idea behind Six Sigma is that if you can measure how many "defects" you have in a process, you can system-atically figure out how to eliminate them and get as close to "zero defects" as possible. To achieve Six Sigma Quality, a process must produce no more than 3.4 defects per million opportunities. As explained earlier, an "opportu-nity" is defined as a chance for nonconformance, or not meeting the required specifications. This means we need to be nearly flawless in executing our key processes."

In a nutshell, Six Sigma revolves around a few key concepts.

- *Critical to quality*: Attributes most important to the customer
- *Defect*: Failing to deliver what the customer wants
- *Process capability*: What your process can deliver
- *Variation*: What the customer sees and feels
- *Stable operations*: Ensuring consistent, predictable processes to improve what the customer sees and feels
- *Design for Six Sigma*: Designing to meet customer needs and process capability..."

24.8   THE VALUES OF THE DEFECT PERCENTAGES

In Table 24.1 below, the last column indicates the percentage of values that lie within the control limits. The more popular measure, the number of defects per million opportunities (1-acceptable specifications per million opportunities), is indicated in the second column. Most of the popular Japanese companies, as well as several Indian industries like the TVS group, have achieved 6σ level and are striving to achieve the 7σ level.

TABLE 24.1 The Values of the Defect Percentages Under Various Sigma Levels

Sigma Level (Note the Word "Sigma" is Used Instead of the Symbol σ)	Defects per Million Opportunities (DPMO)	Percent Defective (%)	Percentage Yield (%)
1 Sigma	691,462	69	31
2 Sigma	308,538	31	69
3 Sigma	66,807	6.7	93.3
4 Sigma	6210	0.62	99.38
5 Sigma	233	0.023	99.977
6 Sigma	3.4	0.0034	99.99966
7 Sigma	0.019	0.000019	99.9999981

When practiced as a management system, Six Sigma is a high performance system for executing business strategy. Six Sigma is a top-down solution to help organizations to

- Align their business strategy to critical improvement efforts
- Mobilize teams to attack high impact projects
- Train its people to focus on key performance areas
- Accelerate improved business results
- Govern so the efforts to ensure improvements are sustained
- Understand where the organization wants to go (its strategy, related to its market-place)
- Understand the services that the organization's customers need most
- Understand and better organize main business processes that deliver these customer requirements
- Measure (in considerable detail) and improve the effectiveness of these processes.

It may be noted that the Bombay Dabbawalas have provided a classic example of achieving 6σ level in their lunch box distribution. The case study in Section 24.14 is a striking example. The other striking illustration is the note-counting machines used in banks, and other cash handling offices in achieving the 7σ level. Several banks have reported that during their use for the past over 5 years, there has not been a single case of wrong counting.

24.9 METHODOLOGIES FOR SIX SIGMA

Deming's PDCA cycle, one of the earliest principle of total quality management, forms a classic illustration of how the design stage forms the first step in the journey towards achieving six sigma at the final product stage.

24.10 DMAIC METHODOLOGY FOR SIX SIGMA

Deming's popular TQM methodology of **Plan-Do-Check-Act Cycle** as described more in detail in Chapter 1, can be effectively adapted in an attempt to achieve six sigma levels. This is generally used for improving an existing product or process and is further developed into DMAIC (Define, Measure, Analyze, Improve, and Control) the methodology described in Section 24.12 and shown in Fig. 24.3.

24.11 DMADV

While DMAIC emphasizes on the improvement of the existing processes, the improvement should start from the design of the product/process itself for developing the process in order to achieve six sigma. In this case, PDCA methodology is further modified as DMADV (Define, Measure, Analyze, Design, Verify). DMADV as adapted for Design for quality is explained in Chapter 32 more in detail.

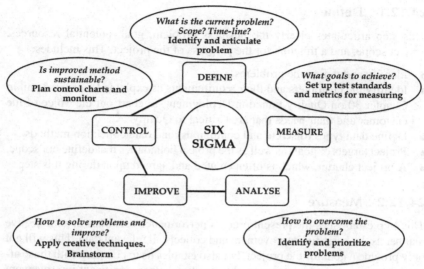

FIG. 24.3   DMAIC methodology.

The other acronyms related to DMAIC/DMADV are

- IDOV (Identify, Design, Optimize, and Validate)
- DCCDI (Define, Customer Concept, Design, and Implement)
- DMEDI (Define, Measure, Explore, Develop, and Implement)

Whatever may be the acronym we adapt, all of these methodologies cited above use techniques similar to Quality Function Deployment, Failure Modes and Effects Analysis, Design of Experiments, methods improvement, value engineering, simulation, statistical optimization, error proofing, Robust Design, etc. The only distinguishing feature is whether they are used for existing product/process improvement (like DMAIC, methods improvement, etc.) or for new products or processes (like DMADV, value engineering, etc.). The term D in other acronyms may mean Design for new products or re-design for existing products.

24.12   DETAILED METHODOLOGY OF DMAIC

Wikipedia details each of these DMAIC steps as below. It may be noted here that though we have numbered them serially, the steps are interrelated and flexible. We may analyze them even during the improvement. We may still be finding scope for new metrics before measuring them. We may be collecting data even after analysis. We may start guessing even while preparing, as emphasized in Chapter 22 on Kaizen.

24.12.1 Define

This step articulates clearly the business problem, goal, potential resources, project scope, and a timeline for the progress of the project. This includes:

- Identify and define the problems.
- Identify the customers and their requirements as explained more in detail in Chapter 30 on Quality Function Deployment, to chart out the Voice of the customer and their needs that are Critical to Quality.
- Define data types (discrete and continuous) and data collection methods.
- Project targets or goals as well as the project boundaries that define our scope.
- A project charter, which is often created and agreed upon during this step.

24.12.2 Measure

This step establishes the present process performance baselines in an objective manner as the basis for improvement and collects all relevant data. This will not only provide a goal for the project, but also enables us to compare with those after the project, so as to determine objectively whether significant improvement has been made or not and if so, by how much. It is essential to plan well ahead and decide the metrics, their suitability and the methods of measuring them. A relevant and substantial data is the heart of the DMAIC process:

- Identify the gap between current and required performance.
- Collect data to create a process performance capability baseline for the project metric, that is, the process output or outputs.
- Draw a **SIPOC** (suppliers, inputs, process, outputs, and customers) chart that summarizes the inputs and outputs in table form.
- Assess the measurement system (for example, a gauge study) for adequate accuracy and precision.
- Establish the following
  - Data collection techniques
  - Concepts of variation
  - QC tools
  - Management tools
  - MSA and measuring process capability
- Calculation of process sigma level
- Calculation of baseline performance
- Establish a high-level process flow baseline. Additional details can be filled in later.

24.12.3 Analyze

The purpose of this step is to identify the root cause of the problem, validate and select the method of its elimination. A large number of potential root causes of the project problem are identified by root cause analysis techniques, such as fish bone diagram, value analysis, and method improvement studies that involve elemental breakdown. The top 3–4 potential root causes are selected using team consensus tools like multi-voting for further validation. Data is collected and analyzed to establish the relative contribution of each root cause affecting the project output metric. This process is repeated until "valid" root causes can be identified. After validating root causes, the following shall be done.

- Listing and prioritizing potential causes of the problem.
- Prioritizing the root causes (key process inputs) to pursue in the Improve step.
- Identifying how the process inputs affect the process outputs. Data is analyzed to understand the magnitude of contribution of each root cause, to the project metric. Cause and Effect Analysis and other Statistical tests using p-values accompanied by Histograms, Pareto charts, and line plots are often used to do this.
- Validating causes and arriving at the root cause.
- Creating detailed process maps to help pinpoint where in the process the root causes reside, and what might be contributing to the occurrence.

24.12.4   Improve

After analyzing as above, the next step is to develop solutions and improve. The purpose of this step is to identify solutions to the problems in part or in whole. This is the step that largely incorporates the concept of creativity explained in Chapter 22 on Kaizen. Special mention can be made here are:

1. Creative methodology (Section 22.9)
2. Brain storming (Section 22.10.5).
3. Value analysis (Chapter 33).
4. Design of experiments with specific reference to the work of Taguchi (Chapter 31).
5. Test solutions using Plan-Do-Check-Act (PDCA) cycle.
6. FMEA which anticipate any avoidable risks associated with the improvement conceived in the above techniques (Chapter 26).
7. After the above indicated validations, create a detailed implementation plan and deploy improvements.

24.12.5   Control

The purpose of this step is to sustain the gains. Monitor the improvements to ensure continued and sustainable success. Create a control plan. Update documents, business processes, and training records as required.

A Control chart is useful during the Control stage to assess the stability of the improvements over time by serving as a guide to continue monitoring the process and to provide a response plan for each of the measures being monitored in case the process becomes unstable. Other steps for control are:

- Assessing the results of process improvement
- Developing process control plan
- Standardization

The flow chart for DMAIC methodology can be represented by Fig. 24.4

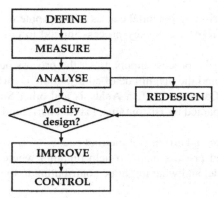

FIG. 24.4 Six sigma process (DMAIC).

24.13 ORGANIZING FOR SIX SIGMA

Several companies abroad and in India have adopted six sigma as their prime tool for maintaining high quality standards. For professionalizing six sigma activities, it is essential to specify the members of the team performing this exercise. Wikipedia details the following classifications.

1 *Executive leader*, including the CEO and other top members of the organization. They are ultimately responsible for the whole project and empower the other role holders with the freedom and resources.
2 *Champions*, take responsibility for Six Sigma implementation across the organization in an integrated manner. The Executive Leader draws them from upper management. Champions also act as mentors to black belts.
3 *Master black belts*, identified by champions, act as in-house coaches on Six Sigma. They devote 100% of their time to Six Sigma. They assist champions and guide black belts and green belts. Apart from statistical tasks, they spend their time on ensuring consistent application of Six Sigma across various functions and departments.
4 *Black belts*, operate under Master black belts to apply Six Sigma methodology to specific projects. They devote 100% of their time to Six Sigma. They primarily focus on Six Sigma project execution, whereas Champions and Master black belts focus on identifying projects/functions for Six Sigma.

5 *Green belts*, the employees who take up Six Sigma implementation along with their other job responsibilities, operate under the guidance of black belts.

6 *Yellow belts*, trained in the basic application of Six Sigma management tools, work with the black belt throughout the project stages and are often the closest to the work.

Unlike the Karate field where the members actually wear the respective cotton belts suiting their respective expertise, the team members of six sigma projects do not wear such belts but their designations are only symbolic with a sole purpose of creating an interest of belonging among the members.

24.14 SOFTWARE USED FOR SIX SIGMA

The following are some of the software popularly used to manage and track a company's six-sigma program. Since the details discussion and explanation of these software is beyond the scope of this book, the readers are advised to refer to their respective websites.

- IBM Web Sphere Business Modeler
- iGrafx
- JMP
- Mentor soft PRO
- Microsoft Visio
- Minitab
- Stat graphics
- STATISTICA
- Telelogic system architect

24.15 THE CASE STUDY OF MUMBAI DABBAWALAS

During the early 1970s, when this author was working at L&T, Powai, Bombay, he was utilizing the services of Mumbai Dabbawalas who were bringing his lunch box from Chembur to Powai. Though much thought was not given to it then, he was pleasantly surprised 40 years later to learn that they have been recognized by international quality gurus, as the best illustration of a supply chain management system in the world, having earned six sigma rating by the Forbes Business magazine in 1998 and found a place in the Guinness Book of World Records. In 2010, the National Institution for Quality and Reliability, a premier quality professionals' institution of India had awarded them the Best Service Organization award (Fig. 24.5).

True, during the 4 years of their day-to-day service, this author did not miss his lunch even for a single day. On a nostalgic chat with the recipient of the award in 2010, the following statistics have been recapitulated (Fig. 24.6).

FIG. 24.5   Mumbai Dabbawalas.

They were supplying one lakh lunch boxes from homes of different parts of Mumbai to offices/factories located at different places. Today the figure must have reached 2 lakh lunch boxes. Around 9 am, 30 boxes would be collected from the houses by each person and carried to the nearest Railway station like Kurla, where all such groups would be re-sorted destination wise and exchanged among the delivery boys. There would be another regrouping and re-sorting at the destination Railway station and carried to the particular work spot by a third delivery boy. This process would be repeated between 2 pm and 5 pm when the empty boxes would be returned to the houses.

Though they do not have formal education, they use a unique coding system indicating the pickup point, regrouping points, and the destination point. They employ 500 persons, including 300 delivery boys and 200 supervisors to ensure accuracy at every re-sorting point. This shows the importance given to this system of having so many supervisors (one for every 2 pickup/delivery boys) which is a rare case in the industrial scenario.

FIG. 24.6   Codes assigned to each lunch box.

The Wikipedia site en.wikipedia.org/wiki/Dabbawala gives a detailed report on this unique organization.

24.16   CONCLUSION

As detailed in Section 24.6, even though this concept has been named six sigma in a casual manner causing confusion with the statistical parameter of 6s, it has become the most significant tool in the quality movement, having achieved wonders in industries all around the world, and has come to stay as the backbone of total quality management.

On the Lighter Side

A company advertised its new Six Sigma approach. A customer placed an order for a thousand parts saying: "We don't want more than two bad parts per thousand." The Six Sigma company shipped a container with the thousand parts, on time to the customer. Along with the container came a small parcel. The customer called and asked the supplier, "What is in the parcel?" The company answered, "The two bad parts."

FURTHER READING

[1] en.wikipedia.org/wiki/Six_Sigma.
[2] www.isixsigma.com.
[3] Kiran DR. Understanding six sigma—13th NIQR convention. 2012. p. 73–5.

Chapter 25

Lean Management

Chapter Outline

25.1 WHAT IS LEAN MANAGEMENT?

Lean manufacturing is a production practice that considers that the available resources must be expended only for creating the required value to a product for the end customer, and any amount of overexpenditure of these resources otherwise is wasteful, and hence, must be curtailed. The term *Lean Management* originated from the word "leaning"; which refers to the reduction of fats from the body in the process of weight reduction and in making one's body attractively lean and agile. This concept was developed in 1990s mostly from the Toyota Production System, and so is also called *Toyotism* or simply *Lean*, and was successfully applied both in Japan and the Western world. In fact, based on Taiichi Ohno's work, John Krafcik in his 1988 article, "Triumph of the Lean Production System," coined this prefix "Lean."

Total Quality Management: Key Concepts and Case Studies. http://dx.doi.org/10.1016/B978-0-12-811035-5.00025-8
363

Originally applied to their manufacturing situation, the prefix "lean" is now applied to several functions, such as lean maintenance, lean quality control, lean inspection, etc., all of which refer to the single concept of reducing wasteful activities in that function.

Subsequently, a new term Lean management has been coined to encompass leanness in all the management functions, including those cited above. Lean production principles are now referred to as *lean management* or *lean thinking*.

- Lean maintenance
  It refers to the proactive maintenance practices employing planned and scheduled maintenance activities through several modern and optimal trends in maintenance management.
- Lean organization
  A lean organization understands customer value and focuses its key processes to continuously increase it. The ultimate goal is to provide perfect value to the customer through a perfect value creation process that has zero waste.
- Lean six sigma
  Several organizations have integrated six sigma with the concept of lean manufacturing and called the methodology as *Lean Six Sigma*.
- Lean inspection through supplier partnership
  Incoming inspection is a nonvalue adding activity. A properly planned and executed supplier partnership moves away from the traditional sample inspection to a system where the supplier takes the responsibility of inspecting his premises. Here the materials move directly from the supplier to the final assembly section of the purchaser.
- Lean Concept in to Office Management
  This lean concept is also applicable to office management also as illustrated by Fig. 23.5 in the Chapter on 5S (Fig. 25.1).

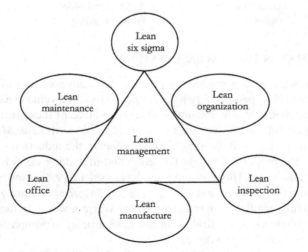

FIG. 25.1   Relationship between various lean management concepts.

In other words, lean management can be said that the thought process in effective coordination of all functional activates so that those products that need to be delivered are manufactured in the right quantity at the right quality, and at the right time with fewer resources. That is to say, it doesn't always have to do with always doing more, that can result in waste, but things that have been sold should be produced as they are needed to be delivered.

25.2 COMPONENTS OF LEAN MANAGEMENT

See Fig. 25.2.

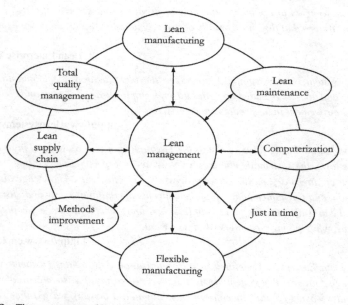

FIG. 25.2   The components of lean management.

25.3 DEFINITIONS ON LEAN MANAGEMENT

Lean manufacturing or lean production, often simply "lean," is a systematic method for the elimination of waste ("Muda") within a manufacturing system. Lean also takes into account waste created through overburdens ("Muri") and waste created through unevenness in workloads ("Mura"). Working from the perspective of the client who consumes a product or service, "value" is any action or process that a customer would be willing to pay for.

Wikipedia

Lean management is an approach to running an organization that supports the concept of continuous; a long-term approach to work that systematically seeks to achieve small, incremental changes in processes in order to improve efficiency and quality.

Whatis.com

Lean production is an assembly-line methodology developed originally for Toyota and the manufacturing of automobiles. It is also known as the Toyota Production System or just-in-time production. Lean production principles are also referred to as lean management or lean thinking.

http://searchmanufacturingerp.techtarget.com

Eliminating waste along entire value streams, instead of at isolated points, creates processes that need less human effort, less space, less capital, and less time to make products and services at far less costs and with much fewer defects, compared with traditional business systems. To accomplish this, lean thinking changes the focus of management from optimizing separate technologies, assets, and vertical departments to optimizing the flow of products and services through entire value streams that flow horizontally across technologies, assets, and departments to customers.

Lean Enterprise Institute

Lean System is a systematic approach to the identification and elimination of waste and non-value added activities through employee development and continuous improvement in all products and services.

http://www.leansystemsinc.com

Lean Manufacturing (also known as the Toyota Production System) is, in its most basic form, the systematic elimination of waste—overproduction, waiting, transportation, inventory, motion, over-processing, defective units, knowledge disconnection, and the implementation of the concepts of continuous flow and customer pull. Lean is about doing more with less: less time, inventory, space, people, and money, while giving customers what they want.

http://www.epplans.com

The Lean Enterprise is defined as an organization that creates customer value through a process that systematically minimizes all forms of waste. Forms of waste include: wasted capital (inventory), wasted material (scrap), wasted time (cycle time), wasted human effort (inefficiency, rework), wasted energy (energy inefficiency), and wasted environmental resources (pollution).

https://www.moresteam.com

25.4   EVOLUTION OF LEAN CONCEPT

It may be noted here that the concept of leaning or the elimination of wasteful elements is not new, but can be traced to the Industrial Revolution era, and has been professed and practiced by industrial engineers all over.

1. In 1785, Benjamin Franklin *Said, "He that idly loses $5 worth of time, loses $5, and might as prudently throw $5 into the river."* He added that avoiding unnecessary costs could be more profitable than increasing sales.
2. In 1911, Frederick Winslow Taylor said in his *Principles of Scientific Management,* "Whenever a workman proposes an improvement, it should be

the policy of the management to make a careful analysis of the new method, and if necessary, conduct a series of experiments to determine accurately the relative merit of the new suggestion and of the old standard. And whenever the new method is found to be markedly superior to the old, it should be adopted as the standard for the whole establishment."

3. Around the same time, Henry Ford continued this focus on waste while developing his mass assembly manufacturing system.

4. In 1915, Frank Gilbreth, who observed the brick laying operation by masons, professed that for better working methods, there should be reduction of wasteful elements. He developed therbligs, a significant tool of the 20th century in identifying and eliminating wasteful micro motions.

5. In 1980 Shigeo Shingo, developed single minute exchange of die and error-proofing or poka-yoke.

6. In 1999, Spear and Bowen identified four rules which characterize the "Toyota DNA":

   a. *Rule 1*: All work shall be highly specified as to content, sequence, timing, and outcome.

   b. *Rule 2*: Every customer-supplier connection must be direct, and there must be an unambiguous yes or no way to send requests and receive responses.

   c. *Rule 3*: The pathway for every product and service must be simple and direct.

   d. *Rule 4*: Any improvement must be made in accordance with the scientific method, under the guidance of a teacher, at the lowest possible level in the organization.

In fact, we can cite several such instances when improvements have been developed based on lean thinking.

25.5 THE HOUSE OF LEAN MANAGEMENT
See Fig. 25.3.

25.6 WHAT CAN LEAN MANAGEMENT ACHIEVE?
Lean management can achieve the following.

- Reduced costs
- Improved morale
- Increased inventory turns
- Reduced lead time
- Creating a learning environment
- Increased profits
- Developing Leadership

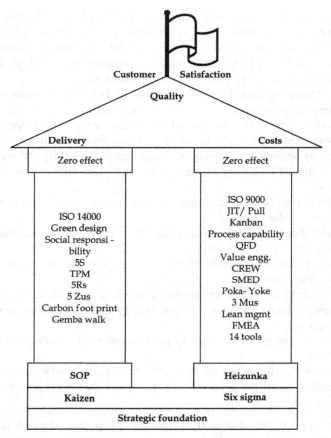

FIG. 25.3   The house of lean management.

- Reducing defects
- Increased customer loyalty and satisfaction
- Gain in market share
- Sustaining long term improvement

25.7   INCREASED RELIABILITY WITH LEAN MANAGEMENT

To become lean, one must reduce maintenance needs and perform the essential maintenance more effectively. By this, the production reliability will increase and increased production reliability will automatically increase the product throughput and reduce the time between incoming raw materials to the finished product. Better reliability is the foundation to a faster and safer manufacturing flow, which will result in lowered losses in delayed deliveries, over production, work in progress, and energy expenditure.

If all three are implemented effectively, the production costs, including maintenance costs and costs for storage, will decrease. If we can eliminate these losses, we would be achieving the lean management.

25.8 THE EIGHT LOSSES IN MANUFACTURING LEADING TO LEAN MANAGEMENT

25.8.1 Manufacturing Reliability

- Loss in quality
- Stop times
- Loss in speed

25.8.2 Partnership Between Operations—Maintenance—Engineering

- Reliability and maintenance related design
- Operator based maintenance

25.8.3 Elimination of Root Cause of the Problem

- Choose problem to eliminate
- Eliminate problems
- Educate and teach

25.8.4 Storage

- Reduce the store value at the same time as you preserve service level to maintenance.

25.8.5 Integration and Application of Increased Knowledge and Skills

- Education and training of crafts people to enable multicraft or multiskills
- Implementation of flexible work systems

25.8.6 Over Manufacturing

- Make more than what has been sold
- Manufacture too early

25.8.7 Over Maintenance

- Perform too much and wrong preventive maintenance
- Perform preventive maintenance before it is needed
- Do corrective maintenance with higher priority than needed

25.8.8   Use of New Technology

- Less need for maintenance
- Better maintainability
- Smart tools and methods

25.9   THE 5 KEY DRIVERS IN LEAN MANAGEMENT SYSTEM

- Apply strong leadership and governance
- Engage staff in daily improvement
- Coaching: the critical difference
- Measure the right things and make results visible
- Embed standard work into the culture

25.10   THE 8 Ps OF LEAN THINKING

Peter Hines in the website http://www.sapartners.com suggests the following 8 Ps for an ideal lean system:

1. Purpose
2. Process
3. Plans
4. People
5. Pull
6. Prevention
7. Partnering
8. Perfection

25.11   LEAN ENTERPRISE IMPLEMENTATION PROCESSES AND TOOLS

Process-oriented practices

- Assure seamless information flow
- Implement integrated product and process development (IPPD)
- Ensure process capability and maturation
- Maintain challenges to existing processes
- Identify and optimize enterprise flow
- Maintain stability in changing environment

Human-oriented practices

- Promote lean leadership at all levels
- Relationships based on mutual trust and commitment
- Make decisions at lowest appropriate level
- Optimize capability and utilization of people
- Continuous focus on the customer
- Nurture a learning environment

25.12   ROAD MAP FOR LEAN MANAGEMENT

Deborah Nightingale of the Massachusetts Institute of Technology suggests the following program for planning and implementing enterprise level lean management:

1. Enterprise strategic planning
   a. Create the business case for lean
   b. Focus on customer value
   c. Include lean in strategic planning
   d. Leverage the extended enterprise
2. Adopt lean paradigm
   a. Build vision
   b. Convey urgency
   c. Foster lean learning
   d. Make the commitment
   e. Obtain senior management buy-in
3. Focus on the value stream
   a. Map value stream
   b. Internalize vision
   c. Set goals and metrics
   d. Identify and involve key stakeholders
4. Develop lean structure and behavior
   a. Organize for lean implementation
   b. Identify and empower change agents
   c. Align incentives
   d. Adapt structure and systems

25.13   ILLUSTRATION OF A PIT SHOP MAINTENANCE SITUATION

Christer Idhammar who initiated the coining of the term *lean maintenance*, offers a nice illustration to signify lean maintenance when it comes to the pit shop maintenance of racing cars.

> The drivers of the car are in constant contact with the pit stop crew. They do not show up suddenly telling the pit crew "I think I have a problem with the right front tire," and the crew answers, "We will go to the store and check if we have any replacement tires." On the other hand, the pit crew can monitor the condition of the car either through radio sensors or through constant communication from the driver. In a car race, there is a strong motivation to win the race, while in a manufacturing plant there are several other factors driving motivation. (As indicated in IDCONo's website, www.idcon.com/resource%2Dlibrary/articles/%E2%80%A6/500%2Dlean%2Dmaintenance)

25.14 CONCLUSION

The power of creativity, starting from Benjamin Frannklin, has proved how simple method improvement ideas have given birth to the concept of leaning as explained in this chapter. It has been proven that this concept can literally be applied in any function and activity of industrial management.

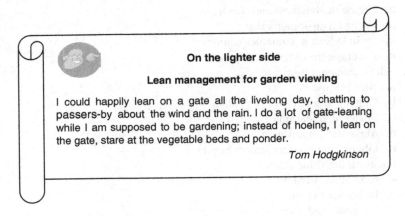

On the lighter side

Lean management for garden viewing

I could happily lean on a gate all the livelong day, chatting to passers-by about the wind and the rain. I do a lot of gate-leaning while I am supposed to be gardening; instead of hoeing, I lean on the gate, stare at the vegetable beds and ponder.

Tom Hodgkinson

FURTHER READING

[1] Nightingale D. Integrating the lean enterprise. Cambridge, MA: Massachusetts Institute of Technology; 2005.
[2] https://home.kpmg.com.

Chapter 26

Failure Modes and Effects Analysis

Chapter Outline

26.1 UNCERTAINTIES DURING DEVELOPMENT

Every product or project undertaken by the engineer is an experiment because each stage of the design or development is experienced for the first time. There are uncertainties at every stage, and the engineer is bound to make presumptions, either from data, books, or from his experience. These uncertainties can be in the form of:

- models used for the design calculations,
- performance characteristics of the materials,
- inconsistencies in the materials purchased,
- nature of the pressure the finished product will encounter,
- size of the product, whether a medium-sized product or a large-sized product,
- volume of production, viz, batch production or mass production,
- specialized materials and skills used in the manufacture.

Apart from the above, the engineer may also experience uncertainties from the viewpoint of several other variables in the development.

26.2   FAILURE MODES AND EFFECTS ANALYSIS

Failure mode and effects analysis (FMEA) is one of the best management tools to analyze the potential failure modes within a system under conditions of uncertainties, as stated above. Its principle is quite basic, and has been practiced since the olden days as the trial and error method. But since learning from each failure is both costly and time-consuming, the modern form of FMEA was developed during the 1940s, as explained in the following section. It emphasizes the probability of occurrence of that failure, and the severity of its effect on the system of every uncertainty. It is used to identify potential failure modes, determine their effect on the operation of the product, and identify actions to mitigate the failures. It analyzes potential reliability problems early in the development cycle, where it is easier to take actions to overcome these issues, thereby, enhancing reliability through design. FMEA should always be done whenever failures would mean potential harm or injury to the user of the end item being designed. According to Besterfield et al., *FMEA is a "before-the-event" action requiring a team effort to easily and inexpensively alleviate changes in design and production.* It is widely used in manufacturing industries in various phases of the product lifecycle and is now being applied in the service industry, too.

26.3   HISTORY OF THE DEVELOPMENT OF FMEA

- FMEA has its origin during the late 1940s for military usage by the US Armed Forces. This is incorporated in the military document MIL-P-1629, dated November 9, 1949, titled *Procedures for Performing a Failure Mode, Effects, and Criticality Analysis*, and subsequently ratified as MIL-STD-1629A.
- Its effectiveness in identifying and reducing any unseen problems encouraged its application for space research and design, specifically in the Apollo Space program and in developing the means to put a man on the moon and return him safely to earth.
- Its industrial application came during the early 1970s when the Ford Motor Company, reeling after the failure of its Pinto car project, introduced FMEA to the automotive industry to improve design, and for safety and regulatory considerations. SAE have documented this as SAE J 1739.
- FMEA methodology is now extensively used in a variety of industries, including semiconductor processing, food service, plastics, software, and healthcare.
- It is integrated into advanced product quality planning (APQP) to provide primary risk mitigation tools and timing in the prevention strategy, in both design and process formats.
- The automotive industry action group requires the use of FMEA in the automotive APQP process and published a detailed manual on how to apply the method. Each potential cause must be considered for its effect on the product or process and, based on the risk, actions are determined and risks revisited after actions are complete.

- Toyota has taken this one step further with its design review based on failure mode (DRBFM) approach. The method is now supported by the American Society for Quality, which provides detailed guides on applying the method.
- The Aerospace industry uses the term "Failure Modes, Effects and Criticality Analysis" (FMECA), to highlight the criticality factor in the application of FMEA is followed by Criticality analysis by which each potential failure is ranked according to the combined influence of severity and probability of occurrence. It identifies single point failures and ranks each failure according to a severity classification of failure effect, helping to identify weak links of design.
- Failure reporting and corrective action system (FRACAS) is another important component of FMEA.

26.4 MULTIPLE CAUSES AND EFFECTS INVOLVED IN FMEA

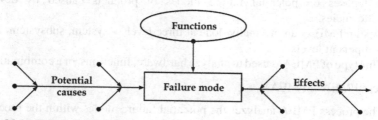

FIG. 26.1 FMEA relationships.

Most real systems do not follow the simple cause and effect model. As explained in the tree diagrams, a single cause may have multiple effects, and a combination of causes may lead to a single or multiple effects. This can also be represented in Fig. 26.1.

26.5 TYPES OF FMEA'S

There can be several forms of FMEA based on which aspect of the activity is analyzed, and can be illustrated as per Fig. 26.2. Other forms of FMEA like System or Service can be interpolated into this illustration.
Concept FMEA (CFMEA)

- The concept FMEA is used to analyze concepts in the early stages before the component is defined (most often at system and subsystem level).
- It focuses on potential failure modes associated with the proposed functions of a concept proposal.
- This type of FMEA includes the interaction of multiple systems and interaction between the elements of a system at the concept stages.

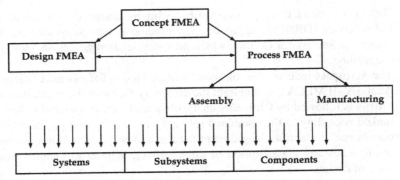

FIG. 26.2   Relationship between the three forms of FMEA.

Design FMEA (DFMEA)

- The design FMEA is used to analyze products before they are released to production.
- It focuses on potential failure modes of products caused by design deficiencies.
- Design FMEAs are normally done at three levels—system, subsystem, and component levels.
- This type of FMEA is used to analyze hardware, functions, or a combination.

Process FMEA (PFMEA)

- The process FMEA analyzes the potential failure modes within the process to identify the severity and frequency, based on past experience with similar processes, enables us to design these failures away from the process system with minimum effort and resource expenditure.
- The process FMEA is normally used to analyze manufacturing and assembly processes at the system, subsystem, or component levels.
- This type of FMEA focuses on potential failure modes of the process that are caused by manufacturing or assembly process deficiencies.

Other forms of FMEA can be

- *System FMEA*, which focuses on global system functions.
- *Service FMEA*, which focuses on service functions.
- *Software FMEA*, which focuses on software functions.
- *Design Review Based on Failure Mode (DRBFM)*, a term coined by Toyota Motor Corporation, as an extension to DFMEA.

26.6   WHEN TO USE FMEA

- Whenever a new product or process is being initiated.
- Whenever changes are made to the product design, process, or the operating conditions. The product and process are interrelated. When the product design is changed, the process is affected and vice-versa.

- Whenever new regulations are being incorporated.
- When the customer feedback indicates problems in the product or process.

26.7   BASIC TERMS OF REFERENCE IN FMEA

26.7.1   Failure Mode

The manner by which a failure occurs and is observed, for example, electrical short-circuiting, corrosion, cracking, or deformation. It may be noted that a failure mode in one component can lead to another failure mode in the same, or another component. Therefore, each failure mode should be listed in technical terms and also giving due consideration for their interrelations. IEC 812-1985 enumerates the generic failure modes as below:

1. Structural failure (rupture)
2. Physical binding or jamming
3. Vibration
4. Failing to remain in position
5. Eccentric rotation
6. Failed interlocking system
7. Failing to open
8. Failing to close
9. Internal leakage
10. External leakage
11. Fails out of tolerance (high)
12. Fails out of tolerance (low)
13. Inadvertent operation
14. Intermittent operation
15. Erratic operation
16. Erroneous indication
17. Restricted flow
18. False actuation
19. Failing to stop
20. Failing to start
21. Failing to switch
22. Premature operation
23. Delayed operations
24. Erroneous input (increased)
25. Erroneous input (decreased)
26. Erroneous output (increased)
27. Erroneous output (decreased)
28. Loss of input
29. Loss of output
30. Shorted (electrical)
31. Open (electrical)
32. Leakage (electrical)
33. Other unique failure conditions as applicable to the system characteristics, requirements, and operational constraints

26.7.2   Failure Cause

The product or process defects or any other quality imperfections would initiate further deterioration leading to a failure. Some failure modes may have more than one cause or mechanism of failure and each of these shall be listed and analyzed separately.

26.7.3   Failure Effect

Failure effect is the immediate consequences of a failure on operation, function or functionality, or status generally, as perceived or experience by the user. Some of the effects can be cited as, injury to the user, inoperability of the product or process, deterioration in product quality, nonadherence to the specifications, emanation of odors, noise, etc. Also the effect of this failure on other systems in immediate contact with the system that failed has to be considered. If a component fractures, it may cause vibration in the subsystem that is in contact with the fractured part. FMEA is the technique used in analyzing the potential failures and their effect on the system.

26.7.4   Severity Factor

A symbolic measure of the failure effect is the severity factor, which is the assessment of the seriousness of the effect of the potential failure. It is noteworthy that the severity represents the seriousness of the failure and not the mode of the failure. Besterfield emphasizes in this connection that no single list of severity criteria is applicable to all designs, and the team should agree on evaluation criteria and on a ranking system that are consistent throughout the analysis. The severity of the effect is given a severity number (S) from 1 (no danger) to 10 (critical), as given in Table 26.1.

TABLE 26.1 Rankings of Severity of Effect

Effect	Severity of Effect	Severity Factor
Hazardous without warning	Very high ranking with potential failure mode affects safe operation and regulation noncompliance. Failure occurs without warning.	10
Hazardous with warning	Very high ranking with potential failure mode affects safe operation and regulation noncompliance. Failure occurs with warning.	9
Very high	Hazardous. Even if the component does not fracture, it becomes inoperable.	8
High	Item is operable, but with loss of performance. Customer is dissatisfied.	7
Moderate	Product is operable but with loss to comfort/convenience. Customer experiences discomfort.	6
Low	Product is operable, but with loss to comfort/convenience. Customer has some discomfort.	5
Very low	Certain item characteristics do not conform to specifications, but noticed by most customers.	4
Minor	Certain item characteristics do not conform to specifications, but noticed by average customers.	3
Very minor	Certain item characteristics do not conform to specifications, but noticed by some discriminating buyers (referred to as dissatisfies in Chapter 3).	2
None	No effect	1

26.7.5   Probability of Occurrence

Probability of occurrence is the chance that one of the specific failure causes will occur. This recoding of probability of occurrence must be done for every cause indicating the probability of occurrence of that cause. This can be done by looking at the occurrence of failures for similar products or processes, and the failures that have been documented for them in technical terms. For FMEA, such an occurrence rate can be assigned a numerical value from 1 to 10, the least frequent being 1 and the most frequent being 10 (Table 26.2). If this value is more than 4, it implies that the actions needed to identify and analyze them shall be more meticulous. Besterfield et al. suggest the following guideline questions for evaluation.

- What is the service history or field experience with similar systems or subsystems?
- Is the component similar to a previous system or subsystem?
- How significant are the changes in the component is a new model?
- Is the component completely new?
- Is the component application any different form the previous?
- Is the component environment any different than before?

TABLE 26.2 Rankings of Probability of Occurrence

Probability of Occurrence	Explanation	Possible Failure Rate	Ranking No.
Very high	Failure is almost inevitable	>1 in 2	10
		1 in 3	9
High	Generally associated with processes similar to previous processes that have often failed	1 in 8	8
		1 in 20	7
Moderate	Generally associated with processes similar to previous processes that have experienced occasional failures	1 in 80	6
		1 in 400	5
		1 in 2000	4
Low	Isolated failures associated with similar processes	1 in 15,000	3
Very low	Only isolated failures associated with almost identical processes	1 in 150,000	2
Remote	Failure is unlikely. No failures ever associated with almost identical processes	<1 in 1,500,000	1

26.7.6   Ease of Detection

This is the ability of the inspecting mechanism and/or design control to detect the potential cause or the subsequent failure mode before the component or the subsystem is completed for production. The proper inspection methods need to be chosen. First, an engineer should look at the current controls of the system that prevent failure modes from occurring, or which detect the failure before it reaches the customer. Hereafter, one should identify testing, analysis, monitoring, and other techniques that can be or have been used on similar systems to detect failures. From these controls, an engineer can learn how likely it is for a failure to be identified or detected. This parameter, too, is given a numerical value between 1 and 10, called the Detection Rating (Table 26.3). This ranking measures the risk that the failure will escape detection. A high detection number indicates that the chances are high that the failure will escape detection.

TABLE 26.3 Rankings of Ease of Detection

Ease of Detection	Explanation	Ranking No.
Absolutely impossible	No known controls available for detection of the failure mode	10
Very remote	Very remote likelihood that the current controls will detect failure mode	9
Remote	Remote likelihood that the current controls will detect failure mode	8
Very low	Low remote likelihood that the current controls will detect failure mode	7
Low	Low remote likelihood that the current controls will detect failure mode	6
Moderate	Moderate remote likelihood that the current controls will detect failure mode	5
Moderately high	Moderately high remote likelihood that the current controls will detect failure mode	4
High	High remote likelihood that the current controls will detect failure mode	3
Very high	Very high remote likelihood that the current controls will detect failure mode	2
Almost certain	Reliable controls are known with similar processes and currant controls almost certain to detect the failure mode	1

Other terminology used in FEMA:

- *Indenture levels*: An identifier for item complexity. Complexity increases as levels are closer to one.
- *Local effect*: The failure effect as it applies to the item under analysis.
- *Next higher level effect*: The failure effect as it applies at the next higher indenture level.
- *End effect*: The failure effect at the highest indenture level or total system.

26.8   RISK PRIORITY NUMBER

Risk priority number (*RPN*) is a function of the three parameters discussed above, viz, the severity of the effect of failure, the probability of occurrence, and the ease of detection for each failure mode. *RPN* is calculated by multiplying these three numbers as per the formula below,

$$RPN = S \times P \times D$$

where S is the severity of the effect of failure, P is the probability of failure, and D is the ease of detection.

RPN may not play an important role in the choice of an action against failure modes, but will help in indicating the threshold values for determining the areas of greatest concentration. In other words, a failure mode with a high *RPN* number should be given the highest priority in the analysis and corrective action. The relationship between the above mentioned parameters of FEMA may be represented as in Fig. 26.3.

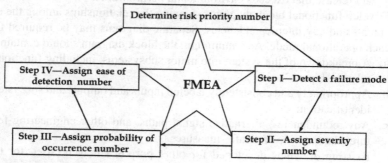

FIG. 26.3   The five basic steps of FMEA.

26.9   PROCEDURE FOR FMEA

In principle, the causes or the specific faults are described in terms of those that can be detected and controlled. Action taken generally should result in a lower severity, lower occurrence, or higher detection rating by adding validation and verification controls

1. Identify the functions.
2. Identify the failure modes.
3. Identify the effects of the failure modes.
4. Determine the probability of occurrence (see Table 12.2).
5. Determine the severity of occurrence (see Table 12.3).
6. Apply this procedure for potential consequences.
7. Identify possible causes.
8. Identify the root cause.
9. Calculate the criticality.
10. Identify special characteristics.
11. Assess the probability that the proposed system detects the potential weaknesses.

Princeton Plasma Physics Laboratory suggests the following basic steps for FMEA:

1. Define the system and its functional and operating requirements;
   a. Include primary and secondary functions, expected performance, system constraints, and explicit conditions that constitute a failure. The system definition should also define each mode of operation and its duration.
   b. Address any relevant environmental factors, such as temperature, humidity, radiation, vibration, and pressure during operating and idle periods.
   c. Consider failures that could lead to noncompliance with applicable regulatory requirements. For example, a failure that could result in a pollutant release that exceeds environmental permit limits.
2. Develop functional block diagrams showing the relationships among the elements and any interdependencies. Separate diagrams may be required for each operational mode. As a minimum, the block diagram should contain:
   a. A breakdown of the system into major subsystems, including functional relationships;
   b. Appropriately and consistently labeled inputs and outputs and subsystem identification;
   c. Any redundancies, alternative signal paths, and other engineering features that provide "failsafe" measures.
      Existing drawings developed for other purposes may be used for the FMEA if the above elements are adequately described.
3. Identify failure modes, their cause and effects.
   a. IEC 812 1985 provides a list of failure modes, reproduced here as table in Section 26.7.1, to describe the failure of any system element.
   b. Identify the possible causes associated with each postulated failure mode. The above list can be used to define both failure modes and failure causes. Thus, for example, a power supply may have a specific failure mode "loss of output" (29), and a failure cause "open (electrical)" (31).

    c. Identify, evaluate, and record the consequences of each assumed failure mode on system, element operation, function, or status. Consider maintenance, personnel, and system objectives, as well as any effect on the next higher system level.

4. Identify failure detection and isolation provisions and methods. Determine if other failure modes would give an identical indication and whether separate detection methods are needed.

5. Identify design and operating provisions that prevent or reduce the effect of the failure mode. These may include:

    a. Redundant items that allow continued operation if one or more elements fail;

    b. Alternative means of operation;

    c. Monitoring or alarm devices;

    d. Any other means permitting effective operation or limiting damage.

6. Identify specific combinations of multiple failures to be considered. The more multiple failures considered, the more complex the FMEA becomes. In many such cases it would be advantageous to perform a FMECA using the guidance of IEC Standard 812 or MIL-STD-1629A. Using the FMECA, the severity of failure effects are categorized, the probability is determined, and the number of redundant mitigating features needed to keep the probability of failure acceptably low are better determined.

7. Revise or repeat, as appropriate, the FMEA as the design changes. Changes may be in direct response results of the previous FMEA or may be due to unrelated factors.

Kenneth Crow, on the website http://www.npd-solutions.com/fmea.html, suggests the following procedure for FMEA, which is quite similar to the above detailed procedure by Princeton Plasma Physics Laboratory, but is more exhaustive.

1. Describe the product/process and its function. An understanding of the product or process under consideration is important to have clearly articulated. This understanding simplifies the process of analysis by helping the engineer identify those product/process uses that fall within the intended function, and which ones fall outside. It is important to consider both intentional and unintentional uses because product failure often ends in litigation, which can be costly and time-consuming.

2. Create a block diagram of the product or process. A block diagram of the product/process should be developed. This diagram shows major components or process steps as blocks connected together by lines that indicate how the components or steps are related. The diagram shows the logical relationships of components and establishes a structure around which the FMEA can be developed. Establish a coding system to identify system elements. The block diagram should always be included with the FMEA form.

3. Complete the header on the FMEA form worksheet: Product/System, Subsys./Assy., Component, Design Lead, Prepared By, Date, Revision (letter or number), and Revision Date. Modify these headings as needed.
4. Use the diagram prepared above to begin listing items or functions. If items are components, list them in a logical manner under their subsystem/assembly, based on the block diagram.
5. Identify failure modes. A failure mode is defined as the manner in which a component, subsystem, system, process, etc., could potentially fail to meet the design intent. Examples of potential failure modes include:
   a. Corrosion
   b. Hydrogen embrittlement
   c. Electrical short or open
   d. Torque fatigue
   e. Deformation
   f. Cracking
6. A failure mode in one component can serve as the cause of a failure mode in another component. Each failure should be listed in technical terms. Failure modes should be listed for the function of each component or process step. At this point, the failure mode should be identified whether or not the failure is likely to occur. Looking at similar products or processes and the failures that have been documented for them is an excellent starting point.
7. Describe the effects of those failure modes. For each failure mode identified, the engineer should determine what the ultimate effect will be. A failure effect is defined as the result of a failure mode on the function of the product/process as perceived by the customer. They should be described in terms of what the customer might see or experience should the identified failure mode occur. Keep in mind the internal as well as the external customer. Examples of failure effects include:
   a. Injury to the user
   b. Inoperability of the product or process
   c. Improper appearance of the product or process
   d. Odors
   e. Degraded performance
   f. Noise
      Establish a numerical ranking for the severity of the effect. A common industry standard scale uses 1 to represent no effect and 10 to indicate very severe with failure affecting system operation and safety without warning. The intent of the ranking is to help the analyst determine whether a failure would be a minor nuisance or a catastrophic occurrence to the customer. This enables the engineer to prioritize the failures and address the real big issues first.
8. Identify the causes for each failure mode. A failure cause is defined as a design weakness that may result in a failure. The potential causes for each

failure mode should be identified and documented. The causes should be listed in technical terms and not in terms of symptoms. Examples of potential causes include:

   a. Improper torque applied
   b. Improper operating conditions
   c. Contamination
   d. Erroneous algorithms
   e. Improper alignment
   f. Excessive loading
   g. Excessive voltage

  9. Enter the probability factor. A numerical weight should be assigned to each cause that indicates how likely that cause is (probability of the cause occurring). A common industry standard scale uses 1 to represent not likely and 10 to indicate inevitable.

10. Identify current controls (design or process). Current controls (design or process) are the mechanisms that prevent the cause of the failure mode from occurring or which detect the failure before it reaches the customer. The engineer should now identify testing, analysis, monitoring, and other techniques that can or have been used on the same or similar products/processes to detect failures. Each of these controls should be assessed to determine how well it is expected to identify or detect failure modes. After a new product or process has been in use, previously undetected or unidentified failure modes may appear. The FMEA should then be updated and plans made to address those failures to eliminate them from the product/process.

11. Determine the likelihood of detection. Detection is an assessment of the likelihood that the current controls (design and process) will detect the cause of the failure mode or the failure mode itself, thus preventing it from reaching the customer.

12. Review *RPN*s. The *RPN* is a mathematical product of the numerical severity, probability, and detection ratings: $RPN = (\text{severity}) \times (\text{probability}) \times (\text{detection})$

    The *RPN* is used to prioritize items than require additional quality planning or action.

13. Determine recommended action(s) to address potential failures that have a high *RPN*. These actions could include specific inspection, testing or quality procedures; selection of different components or materials; derating; limiting environmental stresses or operating range; redesign of the item to avoid the failure mode; monitoring mechanisms; performing preventative maintenance; and inclusion of back-up systems or redundancy.

14. Assign responsibility and a target completion date for these actions. This makes responsibility clear-cut and facilitates tracking.

15. Indicate actions taken. After these actions have been taken, re-assess the severity, probability, and detection and review the revised *RPN*s. Are any further actions required?
16. Update the FMEA as the design or process changes, the assessment changes or new information becomes known.

26.10 RESPONSIBILITY FOR ACTION

FMEA is a team operation. Everyone should feel fully involved in the process and in moving towards the goal. Nevertheless, it is always advisable to delegate certain responsibilities to specified persons, so that the monitoring and reporting can be effective. It is suggested that some of the responsibilities be allocated between the line manager, the analyst, and the reviewer as follows:

Line manager
1. Assign individuals to perform FMEA (analyst) and another individual to review it (reviewer). The reviewer shall have as much expertise and technical experience as the analyst.

Analyst
2. Describe the system under analysis, prepare system diagrams, and use existing documentation to depict all major components and their performance criteria. The level of assembly may vary with the level of the analysis.
3. Perform FMEA as per the procedure described earlier.
4. Sign FMEA and provide it to the reviewer.

Reviewer
5. Review FMEA for technical content and sign if no significant problems are identified. Otherwise discuss the FMEA with the analyst.
6. Ensure that the full FMEA documents are filed in the Operations Center.

26.11 BENEFITS OF FMEA

- Effective prevention planning program
- Identification of change requirements
- Cost reduction
- Increased throughput
- Decreased waste
- Decreased warranty costs
- Reduction of nonvalue added operations
- Improvement in the quality, reliability and safety of a product/process
- Improvement in company image and competitiveness
- Increased user satisfaction
- Reduced system development timing and cost
- Data collection (expert systems) for reduced future failures.

- Reduce warranty concerns
- Early identification and elimination of potential failure modes
- Minimal late changes and associated cost
- Catalyst for teamwork and idea exchange between functions
- Reduction in the possibility of same kind of failure in future

While the general benefits of FMEA can be listed as above, the category-wise benefits can be summarized as under:

Concept FMEA
- Helps selecting the optimum concept alternatives, or determine changes to design specifications.
- Identifies potential failure modes caused by interactions within the concept.
- Increases the likelihood all potential effects of a proposed concept's failure modes are considered.
- Identifies system level testing requirements.
- Helps determine of hardware system redundancy may be required within a design proposal.

Design FMEA
- Aids in the objective evaluation of design requirements and design alternatives.
- Aids in the initial design for manufacturing and assembly requirements.
- Increases the probability that potential failure modes and their effects have been considered in the design/development process.
- Provides additional information to help plan thorough and efficient test programs.
- Develops a list of potential failure modes ranked according to their effect on the customer. Establishes a priority system for design improvements.
- Provides an open issue format for recommending and tracking risk reducing actions.
- Provides future reference to aid in analyzing field concerns.

Process FMEA
- Identifies potential product-related process failure modes.
- Assesses the potential customer effects of the failures.
- Identifies the potential manufacturing or assembly process causes and identifies process variables on which to focus controls or monitoring.
- Develops a ranked list of potential failure modes, establishing a priority system for corrective action considerations.
- Documents the results of the manufacturing or assembly process.
- Identifies process deficiencies.
- Identifies confirmed critical characteristics and/or significant characteristics.
- Identifies operator safety concerns.
- Feeds information on design changes required and manufacturing feasibility back to the designers.

26.12   FMEA SOFTWARE

The following software have been developed for industrial and dedicated application as per information available in internet. It may be remembered that FMEA software refers to the software available for FMEA solutions, whereas Software FMEA refers to the process of applying FMEA so solve problems in software development.

1. *ASENT FMEA Software*—Raytheon's premiere reliability and maintainability tool suite. Includes a very powerful FMECA tool that combines FMECA, RCM analysis, and testability analysis.
2. *Byteworx*—Powerful, cost-effective software for FMEA. It is the global choice of the Ford Motor Company. Byteworx FMEA is fully compliant with SAE J-1739 Third Edition.
3. *FMEA-Pro*—FMEA/FMECA software from Dyadem. An all-in-one software solution provides corporate consistency and assists with corporate compliance.
4. *Isograph Software*—Their Reliability Workbench contains a FMEA/FMECA tool.
5. *Item Software*—FMEA/FMECA/FMEDA—Failure Mode Effects Analysis tool.
6. *Quality Plus*—FMEA software from Harpco Systems, Inc. Performs both Design and Process FMEAs.
7. *RAM Commander Software*—ALD's integrated FMEA/FMECA modules have been adopted by many civil, military, aerospace, energy and pharmaceutical organizations worldwide.
8. *Relex Software*—Offers FMEA tools and FMEA software to process FMEA and meet all functional FMEA standards for criticality matrix.
9. *XFMEA*—FMEA software from ReliaSoft. Provides expert support for all types of FMEA.

26.13   CONCLUSION

As seen in this chapter, FMEA helps us in anticipating unexpected failures and providing for their corrective action during the design stage itself. Right from the days of its conception in the 1940s, it has today become a must for the designers.

On the Lighter Side

Mouse Potato – An amusing modern slang term for a person who sits for long periods in front of a computer, especially using the internet, instead of engaging in more active and dynamic pursuits. Mouse Potato is an adaptation of the older 1970's slang 'couch potato', referring to a person who spends too much time sitting on sofa, watching TV, eating and drinking. Both terms originated in the USA, although these lifestyles are now worldwide.

Clicklexia - Ironic computing slang for a user's tendency to double-click on items when a single click is required, often causing the window or utility to open twice.

-Both from Business Dictionary

FURTHER READING

[1] Langford JW. Logistics—principles and applications. New York, NY: McGraw Hill; 1995.

[2] IEC Standard 812. Procedure for failure mode and effects analysis (FMEA), November 16, 2014.

[3] MIL-STD-1629. A procedures for performing a failure mode, effects and criticality analysis, August 4, 1998.

[4] Stamatis DH. Failure mode and effect analysis—from theory to execution. Milwaukee, WI: ASQ Publications; 1997.

[5] www.pppl.gov/eshis/procedures/eng000.

[6] FMEA, by Kenneth Crow from the website http://www.npd-solutions.com/fmea.html.

[7] www.weibull.com/basics/fmea.

[8] www.qualitytrainingportal.com/resources/fmea.

[9] www.pppl.gov/eshis/procedures/eng008 of Princeton Plasma Physics Laboratory.

[10] www.reliasoft.com/xfmea of Relia Soft Corporation for Xfmea interface.

[11] https://en.wikipedia.org/wiki/Strategic_planning for strategic planning.

[12] www.skymark.com/resources/tools/affinity_diagram for affinity diagram and cause and effect diagram.

[13] www.qualitytrainingportal.com/resources/fmea for severity rating scale.

Chapter 27

Reliability Engineering

Chapter Outline

27.1 FUNCTIONAL RELIABILITY

Reliability engineering is an engineering discipline for applying scientific know-how to a component, product, plant, or process in order to ensure that it performs its intended function, without failure, for the required time duration in a specified environment. It emphasizes dependability in the lifecycle management of a product, which is the ability of a system or component to function under stated conditions for a specified period of time. In other words, reliability has two significant dimensions, the time and the stress. A product has to endure for several years of its life and also perform its desired function, despite all the threatening stresses applied to it, such as temperature, vibration, shock, voltage, and other environmental factors.

In quality management, this principle is applied to a component, product, plant, or process in order to assure that it performs its intended function, without failure, for the required time duration in a specified environment. This is called functional reliability and the application of these principles to achieve high product life is called reliability engineering.

Earlier, we said quality conformation and customer satisfaction are essential for companies to survive in their business. Reliability assumes a major factor in sustaining quality and we can say that, the only companies left in business will be those that are able to control the reliability of their products. Increase in the complexity of the product, as well as in the equipment, has led to an increasing demand for higher reliability.

Reliability engineering is an engineering framework that enables the definition of a complete production regime and deals with the study of the ability of the product to perform its required functions under stated conditions for a specified period of time. It characterizes measures and analyzes the failure and repair of the systems to improve their use by increasing their design life, mitigating defect risks, and reducing the likelihood of failures.

27.2   GENERAL CAUSES FOR POOR RELIABILITY

1. Increasing product complexity,
2. Overemphasis on the "state of the art" factor for the performance,
3. Too many features included in the design that would affect reliability,
4. More complex and severe environmental changes, field stresses, and interactions,
5. Short-circuited development cycles in order to be "the first in the market,"
6. Rapid product obsolescence,
7. Rising customer expectations for guaranteed performance and endurance, and
8. Lack of financial incentives or penalties for reliability in performance.

27.3   DISTINGUISHING BETWEEN QUALITY AND RELIABILITY

Quality is:
- Independent of time.
- Patent failures are removed by quality control methods.
- Lot dependent.

Reliability is:
- Time-dependent.
- Latent failures can be detected.
- Numerical estimates like mean time between failures (MTBF), failure rates are possible.
- These numerical estimates help us to compare between two different designs at the proposal stage itself.

27.4   WHAT IS RBM?

If the principles explained in Section 27.1 are applied to the machinery and equipment, we call it the probability of failure free operation of the system for a given period of time, under specific conditions, and the ability of equipment to perform a required function under stated conditions for a stated period of time, and in other words, how often the breakdown occurs of the equipment or the failure of the components as measured in time units, say, hours. In this context, reliability is as significant to quality management as it is for maintenance management.

Since the quality of a product depends to a large extent on the reliability of the equipment with respect to their breakdown-free performance, it is essential for us to understand reliability with specific reference to the reliability based machine maintenance (Fig. 27.1). Reliability based maintenance; originally called by its synonym *reliability centered maintenance* incorporates sound guidance for managers who wish to attain high standards of maintenance at their operating plants. The amount and type of maintenance which is applied depend strongly on:

- the age of the machine or components
- its replacement cost and
- the cost and safety consequences of system failure

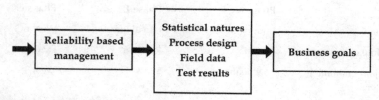

FIG. 27.1   Factors for RBM.

27.5   BATH TUB CHARACTERISTICS

The failure characteristics of a majority of the equipment follow the pattern shown in Fig. 27.2, sometimes called a bath tub pattern, which has three distinct phases:

Phase A or the burning in period: The major contributing factor for this failure is the poor component quality. When the equipment is given initial trials, there might be many initial failures due to poor design, workmanship, assembly errors, etc. Damaged components and poor joints or connections also contribute to this failure. These are tested and replaced generally at the manufacturer's premises to improve the reliability.

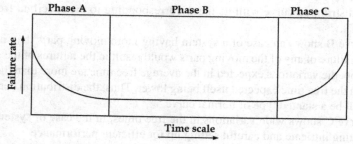

FIG. 27.2   Failure rate of equipment.

Phase B or the useful life period: Here the failure rate is low, but may occur unexpectedly and at random intervals. They are known as random failures or normal failures. It is during this period, that all our availability reliability analysis is based on. The major contributing factor is the stress to which the equipment or products are subjected to and could be due to operating stresses, poor maintenance, operator abuse, and accidents.

Phase C or the wear out period: Beyond the useful period, the wear rate is the major contributing factor because of aging or wear of the components of the system, and could be due to weak design, poor lubrication, wear, fatigue failure, corrosion, and insulation breakdown. In short, Table 27.1 illustrates the contributing factors for each phase.

TABLE 27.1 Contributing Factors for Failures

	Phase A	Phase B	Phase C
Period	Burning in	Useful life	Wear out
Failure occurrence	Trial	Random	Excessive
Major contributing factor	Low quality	Stress	Wear
Other contributing factors	Weak design, assembly errors, damaged components, poor joints/connections	Operating stresses, poor maintenance, operator abuse, accidents	Weak design, wear, fatigue, corrosion

A statistical representation of the probability that a product or system can have maintenance-free performance for a given number of operating hours is given in Fig. 27.3.

Curve A shows a case where the system is not subjected to severe conditions of services and tend to breakdown at nearly constant intervals following the last repair. The statistical variation of these intervals is given by a normally distributed curve with its mean corresponding to the specified free run time (T_a).

Curve B shows an ease of a system having more moving parts than in case A, A's failure of any of the moving parts would result in the failures of the whole machine, the variations expected in the average free time are more than the case A, with the free time expected itself being lower. Thus, the distribution curve for this will be a slanted type of normal curve.

Curve C shows wide variations in the free times, in the case of systems necessitating intricate and careful setting up for efficient performance.

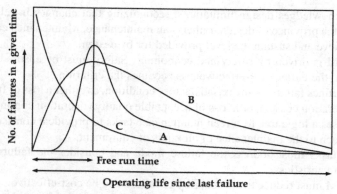

FIG. 27.3   Maintenance-free performance curves.

27.6   BASICS OF RBM

SAE JA1011, *Evaluation Criteria for RBM Processes* sets out the minimum criteria that any process should meet, starting with the seven questions below:

1. What is the item supposed to do and what are its associated performance standards?
2. In what ways can it fail to provide the required functions?
3. What are the events that cause each failure?
4. What happens when each failure occurs?
5. In what way does each failure matter?
6. What systematic task can be performed proactively to prevent, or to diminish to a satisfactory degree, the consequences of the failure?
7. What must be done if a suitable preventive task cannot be found?

27.7   PRINCIPLES OF RELIABILITY ENGINEERING

The EBME website (http://www.ebme.co.uk) lists the following principles upon which reliability based management (RBM) is based:

1. RBM is function-oriented and seeks to preserve system or equipment function.
2. It is group focused and is concerned with maintaining the overall functionality of a group of devices, rather than an individual device.
3. It uses failure statistics in an actuarial manner to look at the relationship between operating age and the failures. However, RBM is not overly concerned with simple failure rate; it seeks to know the probability of failure at specific ages.

4. Acknowledges design limitations, recognizing that changes in reliability are the province of design, rather than maintenance. Maintenance can only achieve and sustain the level provided for by design.
5. RBM is driven by safety and economics. Safety must be ensured at any cost; thereafter, cost-effectiveness becomes the criterion.
6. It defines failure as any unsatisfactory condition, as either a loss of function (operation ceases), or a loss of acceptable quality (operation continues).
7. It uses a logic tree to screen maintenance tasks to provide a consistent approach to the maintenance of all kinds of equipment.
8. Its tasks must address the failure mode and consider the failure mode characteristics.
9. RBM must reduce the probability of failure and be cost-effective.
10. RBM tasks are interval (time- or cycle-)-based and condition-based. Here run-to-failure, is a conscious decision and is acceptable for some equipment.
11. It is dynamic and gathers data from the results achieved, which is fed back to improve future maintenance. This feedback is an important part of the proactive maintenance element of the RBM program.

27.8 HOUSE OF RELIABILITY

The strength provided by Reliability Engineering to an organization can be illustrated in Fig. 27.4.

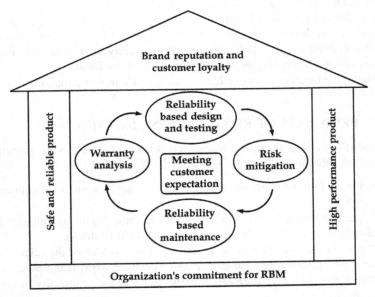

FIG. 27.4 House of reliability.

27.9   TYPES OF FAILURES

Failures can be grouped into the following three categories. Understanding these categories is critical when assigning maintenance tasks.

- Induced
- Intermittent
- Wear out

Induced failures are a result of an outside force causing the failure mode, like a soft foot condition on an equipment train causing coupling misalignment, eventually leading to an inboard bearing failure. Soft foot condition implies improper contact between a machine casing and the baseplate used to support it. In case of rotating machines like the motors, such soft foot condition causes heavy vibrations leading to major breakdowns and accidents.

It is important to understand that induced failure must be recognized and analysis performed to determine the root cause, as explained further by the failure mode and effects analysis (FMEA) concept.

Intermittent failures can happen at any time at random, and the MTBF cannot be predetermined, and the repair cannot be effectively planned and scheduled. A plant can best detect these failure modes through process monitoring and predictive maintenance to some extent.

Wear-out failures have a known MTBF and they occur when the useful life of a component is expended. These types of failure modes are often detectable through process monitoring and predictive maintenance. However, time-based refurbishment or preventive maintenance sometimes could prove to be an effective maintenance strategy.

27.10   SEVERITY OF FAILURES

DOD-STD-2101 defines the characteristics of a component and system defects as:

- Critical, if the failures will have adverse impact on safety,
- Major, if the defective characteristics will degrade with age, and
- Minor, if the defects do not a have significant impact on the performance

This characteristic of severity of failure is dealt in more detail in Table 26.1 of Chapter 26.

27.11   STATISTICAL DISTRIBUTION CURVES OF FAILURES

While the previous chapter gives physical illustrations and computed the failure rate, etc., by simple arithmetic computations, this chapter briefs the several distribution curves that the failures conform to. While the detailed statistical explanation is beyond the scope of this book, the basic explanation of their concepts to the extent an engineer should know is discussed in this chapter.

(a) *The normal distribution* is a probability distribution that associates the normal random variable around central value, called the mean. This is generally applied to analysis of variations, also known as Anova as a special abbreviation around a nominal fixed value, like the variations in the machining of a bar to say, 50 mm diameter. It is also called a bell curve, since it looks like a bell with a central peak as in Fig. 27.5.

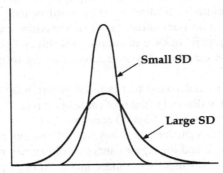

FIG. 27.5 Normal distribution.

(b) *The Poisson distribution* is a curve expressing the probability of a given discreet number of events occurring in a fixed interval of time, or certain sample sizes like the number of defects found in each of the several samples picked up from a lot. This figure can be 0, 1, or 2, etc., unlike the normal distribution which the variations cluster around a central figure. Here, like the σ of normal distribution, the variation factor is represented by λ (lambda). As $\lambda = 1$, it is a simple c-shaped curve, reaching a maximum when the occurrence is zero and smoothing out to a normal curve when the occurrences are high (Fig. 27.6).

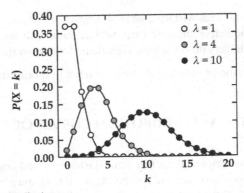

FIG. 27.6 Poisson distribution.

If *X* has a Poisson distribution with a mean of *n* (say 2) failures per year, then the probability that no more than *r* (say 1) failures occur per year is given by $P(X=r)=X!$

If $n=2$ and $r=1$, the $n(X=1)$ would be 0.406 or 40.6% probability.

(c) *The Weibull distribution* (Fig. 27.7) is similar to Poisson, but uses three parameters.

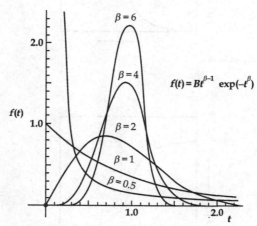

FIG. 27.7   Weibull distribution.

- the shape parameter (β), also known as the Weibull slope
- the scale parameter (η)
- the location parameter (γ)
- the most general expression of the Weibull *pdf* is given by the three-parameter Weibull distribution expression, or:

$$f(T)=\frac{\beta}{\eta}\left(\frac{T-\gamma}{\eta}\right)^{\beta-1} e^{-\left(\frac{T-\gamma}{\eta}\right)^{\beta}}$$

or if $t=\dfrac{(T-\gamma)}{\eta}$, then

$$f(T)=\beta(t)^{\beta-1}\exp\left(-t^{\beta}\right)$$

You notice that when $\beta=0$, it is more or less similar to Poisson distribution, but as β increases, the curve assumes a normal shape and becomes steeper, showing that the determination of failure would be more precise. The Weibull distribution is very useful, not only for the failure analysis in determining the equipment reliability, but also in survival analysis, in the insurance industry, in industrial engineering to represent manufacturing and delivery times, and also in weather forecasting (Fig. 27.8).

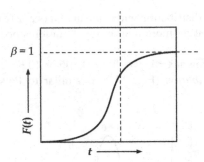

FIG. 27.8   Probability of failure.

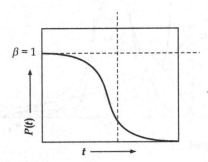

FIG. 27.9   Probability of survival.

More statistical curves can be deduced and represented as follows, indicating the different conditions of failures.

(a) When a fraction of items are expected to fail by a time t that is the probability of failure $f(t)$.

(b) When a fraction of items are expected to survive by a time t that is the probability of survival $f(t)$ (Fig. 27.9).

27.12   PROBABILITY DENSITY FUNCTION

The probability density function (PDF), or density of a continuous random variable, is a function that describes the relative likelihood for this random variable to take on a given value. This PDF is most commonly associated with absolutely continuous univariate distributions and for the random variable to fall within a particular region is given by the integral of this variable's density over the region. In the illustration given alongside, the probability density for the median to fall within the limits of Q_1 and Q_2 is given by the darker shaded area (Fig. 27.10).

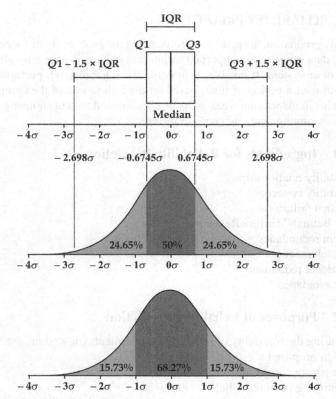

FIG. 27.10 Probability density function. (*Based on Wikipedia*)

27.13 PROCEDURE OF ESTABLISHING RELIABILITY BASED PRODUCT QUALITY

While the application of reliability in the design function is given in more detail in Chapter 32, a brief summary of the steps involved can be as follows.

1. Quantify reliability requirements as design goals or specifications.
2. Allocate and apportion the reliability requirements to specific system components and parts.
3. Apply reliability design methods during the equipment design and development. Perform reliability and maintainability analysis, such as block diagrams, stress-strength analysis, redundancy, etc.
4. Conduct FMEA and criticality analysis.
5. Participate in design reviews.
6. Establish test procedures and conduct reliability testing.
7. Perform reliability prediction and demonstration.
8. Develop a reliability plan.

27.14   RELIABILITY PREDICTION

Reliability prediction, the process of forecasting the probability of success from available data is one of the important techniques in knowing the reliability of an equipment or system. It involves estimating the reliability (ie, performance of the system over a period of time) based on the failure rate of the components. It thus helps in identifying weak areas in a design, and also in choosing the best design from among alternate configurations.

27.14.1   Ingredients for Reliability Prediction

- Reliability relationships
- Reliability concepts
- Constant failure rate
- The "Bathtub" Failure Rate curve
- System redundancy
- Fault tolerance
- Functional redundancy
- Fault avoidance

27.14.2   Purposes of Reliability Prediction

1. Assuring the feasibility of reliability requirements (downtime, etc.) for the design proposed
2. Comparing competing designs
3. Identifying potential reliability problems
4. Planning maintenance and logistic support strategies
5. Reliability predictions can be used to assess the effect of product
6. Reliability on the maintenance activity and on the quantity of spare
7. Units required for acceptable field performance of any particular system. For example, predictions of the frequency of unit level maintenance can be estimated
8. Estimating unit and system lifecycle costs
9. Provide necessary input to system level reliability models
10. Assist in deciding which product to purchase from a list of competing products
11. Useful in setting standards for factory reliability tests and field performance

The failure rate of all the cards in the system are evaluated as per "QM115A Quality Manual on Guidelines to calculate theoretical reliability failures for telecom equipment" issued by Telecom QA circle, DOT, Issue 2, Jan. 1997.

In his address on *Prevention of Problems on Reliability and Safety* at NIQR, Chennai, in January 2015, Professor Kazuyuki Suzuki of University of Electro-Communications, Tokyo, emphasized that events that cannot be predicted, cannot be prevented. But careful consideration of the following would provide an inductive approach to understand the situation for more accurate prediction.

- Sharing of problem information beyond organization
- Abstraction and generalization of individual problems
- Implementation of PDCA cycle
- Practical use of incident information

27.15   MONTE CARLO SIMULATION

The *Monte Carlo method* is a broad class of computational algorithms that relies on repeated random sampling to obtain numerical results, letting us account for risk in the quantitative analysis and decision-making. Just as we keep playing and recording our results in a real casino situation, the simulation is run many times over in order to calculate the probabilities. Hence, the name Monte Simulation. It is specifically useful in systems with several degrees of freedom with uncertainty in the inputs, such as the calculation of risk in purchasing shares.

In fact, this author nostalgically remembers his visit to Monte Carlo Casino in Monaco in 1983, when he practically saw, what he was taught in 1968, how the players in the roulette table kept on recording their and others' bets and results, and used them to decode their next bets.

When Monte Carlo simulations have been applied in space exploration and oil exploration, their predictions of failures, cost overruns, and schedule overruns are routinely better than human intuition or alternative "soft" methods.

Because testing of statistical significance of any activity or parameter is time-consuming, simulation techniques can effectively be adapted for known probabilistic models to predict the outcome. Monte Carlo simulation is one such technique. The steps involved are:

1. Define a domain of possible inputs.
2. Generate random numbers between 0 and 1 to represent the reliability of a component.
3. Replace the reliability as a function of time into the model and calculate the corresponding time to failure by deterministic computation.
4. Generate another random number and repeat the process until a sufficient number of trials are made.
5. Summarize the results.

Monte Carlo simulation can be successfully applied, not only in engineering situations such as microelectronics engineering, geo metallurgy, aerospace engineering, etc., but also in medical situations like biological systems, such as proteins, membranes, images of cancer, etc.

27.16 MARKOV ANALYSIS

Markov Analysis is an analytical method to determine the reliability and availability of a system whose components exhibit strong dependencies or constraints, similar to the tree analysis discussed elsewhere in this book.

This method proposed by Russian mathematician Andrei Andreyevich Markov is used to forecast the value of a variable whose future value is independent of its past history, and is a sure method for forecasting random variables.

Some typical constraints that can be considered for using Markov models are:

- Components in cold or warm standby
- Common maintenance personnel
- Common spares win limited in-site stock

27.17 CONCLUSION

In general, the concept of quality is related to control of lower-level *product* specifications and manufacture, whereas the concept of reliability is related to the systems engineering, manifested by the day-by-day operation for many years. Quality is therefore related to Manufacturing, and Reliability is more related to the validation of systems or sub-systems inherent to design and lifecycle solutions.

On the Lighter Side

Adjustmentality is a word coined by this author to illustrate the philosophy to accept whatever is the outcome of your actions, without getting unduly worked up. This mentality of adjusting oneself to the situation would reduce the blood pressure of a person, but of course is in contradiction to most of the quality management principles professed in this book!

FURTHER READING

[1] Reshetov D, Ivanov A, Fadeev V. Reliability of machines. Moscow: Mir Publishers; 1990.

[2] Ebeling CE. Reliability and maintainability engineering. Noida: Tata McGraw Hill; 2000.

[3] Murthy MN. Excellence through quality & reliability. Chennai: Applied Statistical Centre; 1989.

[4] http://en.wikipedia.org/wiki/Reliability_engineering.

[5] www.iitb.ac.in.

[6] www.weibull.com/basics/reliability.htm.

Chapter 28

Business Process Reengineering

Chapter Outline

28.1 HISTORY OF BUSINESS PROCESS REENGINEERING

The term business process reengineering (BPR) was first coined in 1990 by Michael Hammer of Boston's Massachusetts Institute of Technology (MIT), who observed in his paper published in the Harvard Business Review, "Most of the work being done in the company does not add any value for the customers, and this work should be removed instead of being accelerated by automation. On the other hand, companies should redesign their processes in order to maximize their customer value, while minimizing the consumption of resources for delivering their products or services." Shigeo Shingo, who died in 1990, is also credited with having worked with Hammer in developing this concept. Almost simultaneously, Devonport and Short raised a similar concept, in their paper published in 1990 by Sloan Management Review, also from MIT, Boston.

Subsequently, Michael Hammer together with James Champy published a book *Reengineering the Corporation* in 1993, highlighting the radical design of technology, human resources, time management, and organizational goals with the use of IT tools, such as Oracle Application Software. By 1995, almost half of the Fortune 500 companies declared to have either initiated BPR, or had plans to do so. This trend was fueled by the fast adoption of BPR more by the European industry, as the study *Made in America*, conducted by MIT, showed how companies in many US industries had lagged behind their foreign counterparts in terms of competitiveness, time-to-market, and productivity.

This concept reached India, also, so fast that the Indian Institute if Industrial Engineering, during their National Convention of 2000 at Chennai, named the two halls of the conference Hammer Hall and Champy Hall.

This concept that called for radical changes in organizational thinking won its nickname of "Neo-Taylorism," after Frederick Winslow Taylor, who brought radical changes in Management thinking by his Scientific Management during the latter half of the 19th century. Even Peter Drucker, the great Management Guru who died in 2005, had supported this theory in his 1998 paper on *The Profession of Management*, published in Harvard Business Review.

28.2 DEFINITIONS OF BUSINESS PROCESS REENGINEERING

The explanatory definition offered by Wikipedia on BPR is that it is a management approach aiming at improvements by means of elevating efficiency and effectiveness of the processes that exist within and across organizations. The key to BPR is for organizations to look at their business processes from a "clean slate" perspective and determine how they can best construct these processes to improve how they conduct business.

Let us review some other definitions on BPR as reproduced below:

Business process reengineering is the fundamental rethinking and radical redesign of business processes to achieve dramatic improvements in critical contemporary measures of performance, such as cost, quality, service, and speed.

Hammer and Champy (1993)

Business process reengineering encompasses the envisioning of new work strategies, the actual process design activity, and the implementation of the change in all its complex technological, human, and organizational dimensions.

Thomas H. Davenport (1993)

BPR derives its existence from different disciplines, and four major areas can be identified as being subjected to change in BPR—organization, technology, strategy, and people—where a process view is used as common framework for considering these dimensions.

Leavitt (1965).

28.3 BUSINESS PROCESS REENGINEERING AS A TQM TECHNIQUE

Another definition of BPR relating it with other process-oriented views, such as Total Quality Management (TQM) and Just-in-time (JIT), as offered by Johansson et al. (1993), can be stated as follows:

Business Process Reengineering, although a close relative, seeks radical rather than merely continuous improvement. It escalates the efforts of JIT and TQM to make process orientation a strategic tool and a core competence of the organization. BPR concentrates on core business processes, and uses the specific techniques with in the JIT and TQM 'toolboxes' as enablers, while broadening the process vision.

In the words of Thomas H. Davenport, who explains the major difference between BPR and other approaches to organization development, especially the continuous improvement or TQM movement, when he states, *"Today firms must seek not fractional, but multiplicative levels of improvement—10× (ten times) rather than 10%."*

In order to achieve the major improvements, BPR seeks not only a change of structural organizational variables, but the use of several work improvement methods related to TQM is conceived as a major contributing factor. With the advancement of information technology and computerized data management facilities, the decisions towards BPR is facilitated with IT playing a role as enabler of new organizational forms, and patterns of collaboration within and between organizations.

In a nutshell, the following comparison (Table 28.1) can better explain the conceptual distinctions between TQM and BPR.

TABLE 28.1 TQM vs BPR

Sl. No	Parameter	TQM	BPR
1	Goals	Small scale improvements at all levels of management with cumulative effect	Outrageous and complete turn around
2	Case for action	Necessary	Compelling
3	Scope and focus	Attention to tasks, steps, and processes across the board	Select, but broad business processes
4	Degree of change	Incremental, evolutionary, and continual	High order of magnitude periodic and revolutionary
5	Role of IT	Incidental	Cornerstone
6	Senior management involvement	Important and up front	Intensive throughout

28.4   THE ROLE OF INFORMATION TECHNOLOGY

Wikipedia summarizes the role played by IT in BPR as follows:

- The databases can be shared and information made available to several departments located at different places.
- Expert systems enable generalists to perform specialist tasks.
- Telecommunication networks enable centralized organizations to simultaneously perform decentralized tasks and vice versa.
- IT enables all employees to be easily trained in decision-making.
- The technological advancement of IT has enabled extensive use of laptops or smart mobiles with wireless data communication, making it possible for field personnel to work independently and make quick decisions.
- Interactive videos and other software have enabled businesses to get in immediate contact with potential buyers.
- GPS enablement provides automatic identification and tracking, allowing easy location of the caller and takes immediate decisions.
- High performance computing enables on-the-spot planning and revisions.

Besides IT, the following strategies play an important role in facilitating BPR:

1. Business strategy and the other dimensions are governed by strategy's encompassing role.
2. The organizational dimensions, such as hierarchical levels, the composition of organizational units, and the distribution of work between them.
3. The human resources dimension dealing with aspects, such as education, training, motivation, and reward systems.
4. The concept of business processes—interrelated activities aiming at creating a value added output to a customer—is the basic underlying idea of BPR.
5. These processes are characterized by a number of other attributes: Process ownership, customer focus, value-adding, and cross-functionality.
6. BPR derives its existence from different disciplines, and four major areas can be identified as being subjected to change by BPR—organization, technology, strategy, and people—where a process view is used as a common framework for considering these dimensions. This is called Leavitt's diamond, which is further explained in Section 7.16 of Chapter 7.

28.5   METHODOLOGY FOR BPR (FIG. 28.1)

1. *Develop vision and objectives*, based on the feedback received from different departments, such as marketing, customer relations, and quality assurance apart from the prototype designs details from R & D.
2. Be thoroughly conversant with the existing process and distinguish between the defective elements from the efficient elements.
3. *Identify processes* which are defective, time-consuming, and need elimination or improvement for reengineering, by applying SWOT analysis. It may

be noted from Section 28.7 that because the original concept of radical change by BPR became unpopular due to misuse, apply the new concept of business process management (BPM) by integrating the aspects customer needs with not such radical changes.

4. *Develop the change levers* and methodologies to replace the existing ones and the change levers like automation, new technology-driven solutions, together with the latest enterprise resource planning, customer relationship management, etc., that are to be integrated into the company's overall strategy. Discuss these with all concerned departments. Improve your methodology by brainstorming.

5. *Implement the new process* with the cooperation of all concerned.

6. *Sustain the new process*: In view of the radical change concept, the implementation and reaping of the benefits takes a longer time and hence, it is essential to sustain these operational changes despite the initial resistances. Tiding over the initial teething troubles, companies should integrate BPR into their day-to-day life and strategy.

7. *Evaluate the new process*: In view of the long-term result yielding, you should constantly apply SWOT analysis throughout the implementation

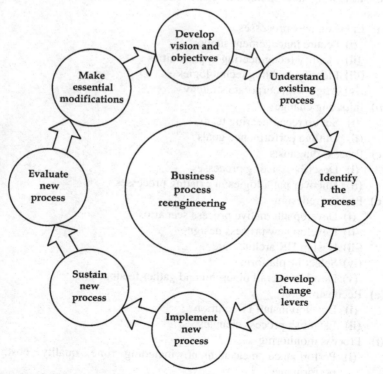

FIG. 28.1 BPR methodology.

period. Regular feedback from the departments involved with the change should be obtained. Even after the product via the new process reaches the customer, the information on its acceptance and popularity should be collected and analyzed.

8. *Make essential modifications*: There is always room for improvement and as such, make necessary modifications periodically, without deviating much from the original goal. Keep getting feedback from all concerned and use it for the evaluation. Remember the whole BPR process is cyclic. Even after making the modifications, revisit the vision and objectives and modify if necessary.

28.6 PROCESS REENGINEERING LIFE CYCLE APPROACH FOR BPR

Guha et al. (1993) have suggested the following IT-centric BPR model, which they called as the process reengineering life cycle approach:

1. Structural organization with functional units
2. Introduction of New Product Development as cross-functional process
3. Re-structuring and streamlining activities, removal of non-value adding tasks
    (a) Envision new processes
        (i) Secure management support
        (ii) Identify reengineering opportunities
        (iii) Identify enabling technologies
        (iv) Align with corporate strategy
    (b) Initiating change
        (i) Set up reengineering team
        (ii) Outline performance goals
    (c) Process diagnosis
        (i) Describe existing processes
        (ii) Uncover pathologies in existing processes
    (d) Process redesign
        (i) Develop alternative process scenarios
        (ii) Develop new process design
        (iii) Design HR architecture
        (iv) Select IT platform
        (v) Develop overall blueprint and gather feedback
    (e) Reconstruction
        (i) Develop/install IT solution
        (ii) Establish process changes
    (f) Process monitoring
        (i) Performance measurement, including time, quality, cost, IT performance
        (ii) Link to continuous improvement

Benefiting from lessons learned from the early adopters, some BPR practitioners advocate a change in emphasis to a customer-centric, as opposed to an IT-centric, methodology. One such methodology, that also incorporates a Risk and Impact Assessment to account for the impact that BPR can have on jobs and operations, was described by Lon Roberts (1994). Roberts also stressed the use of change management tools to proactively address resistance to change, a factor linked to the demise of many reengineering initiatives that looked good on the drawing board.

28.7   CRITICISM AGAINST BPR

During the 1990s, many companies started using reengineering as a pretext to downsize their companies dramatically, though this was not the intention of reengineering's proponents. Consequently, reengineering earned a reputation for being synonymous with downsizing and layoffs.

Hammer himself admitted that:

> *I wasn't smart enough about that. I was reflecting my engineering background and was insufficiently appreciative of the human dimension. I've learned that's critical.*

Devonport, too, commented,

> *When I wrote about business process redesign in 1990, I explicitly said that using it for cost reduction alone was not a sensible goal. And consultants Michael Hammer and James Champy, the two names most closely associated with reengineering, have insisted all along that layoffs shouldn't be the point. But the fact is, once out of the bottle, the reengineering genie quickly turned ugly.*

A similar criticism is reflected in Section 22.3 in the chapter on Kaizen. *Other criticisms leveled against BPR include:*

1. Lack of management support for the initiative and thus, poor acceptance in the organization.
2. Exaggerated expectations regarding the potential benefits and the consequent failures in achieving the expected results.
3. Underestimation of the resistance to change within the organization.
4. The so-called best practices and processes do not fit the specific company needs and the implementation becomes difficult.
5. Too much trust in technology solutions.
6. Tendency to perform BPR as a one-time project with limited strategy.

28.8   SATISFACTORY UNDERPERFORMANCE

Why make a fuss if some little things go wrong? Do you realize the privilege you have in dealing with the best? This is the negative attitude of managers who are satisfied with their underperformance, even though they realize within their heart that their performance has to be improved.

Sumantra Ghosal et al. make the following observation in their book, *Managing Radical Change*:

> *Bottlenecks tend to be at the top of the bottle. That implies that the most critical barrier to change lies at the top of the managers in their lack of belief in and passion for change. They all say the right words, publish them in the annual reports and in house journals, but deep in their hearts they do not believe in what they say. That is why companies find it hard to manage changes. Surely there are many other barriers and obstacles, but none of them is as debilitating as the mindset of the senior managers.*

Fig. 28.2 illustrates this attitude in the form of Dynamics of Satisfactory Underperformance.

Gradual decline towards crises
Satisfactory underperformance
Stifling of initiative and innovation
Excessive focus on control
Layers of staff are built to cope up with growth
Managers start believing they are the Best
Competitiveness, growth, and profits
Successful business strategy

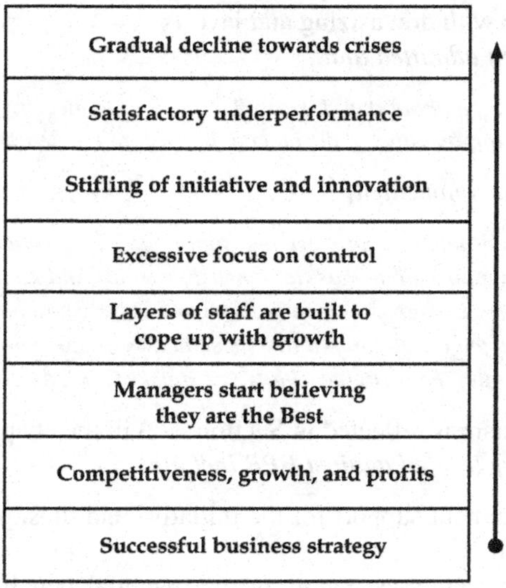

FIG. 28.2   Dynamics of satisfactory underperformance.

The above book also cites the following Illustration for Radical Improvement of Performance (Reengineering):

Japanese competition almost forced Motorola out of its DRAM (Dynamic Random Access Memory) semiconductor business in around 1985. Nevertheless, as a last try, the management came up with a radical reengineering programme and by 1988, the financial position of Motorola has radically improved and established a clear leadership in pagers and cell phones.

28.9 THE SWEET AND SOUR CYCLE

The success of the Reengineering or the Radical improvement process for every business has two components, shown in Fig. 28.3.

● The continuous Improvement process, which the Japanese call Kaizen, provides the resources needed for growth, including money, men, materials, machinery, and management, which we fondly call the 5 Ms of inputs or resources, and
● The continuous revitalization process involving the creation of new processes and exploitation of new opportunities which we call reengineering.

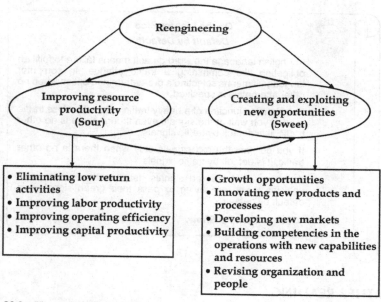

FIG. 28.3   The sweet and sour cycle.

28.10 BUSINESS PROCESS MANAGEMENT

These abuses and misuses of the BPR concept, as stated above, reduced the reengineering fervor in companies world over. This has led to a new concept of BPM which provides a holistic management approach, aligning an organization's business processes with the wants and needs of customers. BPM uses a systematic approach in an attempt to continuously improve business effectiveness and efficiency, while striving for innovation, flexibility, and integration with technology. It can therefore be described as a "Process Optimization Process."

28.11 CONCLUSION

In view of the above stated objections for and criticism of Business process re-engineering, its approach has now evolved into BPM, which focuses on improving corporate performance by managing and optimizing a company's business processes, rather than by radical changes. This approach closely resembles other total quality management tools, especially the Continuous Improvement Process methodologies. A review of the several case studies cited in this chapter emphasizes the fact that BPR is more suitable for small-scale industries producing electronic components, etc., than to majority of large-scale industries.

On the Lighter Side

Default by Default

In English language the word default means failure to fulfil an obligation like committing a traffic offence. In computer language it means selecting a pre-determined option when no other alternative is specified.

In a signaled junction of a heavy traffic road with a less traffic road, if you wait at the signals even though there is no other traffic, it is default by traffic signals.

If you go past the red signals only when there is no other traffic, it is default by traffic signals

If you go past the red signals despite other cross traffic vehicles which are trying to pass their green signal, it is default by default.

-George Bernard Shaw.

FURTHER READING

[1] Ghosal Sumantra, Piramal Gita, Bartlett Christopher. Managing radical change. New Delhi: Penguin; 2002.

[2] http://en.wikipedia.org/wiki/Dynamic_random-access_memory.

[3] onlinelibrary.wiley.com/pdf.

[4] http://en.wikipedia.org/wiki/Business_process_management.

[5] www.aiim.org.

Chapter 29

Benchmarking

Chapter Outline

29.1 WHAT IS BENCHMARKING?

Benchmarking is an ongoing process involving industries from all walks of life and all categories of production. The principle is that no company is 100% perfect, and if you continuously search for better solutions, you will improve your efficiency and become an exceptional company, which can later form a benchmark for similar companies. It essentially compares the business processes and performance metrics including cost, cycle time, productivity, or quality with another company widely considered to be the industry standard benchmark and/or having best practices.

29.2 DEFINITIONS FOR BENCHMARKING

While the dictionary meaning is "anything used as a standard point of reference," the other definitions are cited as:

1. Benchmarking is the process for improving performance by continuously identifying, understanding, and adapting the best practices and processes followed by similar companies, and implementing the results.
   –Wikipedia
2. Benchmarking is the systematic search for best practices, innovative ideas, and highly effective operating procedures.
   –Besterfield et al.

3. Benchmarking is the process used by the management in which the businesses use an industry leader as a model in developing their business practices.

   –ASQ

4. Benchmarking is a continuous systematic process of measuring products, services, processes, or work practices against the toughest competitors or those companies recognized as the industry's best.

   –Xerox Corporation

5. Benchmarking is the search for the industry's best practices and the adoption of such practices to ensure superior performance.

   –Robert Camp

In a nutshell, benchmarking is a systematic method or a popular TQM tool by which the organizations measure themselves against best industry practices. It is a legal and above-board manner of finding out about others' techniques of better performances and using it or improving on it for better results.

29.3 TYPES OF BENCHMARKING

1. *Performance benchmarking*: The company's competitive position is assessed, comparing the products and services with those of other companies.

2. *Product benchmarking*: The basic functional performance and quality features of a company's products are compared and benchmarked with competitors' products, with a view to improve the functional features. It becomes useful in designing products that match precise user expectations, at minimum possible cost, by applying the best technologies available worldwide. The development of Taurus cars by Ford Motors is an illustration.

3. *Process benchmarking*: The firm focuses its investigations with a goal of identifying and observing the best practices adapted in one or more benchmark firms, producing the same or similar products in the cement industry. The study conducted recently by the National Productivity Council of India in the cement industry of South India is an illustration of this.

4. *Strategic benchmarking*: This involves the study of corporate level strategies adapted by successful industries.

5. *Generic benchmarking*: Sometimes one type of industry benchmarks a part of its products with that of other types of industries that use these parts. For example, an automobile manufacturer may want to benchmark their hydraulic systems with those of say, Disneyland.

6. *Functional benchmarking*: Specific functions like billing, distribution network recruitment are compared.

7. *Competitive benchmarking*: Here certain parameters are compared to the competitors' with their cooperation, after obtaining special permission for

getting vital information. While analyzing your competitors, you also identify the best company in the industry, even if it is located elsewhere and is in a different market segment.

8. *Financial benchmarking*: The annual financial results are compared and analyzed in order to assess the overall performance, productivity, and profitability of the concern.

9. *Operational benchmarking*: This embraces everything from staffing and productivity to the office flow, and analyzes the procedures adapted.

10. *Internal benchmarking*: Assessment comparisons are made within departments or within sister concerns.

11. *Collaborative benchmarking*: Sometimes benchmarking is carried out collaboratively by groups of companies either through common consultants or by professional associations or bodies. An example is the earlier cited work done by NPC of India in the cement industry. Another commonly cited illustration is the voluntary collaborative study carried out in 1977 by the Dutch Municipal Water Supply Companies through their industry association.

12. *Metric benchmarking*: This involves using information on cost or production information to identify strong and weak performing units. The two most common forms of quantitative analysis used in metric benchmarking are data envelope analysis (DEA) and regression analysis. DEA estimates the cost level an efficient firm should be able to achieve in a particular market. In infrastructure regulation, DEA can be used to reward companies/operators whose costs are near the efficient frontier with additional profits. Regression analysis estimates what the average firm should be able to achieve.

29.4 SOME OF THE PARAMETERS THAT CAN BE BENCHMARKED

1. Return on investments
2. Return on assets
3. Cost per unit
4. Cost per order
5. Net present worth of the shares
6. Sigma level in quality control
7. Customer satisfaction index
8. Sales cost revenue
9. Service cost/revenue
10. Service response time
11. Distribution cost/Revenue
12. Material overheads

13. Manpower performance ratio
14. Absenteeism
15. Employee morale.

29.5   GENERAL CONCEPT OF BENCHMARKING

While the general concept of benchmarking is shown in Fig. 29.1, the detailed procedure can be as illustrated in Fig. 29.2.

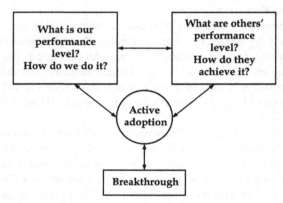

FIG. 29.1   General concept of benchmarking.

29.6   PHASES OF BENCHMARKING

A. *Preparatory phase*: When a preliminary meeting of all concerned people is called for and the details are explained. The planned phases and procedures are subjected to brainstorming based on which of the subsequent phases are planned as explained below. The purpose of this phase is to place bench-marking activities on a firm footing.
B. *Planning phase*: This is the basic phase for conducting a benchmarking investigation. Here the plans are developed by addressing the key questions of
  • *What is to be benchmarked?*
  • *Who will be the benchmark partners?*
  • *What is the method of data collection?*
C. *Analysis phase*: When the data collected is analyzed for the purpose of comparison with other benchmarked companies. The key questions to be addressed here are:
  • *What is the performance of the benchmark partners, for each parameter?*
  • *What is our performance compared to theirs?*
  • *In what way, and in what parameters of their performance are theirs better than ours?*

- *What is the lesson to be learned from them?*
- *How can we apply the lessons to our company?*

D. *Integration phase*: When the goals are developed and integrated into the benchmarked parameter or process so that significant performance improvements are made. The key questions in this phase are:
   - *What are the critical proposals given as a result of the analysis?*
   - *Has management accepted the findings?*
   - *What are the goals that are needed to be modified and what are the finally drawn up goals?*
   - *Have all involved parties been clearly communicated to about the goals?*

E. *Action phase*: When detailed action plans needed to achieve the goals are drawn and developed. The key questions that need to be addressed here are:
   - *Will the plans allow the achievement of the stated goals?*
   - *How will progress be tracked?*
   - *What is the schedule for recalibration of the benchmarks?*

29.7   STAGE OF BENCHMARKING

Wikipedia elaborates the above in the following steps:

1. *Identify your problem areas*: Because benchmarking can be applied to any business process or function, a range of research techniques may be required. They include: informal conversations with customers, employees, or suppliers; exploratory research techniques, such as focus groups; or indepth marketing research, quantitative research, surveys, questionnaires, re-engineering analysis, process mapping, quality control variance reports, or financial ratio analysis. Before embarking on comparisons with other organizations, it is essential that you know your own organization's functions, processes; base-lining performance provides a point against which improvement efforts can be measured.

2. *Identify other industries that have similar processes*: For instance, if one were interested in improving hand-offs in addiction treatment, he/she would try to identify other fields that also have hand-off challenges. These could include air traffic control, cell phone switching between towers, or transfer of patients from surgery to recovery rooms.

3. *Identify organizations that are leaders in these areas*: Look for the very best in any industry and in any country. Consult customers, suppliers, financial analysts, trade associations, and magazines to determine which companies are worthy of study.

4. *Survey companies for measures and practices*: Companies target specific business processes using detailed surveys of measures and practices used to identify business process alternatives and leading companies. Surveys

are typically masked to protect confidential data by neutral associations and consultants.

5. *Visit the "best practice" companies to identify leading-edge practices*: Companies typically agree to mutually exchange information beneficial to all parties in a benchmarking group and share the results within the group.

6. *Implement new and improved business practices*: Take the leading-edge practices and develop implementation plans which include identification of specific opportunities, funding the project, and selling the ideas to the organization for the purpose of gaining demonstrated value from the process.

7. *Repeat the process*: Benchmarking is an ongoing process. Best practices can always be improved upon by constant vigilance and analysis of the competitor practices.

29.8 DIFFERENT APPROACHES TO BENCHMARKING

Robert Camp, one of the early authors on this subject has, in his book, *The Search for Industry Best Practices that Lead to Superior Performance* (1989), summarized the above procedure in 12 steps as:

1. Select the subject ahead
2. Define the process
3. Identify the potential partners
4. Identify the data source
5. Collect data and select partners
6. Determine the gap
7. Establish the process differences
8. Target future performances
9. Communicate
10. Adjust goals
11. Implement
12. Review/recalibrate

Xerox's 10-step procedure
1. Identify what is to be benchmarked
2. Identify comparable organizations
3. Determine data collection methods and collect data
4. Determine current performance gap
5. Project future performance levels
6. Communicate benchmark findings and gain acceptance
7. Establish functional goals
8. Develop action plans
9. Implement specific actions and monitor progress
10. Recalibrate benchmarks

The above can also be represented as in Fig. 29.2.

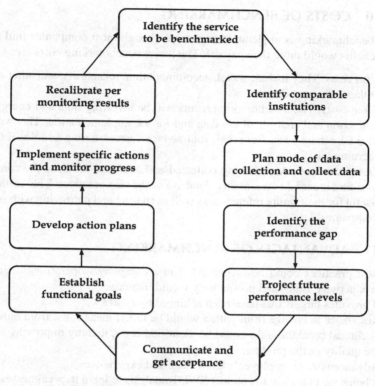

FIG. 29.2 The 10-step approach to benchmarking.

29.9 TIPS FOR THE CONSULTANTS

AT&T had given the following tips to consultants for successful conduct of the benchmarking procedure.

1. Determine who the clients are, who will use the information to improve their processes.
2. Advance the clients from the literacy stage to the champion stage.
3. Test the environment. Make sure that the clients can and will follow through with the benchmarking findings.
4. Determine the urgency. Panic or disinterest indicate little chance for success.
5. Determine the scope and type of benchmarking needed.
6. Select and prepare the team. Overlay the benchmarking process onto the business planning process.
7. Develop the benchmarking plan.
8. Analyze the data.
9. Integrate the recommended action.
10. Take action.
11. Continue improvement.

29.10   COSTS OF BENCHMARKING

This benchmarking is moderately expensive, though most companies find that the benefits would more than pay off. The major benchmarking costs are:

(a) *Visit costs*: They include travel, accommodation, token gifts, and other miscellaneous expenditure.

(b) *Time costs*: The benchmarking teams will be investing time and energy in collection and analysis of the data and for the implementation. This would also take them away from their routine tasks, necessitating additional staff recruitment.

(c) *Database costs*: After having collected and analyzed the data, the companies find it useful to maintain a database of the best practices. This could be useful for their future reference, as well as sharing and partnering with other companies.

29.11   ADVANTAGES OF BENCHMARKING

1. Best practices could be incorporated in a company's operations; thereby productivity, sales, and profitability would increase.
2. It provides targets that have been achieved by others.
3. Resistance to change from within would be less if ideas come from outside.
4. Technical breakthroughs could be identified early, thereby improving upon the quality of the product.
5. Advancement in employee's knowledge and experience.
6. It helps the company to conduct SWOT analysis to learn its weaknesses and the scope of improvement.
7. Customers' requirements can be met in a more systematic manner.
8. Effective goals can be set and achieved.

29.12   LIMITATIONS OF BENCHMARKING

1. It entails subjective judgment.
2. It is difficult to get useful information from the competitors.
3. It requires a thorough understanding of current products and processes.
4. It is purely a creative activity.
5. It is a costly and time-consuming activity.
6. It is still subject to resistance to change from employees to some extent.

29.13   PROFESSIONAL ASSOCIATIONS AND INSTITUTIONS EXCLUSIVELY FOR BENCHMARKING

Internet Service Providers' Benchmarking Association (ISPBA) is a free association of procurement and supply-chain organizations within major corporations, with a mission to identify "Best in class" internet service provision business processes which when implemented would lead member companies for achieving exceptional performance. It conducts benchmarking studies to identify the practices that improve overall performance of the members' organizations.

Its objectives are: To conduct benchmarking studies is important for the internet service provision processes.

- To create a cooperative environment where full understating of the performance and enablers of "best in class" internet service provision processes can be obtained and shared at reasonable costs.
- To use the efficiency of the association to obtain process performance data and related best practices regarding internet service provision.
- To support the use of benchmarking to facilitate internet service provision process improvement and the achievement of overall accuracy, timeliness, and efficiency.

29.14   CONCLUSION

The term benchmark originates from a mark on a permanent object indicating elevation, and serving as a reference in topographic surveys and tidal observations. From this chapter, it can be shown how apt it is to compare our performance with leaders in the industry and improve ourselves to ascend the ladder of success.

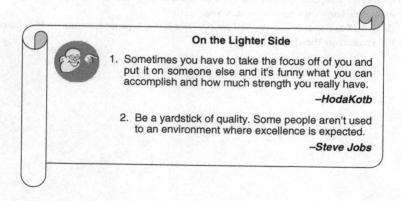

On the Lighter Side

1. Sometimes you have to take the focus off of you and put it on someone else and it's funny what you can accomplish and how much strength you really have.

–HodaKotb

2. Be a yardstick of quality. Some people aren't used to an environment where excellence is expected.

–Steve Jobs

APPENDIX

In one of their course works, NIQR cited the following benchmarking illustrations.

1. Hindustan Lever Limited benchmarked its rural marketing against a beedi manufacturer.
2. Modi Xerox benchmarked its level of Customer Satisfaction Measurement (COSAM) against Rank Xerox, Portugal.

3. Kelvinator India benchmarked its components and product cost against Godrej—GE applications.
4. Gabriel India Auto Ancillary Maker benchmarked its customer delivery times with UK-based British Filters.
5. Arvind Mills regularly benchmarks its compensation strategy against Proctor & Gamble as well as Hindustan Lever.
6. 35 Companies spanning 7 different businesses of RPG group regularly benchmark against each other on cost management, value engineering, purchase management, inventory management, sales forecasting, etc.
7. L & T have benchmarked engineering and project management with that of Bechel Corporation of the US.
8. Citibank has benchmarked five key processes against a courier company.
9. Matrix Applications has benchmarked customer satisfaction against its archrivals Carrier Aircons and National Panasonic.

FURTHER READING

[1] Wikipedia—http://en.wikipedia.org/wiki/Benchmarking.
[2] Boxwell Jr. R.J.. Benchmarking for competitive advantage. New York: McGraw-Hill; 1994. http://www.amazon.com/Benchmarking-Competitive-Advantage-Robert-Boxwell/dp/0070068992.
[3] Beating the competition: a practical guide to benchmarking (1988). Washington, DC: Kaiser Associates. http://www.kaiserassociates.com.
[4] Camp R.. The search for industry best practices that lead to superior performance. Milwaukee, WI: Productivity Press; 1989.

Chapter 30

Quality Function Deployment

Chapter Outline

30.1 WHY QUALITY FUNCTION DEPLOYMENT?

In his book, *Mega Mistakes*, Steven Schnaars states that the technology-bound forecasting methods developed during the onset of Taylor's Scientific Management missed their mark, not by a matter of degrees, but completely, and without regard for what actually is needed by the customer, who is the ultimate user of these technologies or the forecasts. The primary reason can be attributed to the fact the scientific forecasters were themselves seduced by the technological wonders on which to base their predictions and presumed that the customers, too, would find the new technology equally enticing and irresistible. At the same time, these forecasters wanted the changes at "breathtaking speeds," the result being that the economic considerations have been overlooked and the market the industry is expected to serve is ignored.

Under these arguments, understanding customers' means, identifying who they are, and how their needs and wants can be met by the organization that serves them, is the first and foremost need for any corporate before going into production. It assures complete customer satisfaction and keeps frequent design and other engineering changes to the barest minimum. Once a product is defined, quality function deployment (QFD) enables the design phase to focus on the key customer requirements. By addressing these elements, the design phase is shortened to focus on items that the customer really wants. And because of

Total Quality Management: Key Concepts and Case Studies. http://dx.doi.org/10.1016/B978-0-12-811035-5.00030-1

this, seeing a better picture of what features are preferred by the customer, the whole design process can be accelerated. This is the basic reason for the significant reduction in the startup costs as cited earlier.

With this concept in mind, Akao and Mizuno developed the technique of QFD, which is described in more detail in subsequent paragraphs.

As Barb Cleary puts it, "Shaking someone's hand while giving them a defective product represents a bad balance and a lack of understanding of the ways that customers' needs are met." A system, defined by Deming, includes suppliers, inputs, processes, outputs, and above all, the customers.

30.2 DEFINITIONS OF QFD

QFD is basically a planning process with a quality approach to new product design, development, and implementation driven by customer needs and values. QFD has been successfully used by many world-class organizations in automobiles, ship building, electronics, aerospace, utilities, leisure and entertainment, financial, software, and other industries.

Quality Function Deployment (QFD) is a structured approach to defining customer needs or requirements and translating them into specific plans to produce products to meet those needs. The "voice of the customer" is the term to describe these stated and unstated customer needs or requirements. The voice of the customer is captured in a variety of ways: direct discussion or interviews, surveys, focus groups, customer specifications, observation, warranty data, field reports, etc. This understanding of the customer needs is then summarized in a product planning matrix or "house of quality". These matrices are used to translate higher level "what's"[sic] or needs into lower level "how's"[sic] - product requirements or technical characteristics to satisfy these needs.

–Kenneth Crow

Quality function deployment (QFD) is a method to transform qualitative user demands into quantitative parameters, to deploy the functions forming quality, and to deploy methods for achieving the design quality into subsystems and component parts, and ultimately to specific elements of the manufacturing process.

–Wikipedia

QFD is a comprehensive quality system that systematically links the needs of the customer with various business functions and organizational processes, such as marketing, design, quality, production, manufacturing, sales, etc., aligning the entire company toward achieving a common goal.

–QFD Institute.

In QFD, quality is a measure of customer satisfaction with a product or service. QFD is a structured method that uses the seven management and planning tools to identify and prioritize customers' expectations quickly and effectively.

–Jack B. ReVelle

Quality professionals refer to QFD by many names, including matrix product planning, decision matrices, and customer driven engineering. Whatever you call it QFD focuses methodology for carefully listening to the voice of the customer and effectively responding to those needs and expectations.

–ASQ

The Quality Function Deployment (QFD) process is probably the most effective methodology available for capturing and responding to the "voice of the customer".

–Joseph P. Merts

Quality Function Deployment is a systematic approach to design based on a close awareness of customer desires, coupled with the integration of corporate functional groups. It consists in translating customer desires (for example, the ease of writing for a pen) into design characteristics (pen ink viscosity, pressure on ballpoint) for each stage of the product development.

–Creative Industries research Institute

30.3   HISTORY OF QFD

During the late 1960s, Yoji Akao and Shigeru Mizuno, both of Tokyo Institute of technology, while combining their work in quality assurance and quality control points with function deployment used in Value Engineering, came up with this idea of QFD which they described as a method to transform user demands into design quality, to deploy the functions forming quality, and to deploy methods for achieving the design quality into subsystems and component parts, and ultimately to specific elements of the manufacturing process.

Shigeru Mizuno first applied this principle in the design of oil tankers for the Kobe Shipyard of Mitsubishi Heavy Industries in 1972, and later in Tokyo Motor Corporation in 1977. He achieved the following astounding results.

- 20% reduction in start-up costs by 1979
- 38% reduction by 1982
- 61% reduction by 1984

The encouraging results of QFD enabled its application in deploying highly controllable factors in strategic planning, also known as Hoshin Kanri (see Chapter 24) The QFD associated Hoshin Kanri somewhat resembles Management by objectives (MBO), and adds a significant element in the goal-setting process.

In the United States

- Xerox Company introduced QFD in 1984.
- Florida Power and Light Co. applied QFD and got the Deming Award in 1990.
- AT&T Power system applied QFD and got the Deming Award in 1994.

- Host Marriot, the Airlines Catering Company applied QFD.
- Other notable US companies that benefited from QFD application were Ford, Chrysler, IBM, GE, Boeing, Lock heed, and a host of other famous companies.

In the International Symposium on QFD 1997 at Linköping, Sweden, Yoji Akao, then at Asahi University, himself stated in his paper "QFD: Past, Present and Future" that

> *QFD was conceived in Japan in the late 1960's, during an era when Japanese industries broke from their post-World War II mode of product development through imitation and moved to product development based in originality. QFD was born in this environment as a method or concept for new product development under the umbrella of Total Quality Control. A sub-title 'An approach to Total Quality control' was added to the book Quality Function Deployment, the very first book on QFD written by the late Dr. Shigeru Mizuno and myself illustrates this relationship.*

QFD gained such momentum in industry that an International Council for Quality Function Deployment (ICQFD) was established in 1997 during the International Symposium on QFD (ISQFD'97), to provide a unified body to coordinate the many local QFD organizations, efforts, and events around the world. An International QFD Akao Prize is being awarded during every ISQFD. The recent Symposia being held were:

- The 20th ISQFD'14 in Istanbul, Turkey, on September 3rd and 5th, 2014,
- The 21st ISQFD'15 in Hangzhou, China on August 22nd, 2015,
- The next 22nd ISQFD'16 is scheduled in Boise, Idaho, USA in September 2016.

Besides, several countries have started QFD institutes, some of which are cited below:

- QFD Institute (North America)
- QFD Institute Deutschland (Germany)
- Latin America QFD Association (Mexico)
- Union of Japanese Scientists and Engineers (Japan)
- Dokuz Eylul University—QFD Turkiye Research Group (Turkey)
- Institute de GDP do Brasil (Brazil)
- Hong Kong QFD Association/China Association of Quality (China)
- Mac Quarie University (Australia)
- Linköping University (Sweden)

As www.qfdi.org puts it,
QFD is:

1. Understanding "true" customer needs from the customer's perspective
2. What "value" means to the customer, from the customer's perspective

3. Understanding how customers or end users become interested, choose, and are satisfied
4. Analyzing how we can know the needs of the customer
5. Deciding what features to include
6. Determining what level of performance to deliver
7. Intelligently linking the needs of the customer with design, development, engineering, manufacturing, and service functions
8. Intelligently linking Design for Six Sigma (DFSS) with the front end Voice of Customer analysis and the entire design system.

30.4   ISSUES THAT WOULD BE ADDRESSED BY QFD

The *website of* QFDE Capture cites the issues that would be addressed by QFD as:

- *Competitive product positioning*: QFD enables specific competitive positioning of targets that are communicated throughout the organization and provides a shared focus for management and project teams.
- *Product portfolio management*: Application of QFD provides a unique opportunity to not only define what the current new product should be all about, but also what constitutes better as technologies improves.
- *Technology planning*: QFD provides visibility to technology shortcomings and focus for future technology needs.
- *Timely progress communication (to support stage-gate process)*: Each step in the QFD process produces the exact data needed for decisions to be made by management at each gate in the product development process.
- *Meaningful definition of critical parameters for DFSS process*: The QFD process enables the project team to identify the most critical parameters needed to obtain the competitive positioning that will be critical for the success of the project.
- *Data-driven decisions*: The QFD process limits the reliance on subjective opinions to make key decisions by letting decisions be driven by data collected from customers.
- *Traceability of decisions and intent (requirements management)*: The proper execution of QFD leaves an organization with traceability of requirements and design decisions all the way back to the initial targeted customer and business needs.

30.5   THE FOUR PHASES OF QFD

QFD basically is performed in four phases, which are explained and discussed in Table 30.1:

1. Product Planning
2. Design Deployment
3. Process Planning and
4. Production Planning

TABLE 30.1 The Four Phases of QFD

No.	Phase	Activity	Tools
1	Phase I: Product planning	1. Determine customer requirements 2. Translate customer requirements into design requirements	• Direct data collection such as Surveys • Indirect data collection such as Contract or order • Regulatory requirements • Competitive benchmarking • Ability of competing • Cause and effect diagrams
2	Phase II: Design deployment	1. Select process concept 2. Develop alternative processes and concept 3. Evaluate 4. Analyze the relationship between the design requirements for each product feature 5. Identify critical part characteristics	• Engage assessment of competitor's products • Tanaka's functional evaluation system • Value engineering analysis of the design • Fault tree analysis • FMEA
3	Phase III: Process planning	1. Analyze and evaluate alternative designs for processes 2. Compare relationship between process parameters and critical part characteristics 3. Identify critical part characteristics 4. Apply method study techniques to identify and eliminate non-value adding elements	• Work study • Taguchi loss function • Event analysis • Value engineering • Flow diagrams and process sheets
4	Phase IV: Production planning	1. Develop specific process controls 2. Set up Production Planning and Controls 3. Prepare visuals of the critical process parameters for everyone to understand (Seiketsu) 4. Train workers and ensure on the job guidance and supervision	• Master flow diagram • Operating instruction charts • Production layout • PP&C • Gantt chart • Process charts • Poka-yoke • Maintenance plans and schedules

These phases can also be illustrated by Fig. 30.1.

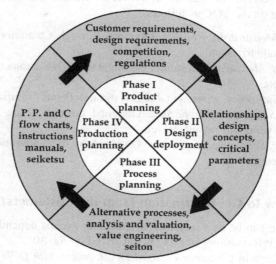

FIG. 30.1 The four QFD phases for product development.

30.6 BUILDING A HOUSE OF QUALITY

- List Customer Requirements (What's)
- List Technical Descriptors (How's)
- Develop Relationship (What's & How's)
- Develop Interrelationship (How's)
- Competitive Assessments
- Prioritize Customer Requirements
- Prioritize Technical Descriptors

30.7 VOICE OF THE CUSTOMER

In QFD terminology, the process of identifying the customer requirements is called the Voice of the customer (VOC). This forms the driving force behind QFD. Though the customers often speak about what features they want for a product or service, they do not know or cannot explain fully what they want from the design point of view. It is the organization's job to understand why the customers want these features and then translate them into innovative or modified deigns. Understanding the customers' needs is of primary importance to develop the solutions before the competitors do. Kano's classification of the types of customers and his model of customer satisfaction explains this situation in more in detail Chapter 10.

It may also be noted as stated therein that while Kano classifies the customer based on their satisfaction detailed as above, Besterfield classifies information collected based on the VOC as follows.

1. *Solicited, Measurable, Routine*: Obtained through Customer and Market Surveys, trade trials, etc.
2. *Unsolicited, Measurable, Routine*: Obtained through Customer Complaints, Lawsuits, etc.
3. *Solicited, Subjective, Routine*: Obtained through Focus Groups
4. *Solicited, Subjective, Haphazard*: Obtained through Trade & Customer Visits, Independent Consultants, etc.
5. *Unsolicited, Subjective, Haphazard*: Obtained through conventions, Vendors, Suppliers, etc.

30.7.1   How to Get Information From the Customers?

Basically there can be two segments of customer groups depending upon the quantum. The information gathering is illustrated in Fig. 30.2.

Kenneth Crow in *Customer Focused Development with QFD* recommends the following steps to facilitate initially using QFD.

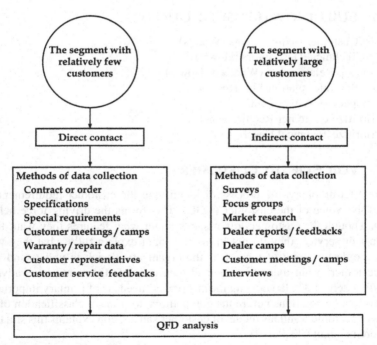

FIG. 30.2   Flow of information for QFD.

- Obtain management commitment to use QFD.
- Establish clear objectives and scope of QFD use. Avoid first using it on a large, complex project if possible. Will it be used for the overall product or applied to a subsystem, module, assembly, or critical part? Will the complete QFD methodology be used or will only the product planning matrix be completed?
- Establish a multi-functional team. Get an adequate time commitment from team members.
- Obtain QFD training with practical hands-on exercises to learn the methodology and use a facilitator to guide the initial efforts.
- Schedule regular meetings to maintain focus and avoid the crush of the development schedule overshadowing effective planning and decision-making.
- Gathering only the obvious data is not enough. Many times, significant customer insights and data exist within the organization, but they are in the form of hidden knowledge—not communicated to people with the need for this information. On the other hand, it may be necessary to spend additional time gathering the VOC before beginning QFD. Avoid technical arrogance and the belief that company personnel know more than the customer.

30.8   VOICE OF THE ORGANIZATION

Voice of the organization (VOO) is the other side of the coin of QFD, one side being the VOC. This voice analyzes the capabilities of the company by a gap analysis of the process and the team to determine whether the customers' requirements can be fulfilled by the company in the present circumstances. This analysis enables the company to fill these gaps so as to fulfill the VOC. In fact, the name "voice of organization" has no other significance, except to indicate that this is the other side of the coin.

The VOO needs to analyze the gaps by asking the following questions:

- Who are the customers?
- Determine the VOC that is the customer requirements.
- Survey customers for determining the relative importance of requirements.
- Identify and evaluate competition. How satisfied is the customer now?
- Generate product specifications. How will the customer's requirements be met?
- Translate customer requirements into measurable technical requirements.
- Identify relationships between customer and technical requirements.
- Set targets for design. How much is good enough?

30.9   FRAMEWORK FOR HOUSE OF QUALITY

The next step in the QFD process is forming a planning matrix. The main purpose of the planning matrix is to compare how well the team met the customer requirements compared to its competitors. The planning matrix shows the weighted importance of each requirement that the team and its competitors are attempting to fulfill. Because the top portion normally fits in to a triangle, the whole matrix looks like a house. Thus, it gets the name House of Quality.

30.10 BUILDING UP OF HOUSE OF QUALITY

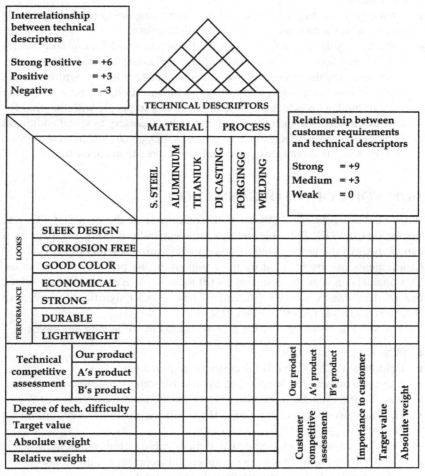

FIG. 30.3 An illustration of comparing VOC and VOO for building up the house of quality.

Fig. 30.3 on the facing page illustrates how the VOC and the VOO can be compiled by building up the house of quality in line with the above framework illustrated in Fig. 30.4.

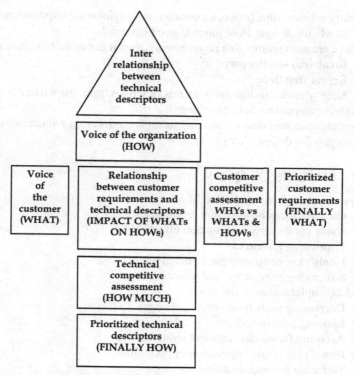

FIG. 30.4   Framework for the house of quality.

30.11   PROCEDURE FOR QFD

ITI website gives the following procedure:

1. Identify the customer
   (a) Use Kano model
   (b) Determine customer requirements
   (c) Specify information needed
2. Determine type of data collection
   (a) Determine content of questions
   (b) Design questions
   (c) Order questions
   (d) Take data
   (e) Reduce data
3. Determine relative importance of requirements
   (a) Identify and evaluate competition: How satisfied is the customer now?
   (b) Generate product specifications: How will customers' requirements be met?
   (c) Translate into measureable engineering requirements

4. Identify relationships between customer and engineering requirements
5. Set targets for design: How much is good enough?
6. If there are not measureable requirements, then it is not well understood
    (a) Break into smaller parts
    (b) Repeat step three
    (c) State whether desires are demands or wishes, rank the wishes
7. Conduct competition benchmarking
8. Translate customer desires into measureable engineering requirements
9. Set targets for design: dates

30.12   BENEFITS OF QFD

1. Improves customer satisfaction by:
    (a) Creating focus on customer requirements
    (b) Using competitive information effectively
    (c) Prioritizing resources
    (d) Identifying items that can be acted upon
    (e) Structuring experience and information
2. Reduce implementation time by:
    (a) Decreasing mid-stream design changes
    (b) Limiting post introduction problems
    (c) Avoiding future development redundancies
    (d) Identifying future application opportunities
    (e) Surfacing missing assumptions
3. Promotes team work by:
    (a) Basing the procedure on consensus
    (b) Creating communication interfaces
    (c) Identifying actions at interfaces
    (d) Creating global view out of details
4. Provides documentation by:
    (a) Providing documents rational for the design
    (b) Making it easy to assimilate
    (c) Adding structure to the information
    (d) Adapting to changes by dynamic documentation
    (e) Providing framework for sensitivity analysis.

30.13   CONCLUSION

The history of QFD, how it evolved and how it gained the global importance as a necessary tool for the designers as detailed in paragraph 30.3, and also the numerous websites created on QFD, provide an interesting study. This chapter may be treated as a supplementary extension to Chapter 10 on Customer Satisfaction.

On the Lighter Side

It's funny. All you have to do is say something nobody understands and they'll do practically anything you want them to.

—J. D. Salinger

FURTHER READING

[1] en.wikipedia.org/wiki/Quality_function_deployment.

[2] www.qfdi.org/what_is_qfd/what_is_qfd.htm.

[3] www.webducate.net/qfd/qfd.html.

[4] www.qfdonline.com/.

[5] asq.org/learn-about-quality/qfd-quality-function-.

[6] www.qfdcapture.com/.

[7] Quality Function Deployment—www.ciri.org.nz/downloads/.

[8] www.npd-solutions.com/qfd.html.

[9] www.eng.usf.edu/~besterfi/class/capDOCS/qfd.ppt.

[10] intra.itiltd-india.com/quality/QulandRelTools\QFD_History.pdf.

Chapter 31

Quality Loss Function

Chapter Outline

31.1 WHAT IS QUALITY LOSS?

The quality loss function as defined by Taguchi is the loss imparted to the society by the product from the time the product is designed to the time it is shipped to the customer. In fact, he defined quality as the conformity around a target value with a lower standard deviation in the outputs. It is a graphical representation of a variety of non-perfect parts that can each lead to losses, and these losses can be measured in rupee value. These losses basically originate from:

- Cost to produce,
- Failure to function,
- Maintenance and repair cost,
- Loss of brand name leading to customer dissatisfaction,
- Cost of redesign and rework.

Wikipedia defines *Taguchi loss function* as the graphical depiction of loss to describe a phenomenon affecting the value of products produced by a company. It emphasizes the need for incorporating quality and reliability at the design stage, prior to production. It is particularly significant in products not involving close tolerances, but with wider than usual tolerances, or in other words, where the design specifies larger variations between the upper and lower control limits to suit the manufacturing facilities, rather than as required for the matching between two mating components in an assembly components.

The pre-1950 concept cost of quality as discussed in Chapter 8 is based on general non-statistical aspects of the quality costs to provide the basic understanding of how product costs are built up from the materials and labor costs point of view. On the other hand, the cost due to poor quality as emphasized by Taguchi's loss function (post-1950) and discussed in this chapter has a different approach and treatment involving statistical analysis and graphical depiction. It emphasizes precision more than accuracy within the production standards.

31.2   PRECISION VS. ACCURACY

The two terms precision and accuracy are more or less synonymous in general sense, but in scientific measurement especially related to quality management, they are significantly different. Accuracy of a measurement or a system is the degree of closeness of measurements of a quantity to that quantity's true value. On the other hand, precision is the degree to which repeated measurements under unchanged conditions show the same results. These are illustrated in Fig. 31.1. Interestingly, measurements can be accurate but not precise as in Fig. 31.1A, or can be precise but not accurate as in Fig. 31.1C. The ideal situation is when the measurements are both accurate and precise as in Fig. 31.1E, and as emphasized by Taguchi. If the measurements vary too much and also far away from the actual values, the measurements are neither accurate nor precise.

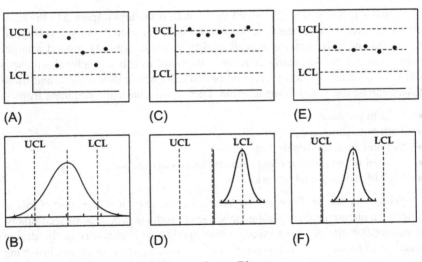

    (A) Accurate but not precise—Scatter Diagram
    (B) Accurate but not precise—Normal curve
    (C) Precise but not accurate—Scatter Diagram
    (D) Precise but not accurate—Normal curve
    (E) Accurate and precise—Scatter Diagram
    (F) Accurate and precise—Normal curve

FIG. 31.1   Illustrations of precision vs. accuracy.

31.3   HISTORY OF THE DEVELOPMENT OF THE CONCEPT OF THE LOSS FUNCTION

During his work at the Japanese Electrical telecommunications lab in the 1950s and 1960s, Dr. Genichi Taguchi observed that for developing new products, a great deal of the engineers' time and energy was spent in experimentation and testing. This, in fact, affected the final quality output to a large extent.

Taguchi argued that the performance requirements of the system are generally underspecified, that is, given too loose tolerances to allow for the process variations resulting in the quality loss function as described earlier.

He also deduced that 85% of the poor quality can be attributed to manufacturing process defects and only 15% to the operative. This led him to believe in the philosophy that instead of attributing the poor quality to the operative, the process and the product should be designed perfectly by building the quality into the design. It should start from the very beginning, that is, the product conception stage and continue during the design stage, the process development stage, and also the production stage, when care should be taken to eliminate variation.

The traditional method of calculating the losses are based on the number of parts rejected and reworked in a production facility. This method does not distinguish between two samples, both being within the specifications, but with different nominal values within those limits, as explained in the Fig. 31.1. It is thus to be understood that any item manufactured away from nominal would result in some loss to the customer, or the wider community through early wearout; difficulties in interfacing with other parts, etc., even though they may be within the acceptable limits. These losses may be minimal, as Edwards Deming put them—unknown and unknowable. But Taguchi argued that such losses would inevitably find their way back to the originating corporation, like the saying, "Little drops of water make a mighty ocean."

Taguchi, in his series of lectures emphasized the two related ideas.

- By statistical methods that are concerned with the analysis of variance. Experiments can be designed to enable identification of the important design factors responsible for degrading product performance.
- By meticulously and effectively judging the effectiveness of designs, the degree of degradation or loss as a function of the deviation of any design parameter from its target value can be assessed.

31.4   TAGUCHI PHILOSOPHY

Taguchi's philosophy basically consists of three components which are described in the following paragraphs.

- A specific loss function
- The philosophy of *off-line quality control*; and
- Innovations in the design of experiments.

31.5 QUALITY LOSS FUNCTION

Taguchi loss function or quality loss function is a graphical depiction of the losses accrued by the phenomenon that affects the value of products produced by a company, by variations within the production standards, even though the products themselves are within tolerance limits and hence, with acceptable quality standards.

Deming, the quality guru and Taguchi's contemporary, stated in his book, *Out of the Crisis*, that Taguchi Loss Function shows "a minimal loss at the nominal value, and an ever-increasing loss with departure either way from the nominal value."

Taguchi loss function is largely credited for the increased focus on continuous improvement throughout the business world. It has also been instrumental to the Six Sigma movement and the concept of variation management.

This significance can be understood from Figs. 31.2 and 31.3.

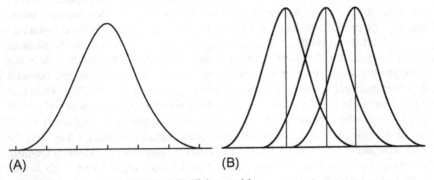

(A)   (B)

FIG. 31.2   (A) Normal precision and (B) Higher precision.

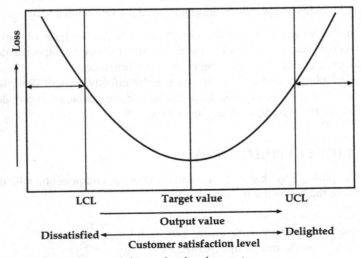

FIG. 31.3   Loss incurred by not being precise, though accurate.

31.6 OFF-LINE QUALITY CONTROL RULE FOR MANUFACTURING

According to Taguchi, more than in the manufacturing stages, it is during the design stage that the variation originates. This is called Taguchi philosophy, which later evolved into Design for Quality and Design for Six Sigma, which are described in detail in Chapter 32.

His strategy for quality engineering that can be used has three stages of design, viz:

- System design
- Parameter design
- Tolerance design

System design is the conceptual and non-statistical stage for engineering, marketing, and gathering customer knowledge and involves creativity and *innovation* with the adaptation of value engineering, as illustrated in Chapter 33.

Parameter design is concerned with the detail design phase of conventional engineering, the nominal values of the various dimensions, and design parameters need to be set. It involves analysis of how the product should perform against defined parameters and an optimal solution of cost-effective manufacturing, irrespective of the operating parameters.

Tolerance design—with a successfully completed parameter design, and an understanding of the effect that the various parameters have on performance, resources can be focused on reducing and controlling variation in the critical few dimensions.

31.7 DESIGN OF EXPERIMENTS

While R. A. Fisher is credited for developing the concept of design of experiments as early as 1954, Taguchi developed a number of innovations in designing these experiments.

31.7.1 Outer Arrays

Because the conventional sampling method is inadequate here, as there is no way of obtaining a random sample of future conditions, Taguchi proposed judgmental sampling, extending each experiment with an outer array that simulates the random environment in which the product would function.

31.7.2 Management of Interactions

The impact of the uncontrolled variations, as explained below, plays a significant role in creating noise and affects the assessment of the end result of an experiment. Hence, Taguchi emphasized on robustification and noise variables as explained below.

31.8 ROBUSTIFICATION

Robustification, as per Wiktionary, is to make designs more robust, that is, more tolerant of unexpected events. Robustification, also called robust parameter design, is the process of determining the settings of the control factors that minimize response variation from uncontrollable factors. This can be done by understanding and controlling the interaction between these variables. While these variables may easily be controlled in an experimental setting, outside of the experimental world, they are very hard and are called noise variables.

31.9 NOISE VARIABLES

Noise variables are those which are difficult or impossible to control at the design and production level, but can be controlled at the analysis or experimental stages, as in the variation in the loading pattern, or material variation.

While the controlled variables in electronic signal transmission perform perfectly, the uncontrolled variables cause audible noise, which is bothersome to the signal receivers. In a similar manner, in TQM, also such unwanted and uncontrolled variables are called "noise," even though they are not audible.

31.10 CASE STUDY

The following illustration explains how specifications of loose tolerances would create a loss to a company in the long run.

> *A company that manufactures parts that require a large amount of machining grew tired of the high costs of tooling. To avoid premature replacement of these expensive tools, the manager suggested that operators set the machine to run at the high-end of the specification limits. As the tool would wear down, the products would end up measuring on the low-end of the specification limits. So, the machine would start by producing parts on the high-end and after a period of time, the machine would produce parts that fell just inside of the specs. The variation of parts produced on this machine was much greater than it should be, since the strategy was to use the entire spec width allowed rather than produce the highest quality part possible. Products may fall within spec, but will not produce close to the nominal. Several of these "good parts" may not assemble well, may require recall, or may come back under warranty. The Taguchi loss would be very high.*

This case study makes us consider if the savings of tool life was worth the cost of poor product?

31.11 CONCLUSION

For those who have been thinking that by specifying loose tolerances at the design stage, the rejection losses of the manufactured goods would be lower, Taguchi philosophy and the explanation is an eye-opener and made them realize that in the long run, the total losses, especially the losses due to the customer dissatisfaction, would be higher. Hence, precision plays a significant role in reducing the total losses, as explained in this chapter.

On the Lighter Side

The words "robust" and "lean" are more or less antonyms in English. Nevertheless in TQM they are more or less synonymous in representing the concept of ensuring that no single factor or variable is neglected at the conceptual and design stages at the same time eliminating all wasteful elements in the operations or functions at post-design stages.

FURTHER READING

The following websites can be referred to for further information and additionally to some text books indicated in the bibliography.

[1] en.wikipedia.org/wiki/Taguchi_methods.
[2] http://www.businessdictionary.com/definition/Taguchi-loss-function.html#ixzz37PnlNpIf.
[3] http://www.scribd.com/doc/73104345/Taguchi-Loss-Function.
[4] www.terninko.com/loss.htm.
[5] It provides an animated graph showing how the loss reduces when the kurtosis increases—http://elsmar.com/Taguchi.html.
[6] www.annauniversity.info/NetLearn/TQM_TAGUCHI.ppt.
[7] https://www.kellogg.northwestern.edu/faculty/dranove/htm/dranove/coursepages/Mgmt%20469/noisy-variables.pdf.

Chapter 32

Design for Quality

Chapter Outline

Quality must be designed into the product not inspected into it.

—Kenneth Crow

32.1 DESIGN FOR QUALITY

Traditionally, quality conformance in the production process can be said to be Quality after design (QaD) viz, applying quality control procedures during the production process only to ensure that the product conforms to the specifications given by the designers.

In his book, *Quality by Design*, Juran propounded his theory of QbD contrary to the traditional QaD, and emphasized that that quality could be planned, and that most quality crises and problems relate to the way in which quality was planned. QbD provides guidance to facilitate design of products and processes that maximize the product's efficacy and safety profile, while enhancing product manufacturability and control. It is defined by Business Dictionary as a systematic process to build quality into a product from the inception to final output.

Juran's emphasis on the above concept, as well as Taguchi's experiments at Toyota as explained in Chapter 31, made several automobile manufacturers rethink the way the design process for a product is conceived, by adapting the principle of Quality by Design (QbD), by which detailed planning and checklists shall be prepared, to highlight all the factors that affect the production.

These factors must be considered and analyzed before embarking upon the design procedure. This detailed planning builds quality into the design process as highlighted by Juran.

This term, QbD, is now replaced by a more popular term, *Design for quality (DFQ)*, which is complementary to DFSS (Design for Six Sigma). It may be noted that while Six Sigma emphasizes the improvement of the process to achieve higher levels of quality, DFSS emphasizes meticulous planning in the design stage itself. Thus, while the former adapts the Define, Measure, Analyze, Improve, and Control (DMAIC) methodology, the latter adapts the Define, Measure, Analyze, Design, and Verify (DMADV) methodology, which is described more in later paragraphs.

It may also be noted that several books, as well as several six sigma practitioners relate the DMAIC methodology to QbD as they do equally to the six sigma process. The following paragraphs describe DMAIC as the methodology for achieving six sigma levels.

While DMAIC is described in detail in Chapter 24, it is cited here again to provide a contrast with DMADV or DFSS.

- *Define* the project goals and customer (internal and external) requirements.
- *Measure* the process to determine current performance.
- *Analyze* and determine the root cause(s) of the defects.
- *Improve* the process by eliminating defect root causes.
- *Control* future process performance.

This can also be represented below.

- Define the problem/defects
- Measure the current performance level
- Analyze to determine the root causes of the problem/defects
- Improve by identifying and implementing solutions that eliminate root causes
- Control by monitoring the performance of the improved process

The process of DFSS can be understood better by some of its explanatory definitions.

- A methodology for designing new products and/or processes.
- A methodology for redesigning existing products and/or processes.
- A way to implement the six sigma methodology as early in the product or service life cycle as possible.
- A way to exceed customer expectations.
- A way to gain market share.
- A strategy toward extraordinary return on investments.

It may be noted that this procedure is broadly similar to any other method improvement procedure followed by industrial engineers, which is explained more in Section 22.9 of Chapter 22 on Kaizen.

32.2 DESIGN FOR SIX SIGMA

As explained in the previous paragraph, the six sigma practitioners have gone a step further by adapting DFSS to design a new product to achieve not only the high quality standards put forth by Juran, Taguchi et al., but to achieve the six sigma quality levels. The full impact the six sigma concept had on the modern manufacturing industries is explained in Chapter 24. This approach is similar to DMAIC methodology, which is explained in an earlier chapter, and consists of the steps detailed in the next paragraph.

Earlier, the design function and quality control function had separate identities and functions. But today it is DFQ (Design for Quality) and DFSS that has become part and parcel of the quality function, at least in coordinating with the design and development departments to ensure for application of DFQ and DFSS concepts for creating new products or process designs.

DFSS uses a process management and performance improvement strategy methodologies DMAIC or DMADV leading to Lean Six Sigma. As discussed in the previous paragraph, DMAIC is adapted to improve the existing process to achieve six sigma, while DMADV is adapted for designing new products and new processes that yield six sigma levels. This process of designing new products for six sigma is generally termed as Design for Quality (DFQ).

Design for Six Sigma is about leaping past incremental improvements by utilizing a rigorous design method to create processes that surpass customer expectations by delivering value and excellence. Design For Six Sigma (DFSS) is a structured approach that designs Six Sigma performance into processes from the start. When applying DFSS, the DMADV methodology is utilized.

website of Acuity Institute (http://www.acuityinstitute.com)

DFSS is not a methodology. It is an approach and attitude towards delivering new products and services with a high performance as measured by customer critical to quality metrics. Just as the Six Sigma approach has the DMAIC methodology (Define, Measure, Analyze, Improve, Control) by which processes can be improved, DFSS also has methodologies such as DMADV by which new products and services can be designed and implemented.

website of design six sigma (http://www.designsixsigma.com)

32.3 ACRONYMS FOR METHODOLOGIES AKIN TO DMAIC

The other acronyms used in DFSS are:

- DMAIC (Define, Measure, Analyze, Improve, and Control)
- DMADV (Define, Measure, Analyze, Design, and Verify)
- IDOV (Identify, Design, Optimize, and Validate)
- DCCDI (Define, Customer Concept, Design, and Implement)
- DMEDI (Define, Measure, Explore, Develop, and Implement)

While six sigma (DMAIC) requires a process to be in place and functioning, DFSS is applied during the design process with an objective of determining the needs of customers and the business, and driving those needs into the product solution duly created. In other words, DFFS is concerned with using tools, training, measurements, and verification so that products and processes are designed at the outset to meet six sigma requirements.

32.4   DMADV

DMADV is a DFQ methodology (Fig. 32.1) for the development of a new service, product, or process, as opposed to improving a previously existing one. The letters indicate:

- Define the problem/defects
- Measure the current performance level
- Analyze to determine the root causes of the problem/defects
- Design meticulously for the selected alternative
- Verify to validate that the design is acceptable to all stakeholders

This approach—Define, Measure, Analyze, Design, and Verify—is especially useful when implementing new strategies and initiatives because of its basis in data, early identification of success, and thorough analysis. Thus, it includes five phases as explained below and illustrated in Fig. *32.1*.

32.4.1   Define Phase

This phase identifies the purpose of the project, process or service, and then sets realistic and measurable goals, as seen from the perspectives of the organization and the stakeholders. The goals shall be consistent with the customer demands as well as the company's capabilities, as determined by the quality function deployment (QFD). This also involves preparation of DFQ charter and assessment of possible risks.

32.4.2   Measure Phase

Here the factors such as the voice of the customer, which is critical to quality (CtQ's), process capability risks, etc., are measured and recorded, as well as preparing the design scorecard. Refer to Chapter 30 on QFD.

32.4.3   Analyze Phase

This phase develops design alternatives by:

- identifying the optimal combination of requirements to achieve value within constraints,
- developing conceptual designs,

- evaluating, then selecting the best components,
- developing the best possible design
- make an estimate of the total life cycle cost of the design.

One of the best traditional management techniques useful in this stage is Value Engineering as explained in Chapter 33. While value engineering is basically applied to the existing designs, it is also used to validate the new product/process designs made.

32.4.4 Design Phase

This phase is generally an integral part of the Analyze phase, when the design process for a new process or product is carried out in accordance with the analysis phase and a corrective step is taken to the existing one to meet the target specification.

32.4.5 Verify Phase

Verify, by simulation or otherwise, the performance of the developed design and validate its ability to meet the target needs, before handing it over to the process department. This phase also includes:

- Setting up pilot runs
- Training and implementation
- Transition and control

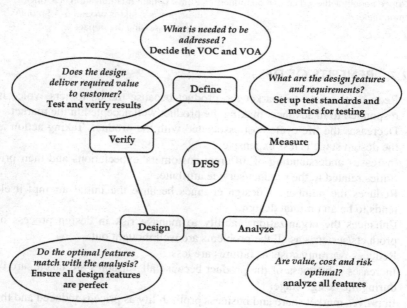

FIG. 32.1 DMADV methodology.

32.5 SCOPE OF DFSS

The scope of DFSS is as follows:

- Designing new products and or processes,
- Redesigning the existing products or processes,
- Developing products that meet the customers' expectations/demands,
- Predicting and improving quality before building prototypes.

We can say that DMAIC : DMADV :: Method Improvement study : Value Analysis.

32.6 SIX SIGMA VERSUS DFSS

We can summarize the differences between traditional six sigma and DFSS as follows:

Six Sigma	Design for Six Sigma
DMIAC (Define-Measure-Analyze-Improve-Control)	DMADV (Define-Measure-Analyze-Design-Verify)
Looks at the existing processes and analyzes to fix problems	Focuses on the new design of the products and processes
More reactive (removing defects)	More proactive (correcting before the event)
Rupee benefit achieved can be quantified immediately	Rupee benefit quantification is a long-time process, taking around 6–12 months before assessing the impact

32.7 BENEFITS OF DFSS

1. Reduces the time to market the product because the customers' voice is responded to in advance, making the product well accepted in the market.
2. Decreases the life cycle cost associated with the product. Taking action at the design stage is less expensive.
3. Increases understanding of different customers' expectations and their priorities related to the product/service attribute.
4. Reduces the number of design changes because the initial attempt itself tends to be an optimal design.
5. Enhances the organizations' ability to manage risk in design process of products/services as all the problems are well thought out.
6. Reduces warranty costs as failures are less.
7. Increases robustness of the product because all variables are taken care of during the design stage.
8. Improves market share and business profitability as price is reduced and the products are accepted in the market.

32.8   CONCLUSION

Whereas several books relate DFSS procedure to DMIAC, some websites like Six Sigma and a few bloggers on the Minitab website say that the DFSS should not be related to DMAIC, but only to DMADV, indicating that the steps of *Design* and *Verify* are more essential to DFSS than merely *Improve* and *Control*. These versions indicate that even today, there is a controversy whether DMAIC or DMADV should be followed for DFSS. However, this author feels the distinction lies not in relation to DFSS, but in what context the procedure is applied. DMAIC focuses on improving an existing product or process, while DMADV focuses on creating a new product or process. The term D in other acronyms may mean Design for new products or redesign for existing products.

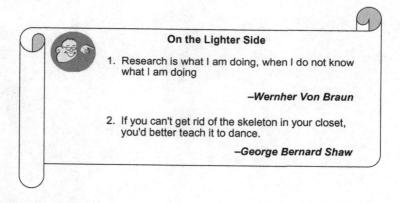

On the Lighter Side

1. Research is what I am doing, when I do not know what I am doing

   –Wernher Von Braun

2. If you can't get rid of the skeleton in your closet, you'd better teach it to dance.

   –George Bernard Shaw

FURTHER READING

[1] http://en.wikipedia.org/wiki/Design_for_Six_Sigma.
[2] www.isixsigma.com/new-to-six-sigma/design-for-six-sigma-DFSS/design-six-sigma-DFSS-versus-dmaic/.
[3] www.designsixsigma.com/whatis_DFSS.htm.
[4] www.npd-solutions.com/DFSS.html.
[5] www.dtic.mil/ndia/2003test/kiemele.pdf.
[6] http://www.businessdictionary.com/%E2%80%A6/quality-of-design.html.
[7] http://www.acuityinstitute.com.

Chapter 33

Value Engineering

Chapter Outline

33.1 WHAT IS VALUE ENGINEERING?

Value engineering is a design engineering technique involving critical examination and analysis of the design of a component with reference to its functional value. It is a systematic and creative method using proven methods akin to the modern design for six sigma (DFSS) tool of define-measure-analyze-improve-control, to obtain the same or better performance at a lower cost, so as to improve the *value* of goods, or products and services by critical examination of its function. Value engineering seeks to optimize value for the money in projects and emphasizes analyzing the functional values of all design features of a component, making it a special among all DFSS tools.

It has been used in almost any kind of application. It helps people creatively generate alternatives to secure essential functions at the greatest worth, referred to as value as opposed to costs. It is also known as Value Analysis, Value Management, Value Planning, and a host of other names.

33.2   DEFINITIONS OF VALUE ENGINEERING

Some definitions of value engineering, which also give an insight into the technique, are:

A discipline action system, attuned to one specific need: accomplishing the functions that the customer needs and wants at the lowest cost.

–Miles (1972)

A proven management technique using a systematized approach to seek out the best functional balance between the cost, reliability, and performance of a product or project.

–Zimmerman (1982)

Systematic analysis that identifies and selects the best value-added alternatives for designs, materials, processes, and systems. It proceeds by repeatedly asking "can the cost of this item or step be reduced or eliminated, without diminishing the effectiveness, required quality, or customer satisfaction?"

–Business Dictionary

Value Analysis/Value Engineering is an intensive, interdisciplinary problem-solving activity that focuses on improving the value of the functions that are required to accomplish the goal, or objective of any product, process, service, or organization.
–Value Engineering: The Forgotten Lean Technique by James R. Wixson

Value engineering is a systematic and organized approach to provide the necessary functions in a project at the lowest cost. Value engineering promotes the substitution of materials and methods with less expensive alternatives, without sacrificing functionality. It is focused solely on the functions of various components and materials, rather than their physical attributes.

–Investopedia

Value engineering can be defined as an organized effort directed at analyzing features, systems, equipment, and material selection for the purpose of achieving the essential functions at the lowest life cycle cost consistent with the required performance, quality, reliability, and safety.

–US General Services Administration

The value methodology (VM) is a systematic and structured approach for improving projects, products, and VM, which is also known as value engineering, is used to analyze and improve manufacturing products and processes, design and construction projects, and business and administrative processes.

–Save

A professionally applied, function- oriented, systematic team approach used to analyze and improve value in a product, facility design, system, or service.

–SJVE

33.3   HISTORY OF VALUE ENGINEERING

Value engineering began at the General Electric Co. when Lawrence Miles conceived Value Analysis in 1945, based on the application of function analysis to the component parts of a production in light of shortages of skilled labor, raw materials, and component parts during World War II. This analysis produced substitutes which were often found to reduce costs and provided equal or better performance. By 1952, value engineering began its growth throughout industry.

The Public Law 104-106 of US states, "Each executive agency shall establish and maintain cost-effective value engineering procedures and processes."

The Federal-Aid Highway Act of 1970, too, has recognized the effectiveness of value analysis, made the first Federal Highway reference to VE, requiring that "in such cases that the Secretary determines advisable plans, specifications, and estimates for proposed projects on any Federal-Aid system shall be accompanied by a value engineering or other cost reduction analysis."

Some of the acronyms related to value engineering:

- VA/VE, value analysis and value engineering
- VM, value management
- VAVE, value-added value engineering
- FA, functional analysis
- FAST, functional analysis system technique
- SAVE, American Society of Value Engineering
- SJVE, Society of Japanese Value Engineering
- INVEST, Indian Value Engineering Society
- INVAVE, Indian Journal of Value Analysis/Value Engineering.
- DARSIRI, data (collect), analyze, record ideas, speculate, innovate, review, and implement

33.4   WHAT IS VALUE?

Here the term value is distinguished from price or cost because value is more of an abstract concept referring to the cost benefit aspect. It is the ratio between a function for customer satisfaction and the cost of that function. Functional worth is the lowest cost to provide a given function.

Value is a perception, hence, every customer will have their own perceptions on how they define value. However, we can relate value to quality, performance, style, and design in comparison to the product cost.

In fact, value is the performance of a product in relation to the function or what the product or service is supposed to do in relation to the cost and the expenditure to create it.

Value = Performance/Cost, or $V = \dfrac{P}{C}$

If the numerator is increased without increasing the denominator, value increases. In other words, a product can be engineered for improved value by either increasing the quality, reliability, availability, maintainability, serviceability, etc., for the same cost or by reducing cost for the same degree of the above factors of quality, reliability, etc. In short, value analysis results in more functions at the same cost or same function performed at lesser cost. The manner in which the thinking is applied gives the concept its name value analysis, and planning the analysis in the design function gives it the name value engineering. We can say that value engineering is the functional aspect of value analysis, which refers to the analytical overall concept. This chapter, hence, uses the term value analysis as a synonym for the term value engineering.

33.5   VALUE ANALYSIS

As stated earlier, value analysis is a methodology to increase the value of an object. Value analysis is a planned, scientific approach to cost reduction which reviews the material composition of a product and process design, so that modifications and improvements can be made which do not reduce the value of the product to the customer or to the user. The object to be analyzed could be an existing or a new product or process, and it is usually accomplished by a team following a work plan.

33.6   OBJECTIVES OF VALUE ENGINEERING

1. To reduce piece cost and total cost as illustrated by the case study below.
2. To improve operational performance.
3. To improve product quality.
4. To reduce the manufacturing costs.
5. To improve customer-supplier relations.
6. Cost avoidance on future programs.
7. To reduce in product variations.

33.7   TYPICAL BENEFITS OF VALUE ENGINEERING PROJECTS

The benefits of value analysis, which fulfill the above objectives, can be summarized in the following section:

- Value analysis aims to simplify products and process, thereby increasing efficiency in managing projects, resolving problems, encouraging innovation, and improving communication across the organization.
- Value analysis enables people to contribute in the value addition process by continuous focus on product design and services.
- Value analysis provides a structure through cost saving initiatives, risk reduction, and continuous improvement.

33.8   FUNCTIONS OF A PRODUCT AS THE CUSTOMER WANTS IT

Value Analysis is based on a study of functions of a product or service. It involves the identification of functions from the knowledge of the customer needs. The first approach to the identification of functions should be focused on basic functions. These functions are those things for which the customers believe they are paying. There are usually only one or two basic functions per product or service.

All the functions can be grouped below, as per their levels of importance,

- The basic function, which is the very purpose of the product or service.
- The secondary functions are those not directly accomplishing the primary purpose, but support it from a specific design approach. These can also be subcategorized as use functions or aesthetic functions.
- Use functions are those which answer the question of how the basic function is achieved. For example, if the primary purpose of a bottle is to contain a liquid, the secondary purpose can be strength to support the contents, even when dropped, or transparency so that the contents can be identified without opening the bottle.
- Aesthetic functions, whose purpose is only to add beauty or esteem value to the product are often associated with feelings. In the above example, the attractive color or shape provides the aesthetic function.

It is generally found that the primary functions are achieved by 20% of the total cost, whereas the secondary functions account for 80% of the cost. This is the crux of value analysis.

Once you identify the functions, they must be written down in two words, a verb and a noun, as further explained in Section 33.10.3.

33.9   FUNCTIONAL VALUE OF A PRODUCT VERSUS OTHER VALUES

Functional value or use value of a product is the purpose the product fulfills and is an attribute that provides the customer with functional utility, which can be distinguished from other values of a product as illustrated in the following section.

- *Cost value*: It is the cost of manufacturing and selling an item.
- *Exchange value*: It is the price a customer is prepared to pay for the product, or service.
- *Esteem value*: It is the prestige a customer attaches to the product.
- *Place value*: Same items may have different values at different places. For example, a glass of water in a desert.
- *Time value*: An item may have a high value at certain point of time. Once the time is passed, it may lose its value. For example, blood transfusion to a patient during an operation.

33.10   METHODOLOGY OF VALUE ENGINEERING

The creative mind required for value analysis makes it comparable to the method study that analyzes and improves the manufacturing operations, while value engineering analyzes and improves the design factors. We can say:

Method Study:Production::Value Engineering:Design

Hence, the methodology for value analysis is similar to the SREDDIM of Method study, as explained in Chapter 22 on Kaizen. Nevertheless, in case of value analysis, it is split into 8 phases, the terminology for each phase as follows:

The 8 Phases of Value Analysis

1. General Phase
2. Information Phase
3. Functional Phase
4. Investigation Phase
5. Creative Phase
6. Evaluation Phase
7. Recommendation Phase
8. Follow-Up Phase

33.10.1   General Phase

After identifying the existing product or the process to be analyzed, its general description is given, indicating the functions and design features, etc., of the product, as well as its components. List the basic functions (the things for which the customer is paying), as identified by the function phase.

33.10.2   Information Phase

Additional data such as the operational sequences or the time standards are recorded. These data would assist in analysis and in the comparison of the proposed process to the existing process.

33.10.3   Function Phase

Identify and list all the functions of the product or process, for which the customer is paying. Here it is necessary to indicate each function in only two words, a noun and a verb. This enables conciseness. By trying to describe a function in a sentence, we may unwittingly combine two or more functions, which would cause confusion in our analysis. Table 33.2 of the case study provides an illustration to this concept.

Again, while identifying a function, specify it so as not to limit the ways in which it can be performed. For example, don't say, "screw nameplate," but say, "attach a nameplate," because the nameplate can be attached, not only by screwing, but also by soldiering, riveting, or gluing, etc. The later specification would help us in thinking of alternative solutions for this function.

Once all the functions are listed, isolate the basic function followed by all the secondary functions. This will help in our analysis if each of the secondary functions is really necessary, or can be done away with. Given below is the criteria to distinguish between the basic and secondary functions:

- Basic function is the primary reason for an item or system. It is the performance feature that must be attained if it has to perform its purpose.
- A secondary function is the features of an item which supports the basic function, and even without that function, the item can perform its functions. For example, the primary function of paint is to protect the surface, while the secondary function is to give a good appearance.

Guidelines for defining the functions:

- The function shall be defined only by two words, a verb and a noun.
- The noun shall be measurable and/or countable.
- The noun shall as much as possible signify the design-based constraint.
- The verb shall be active and affect the noun directly.
- The function shall be verifiable.

33.10.4   Investigation and Creative Phases

While all required data is collected and recorded in the investigation phase, it is the creative phase that is the heart of the methodology. Because these two phases overlap each other, they are discussed together in this paragraph. All the points discussed in Sections 22.8–22.10 including the brainstorming, in the chapter on Kaizen with respect to creativity, are applicable here.

The objective of this phase is to find a better way to do the main function by asking the following questions for each of the identified functions and determining the relative importance of each function, preferably by asking a representative sample of customers.

- Does it contribute value? (Is there something that does not contribute value?)
- Is the cost in proportion to the function realized?
- Does it need all its parts, elements, procedures?
- Is there something else to do the same function?
- Is there a standard part that can do this function?

33.10.5   Evaluation Phase

- Each idea generated should be analyzed and developed in a manner to be more logical and practical, making it function better.
- Identify barriers like mindset concepts opposing the idea and discuss whether the barriers hold strongly against the ideas. Isolate and eliminate them, but after recording them for future reference.
- Choose two to four ideas among them and make a comparative study regarding the cost, as well as performance.

33.10.6   Recommendation and Follow-Up Phases

After all, any analytical study has to be approved by the top management. It is hence imperative that the value analyst team prepares a report detailing the several factors considered as detailed earlier emphasizing the net cost savings, as well as the functional improvements achieved and submit the same to the top management as their recommendation. Once the recommendation is accepted, the operatives and other related personnel will have to be trained and regular follow-up with the implementation has to be maintained. This phase is similar to the steps *Install* and *Maintain* of method study.

33.10.7   DARSIRI Methodology for Value Analysis

Some books cite a 7-step DARSIRI methodology which is similar to the above 8 phases and is as follows:

1. Data (collect),
2. Analyze,
3. Record ideas,
4. Speculate,
5. Innovate,
6. Review, and
7. Implement

33.11   FUNCTION ANALYSIS SYSTEM TECHNIQUE

Function Analysis System Technique (FAST) is a graphical representation of the functions identified during the value analysis program. It builds upon value analysis by linking the simply expressed, verb-noun functions to describe complex systems in a logical sequence, visualize the need for and the role of each major component and prioritizing them.

33.12   CASE STUDY

The JLO division of the Surat Unit of the Ralli group manufactures 25 cc petrol engines used for agricultural sprayers and cycle rickshaws. A major component of this engine is the *clutch base plate* of the transmission assembly whose main function is to absorb the thrust exerted by the clutch plate during the power transmission. This being the costliest component of the engine, a value engineering study was conducted as detailed in the following case study. This will provide a clear understanding of the methodology and benefits of value engineering.

Component: *Clutch Base Plate in the Transmission Assembly of a Petrol Engine*

(i) *General phase*: The general information about the function of the Clutch Base Plate is as follows.

The 25 cc JLO petrol engine, used in agricultural sprayers and cycle rickshaws, has a clutch base plate as a part of the power transmission system. This component is under study, being one of the costliest components in the transmission assembly (Fig. **33.1**), as described briefly in the following section.

The drive from the drive shaft is through the gear (which rotates freely on the clutch base) and the friction plates on one side and the clutch plate, the clutch base, and the clutch shaft on the other side. In the normal running, the friction plates hold the gears tightly against the clutch plate, (and hence on to the clutch base), transmitting motion by friction under pressure. When the clutch pin is released, the friction plates move to the right against the spring pressure, thus releasing the pressure between the gear and the clutch plate, thereby enabling the gear to rotate freely in the clutch base.

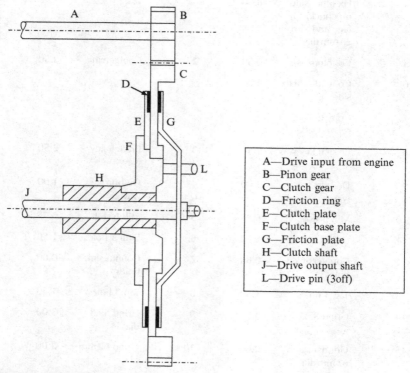

A—Drive input from engine
B—Pinon gear
C—Clutch gear
D—Friction ring
E—Clutch plate
F—Clutch base plate
G—Friction plate
H—Clutch shaft
J—Drive output shaft
L—Drive pin (3off)

FIG. 33.1   Clutch transmission assembly.

(ii) *Information phase*: The clutch base is machined from 80 mm dia EN8 steel bar, the sequence of operations (Table 33.1) being as under:

TABLE 33.1 Operation Sequence

Machined on Bar stock			Machined on Forging		
Op. No.	Operation	Std. Time (min)	Op. No.	Operation	Std. Time (min)
1.	Cut blanks 28 mm thick	2.60	–		
2.	Load on lathe, face one side, drill 17 dia and counter bore	8.00	1	Load on lathe, face one side, drill, and ream 17 dia and counter bore	8.20
3	Bore 17 mm dia	5.80	–		
4	Fix other side on chuck, face and turn 23 mm dia	8.00	–		
5	Countersink	0.50	2	Countersink	0.50
6	Copy turn first side	4.20	–		
7	Copy turn second side	3.93	–		
8	Broach key way	2.50	3	Broach key way	2.50
9	De-burr key way	1.00	4	De-burr key way	1.00
10	Mill 6 slots	6.75	5	Mill 6 slots	6.75
11	Drill 3 holes	1.50	6	Drill 3 holes	1.50
12	Countersink 3 holes	0.80	7	Countersink 3 holes	0.80
13	Tap 3 holes	1.50	8	Tap 3 holes	1.50
14	Grind 55.3 dia	3.00	9	Grind 55.3 dia	3.00
15	Grind 62 mm dia	2.00	10	Grind 62 mm dia	2.00
16	Face the boss	1.00	11	Face the boss	1.00

TABLE 33.1 Operation Sequence—Cont'd

Machined on Bar stock			Machined on Forging		
Op. No.	Operation	Std. Time (min)	Op. No.	Operation	Std. Time (min)
17	De-burr, coat antirust and store	3.00	12	De-burr, coat antirust and store	3.00
	Total time	56.08			31.75

(iii) *Functional phase*: The functions of the clutch base plate and their functional levels are indicated in Table 33.2.

It may be noted that the primary function is specified as "Absorb Thrust," while others like facilitate Drive (from pinion to drive shaft) are categorized as secondary. This conforms to the fact that by fixing the clutch plate in the drive shaft and making the gear rotate freely on the drive shaft, it may be possible to transmit the drive through the clutching and declutching action. But the spring thrust is so high that the clutch plate would fail, and it is the clutch base plate that will absorb the thrust and thereby, enable the friction and the clutch plates to perform the clutching and declutching action, and transmit the power without deforming the plates.

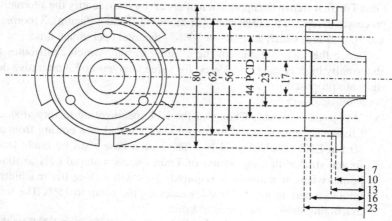

FIG. 33.2   Clutch base plate.

(iv) *Investigation phase*: If we analyze the costs involved in sustaining each of the 5 functions, we notice that the total cost of maintaining the secondary functions amount to as much as 80% of the total cost. Especially, the secondary function of "facilitate drive" can be rated as the costliest, due to the cost of providing the flange and the slots. This should give an idea as to which design features should be questioned more thoroughly to get maximum cost reduction by changing only the minor and secondary design features.

TABLE 33.2 Function Levels

	Functions		Functional Level		
Sl. No	Verb	Noun	Primary	Secondary	Remarks
1	Facilitate	Drive	–	Yes	To pinion
2	Resist	Bending moment	–	Yes	
3	Support	Weight	–	Yes	Of the reduction assembly
4	Absorb	Thrust	Yes	–	From the clutch plate
5	Provide	Location	–	Yes	For drive pin

(v) *Creative and evaluation phases*: While the finished piece of the clutch base plate is 300 gm, the raw material used is 80 mm blank weighing 1200 gm, thus 75% of the material is being lost as scrap.

Table 33.1 pinpoints the operations that take unusually high machining time. The first step in the application of creativity is to identify the alternative processes to replace the high cost operations viz, 2 and 4 (facing), 3 (boring), and 10 (milling) which account for 90% of the total operational time.

Hence the flange feature of the existing design, which contributes to abnormally high cost, is considered for value analysis and 3 alterative designs are proposed.

(vi) *Recommendation phase*:

  (a) *Design change no.1*: Changing the RM specification to forged steel. Change the raw material to forged steel, instead of a cutting from an 80 mm EN8 (BS970) steel bar. The clutch base shall be made from forged steel with a maximum of 1 mm excess material only at those points where machining is required. This will reduce the machining time to a bare minimum besides reducing the scrap to 15%. The total machining time is estimated at 8 min.

  • The quoted weight of the forged blank of specification EN 8 steel is 500 gms and the price of this forged piece is Rs. 7.25 per piece (Rs. 6.25 for IS 226-MS).

  • Table 33.3 below compares the costs between fully machined and forged components, yielding a direct saving of Rs 4.55 per piece by replacing bar stock with forged pieces.

TABLE 33.3 Cost Comparison in Changing to Forged Steel

Sl. No	Cost Element	Fully Machined Component		Forged Component	
		Details	Cost in Rs.	Details	Cost in Rs.
1	Raw material cost	1.2 kg of En8 bar at Rs 5.75/kg	6.90	Vide quotation	7.25
2	Hacksaw cutting	Subcontracting	2.00	Nil	–
3	Machining	56.08 m	7.10	31.75 m	4.20
	Total		16.00		11.45

Note: The prices and costs were as prevalent in the 1970s when this study was undertaken. Nevertheless, the comparison between the existing and the proposed processes is valid today. Fig. 33.2 indicates the dimensions of the clutch base plate after machining. It is to be noted that there are no dimensional variations of the machined piece, whether from bar stock or from forged stock.

(b) *Supplementary design change no.* 2: Integration of the flange portion of the base plate with the clutch plate itself by screwing.

By looking at the component and its assembly with the clutch plate, it can be deduced that the flange portion and its milling is designed purely to hold the clutch plate and to receive positive power transmission. The axial movement of the plate, with respect to the base is not required. In fact, there were initial complaints that the plate tends to slip into the gap between the gear and base when the clutch is released.

Analyzing the available bar sizes and the ideal overlapping, a bar size of 75 mm dia is considered optimal with a potential saving of Rs. 30,000 per annum.

Salient features of this design change are:

1. The bar stock size is reduced from 80 to 75 mm, resulting in a further saving of Rs. 3.45 ore piece in material cost and Rs. 1.50 per piece in machining cost, ie, Rs. 4.95 per piece. The total saving by implementing both the recommendations is hence Rs. 4.55 + Rs. 4.95 or Rs. 9.50 per piece. At 2000 pieces per month, the monthly savings potential is Rs. 19,000.

2. The new design for the clutch plate has circular holes instead of complicated serrated slots, possibly reducing the procurement cost.

3. The 3 pins presently screwed on to the clutch can themselves be used to fix to the clutch pate. By this, the costly milling operation can be eliminated. The proposed design has 6 holes for better strength and power transmission.

4. The provision of nuts may involve slight modification on the crankcase, without effect on any functional design of the latter.

(c) *Supplementary design change no.* 3: *reducing the bar size*

    The bar procured is of 80 mm outside diameter (OD). It is provided with a 19 mm bore to match with the clutch shaft and is provided with two steps at 55 mm dia and 62 mm dia, the former to match with the clutch plate, and the latter to match with the gear. Discussions with the R&D revealed that the additional step has no functional value and that the 62 mm step can be reduced to 58 mm, thereby reducing the OD to 75 mm from 80 mm. Table 33.4 indicates the cost reduction

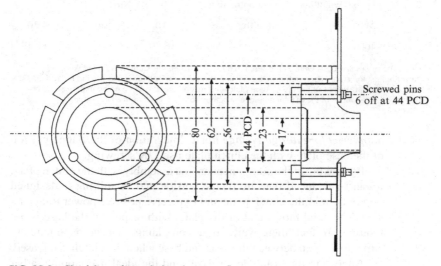

FIG. 33.3   Clutch base plate—design change no. 2.

TABLE 33.4 Cost Reduction Achieved by Reducing the Bar Sizes

Sl. No	Radial Overlap in mm	Bar dia Required in mm	Wt. in kg/m	Reduction in Weight From 80 mm bar in kg/m	%age Reduction in Weight, Thereby in Raw Material Procurement Price
1	9	80	39.5	–	–
2	8	78	37.5	2.0	5.0
3	7	76	35.6	3.9	10.0
4	6.5	75	34.7	4.8	12.2
5	6	74	33.8	5.7	14.5
6	5	72	32.0	7.5	19.0
7	4	70	30.2	9.3	23.6

achieved by reducing the bar sizes, ignoring the reduced machining time (Fig. 33.3).

(vii) *Follow-up phase*:

- Recommendation no. 1, viz to change to forging component, has been accepted and implemented. Net saving is Rs. 4.55 per piece or Rs. 9000 per month.
- Recommendation no. 2 viz, Integration of the flange portion of the base plate with the clutch plate itself by screwing, was accepted in principle, but was deferred due to involvement of a change in the die of the aluminum casting of the crankcase. Correspondence was initiated with the supplier, nevertheless without much positive result in this direction. Net saving is s 4.95 per piece or Rs 25,000 per month.
- Recommendation no. 3, viz, reduction in the OD of the bar stock to an optimal level was withheld due to the design imposition from the principals. Expected savings: Rs 19,000 per month.

It may be noted that though these savings figure appear low, they were substantial during the 1970s.

33.13 CONCLUSION

It can be seen from the case study that value engineering which has been practiced by industrial engineers even when the concept of six sigma did not exist, is still prevalent, and has become a significant tool of DFSS.

The savings illustrated in the case study may appear to be too small to call for a citation. But in 1974, when this study was done, the petrol price was Rs.1.50 per liter and rice was nearly a rupee per kg, compared to 2015 prices of Rs. 60.00 and Rs.40.00, respectively.

On the Lighter Side

India believes very much in education. A professor makes more money in a year than an average cricketer does in a whole week.

–Indianised version of Evan Esar's Quote

FURTHER READING

[1] Otto KN, Wood KL. Product Design. Upper Saddle River, NJ: Pearson Education; 2004.
[2] Maynard HB, editor. Industrial Engineering Handbook. New York: McGraw Hill; 1963.
[3] ILO. Introduction to Work Study. Geneva: ILO; 1979.
[4] Ireson WG, Grant EL. Handbook of Industrial Engineering and Management. Upper Saddle River, NJ: Prentice Hall; 1971.
[5] Hurst Ronald. Industrial Management Methods. UK: Hutchinson; 1970.

[6] Yoga M, Ananthkrishnand B. Project Studies in Industrial Engineering. In: Vol. I. New Delhi: NPC; 1974.
[7] Walker WF. Engineering Productivity. Bombay: BI Publications; 1965.
[8] Eilon Sameul. Elements of Production Planning and Control. McMillan & Maruzan; 1962.
[9] Kiran, D.R. (1977). Value Engineering, A Case Study—Industrial Engineering Journal.
[10] Kiran, D.R. (2015). Value Engineering, NIQR Journal.

WEBSITES

[11] http://www.businessdictionary.com/definition/value-analysis.html.
[12] www.productivity.in/ValueAnalysis.
[13] en.wikipedia.org/wiki/Value_engineering.
[14] www.gsa.gov/portal/category/21589.
[15] www.investopedia.com/terms/v/value-engineering.
[16] www.dfma.com/resources/vave.htm.

Chapter 34

ISO 9000 Quality Systems

Chapter Outline

34.1   NEED FOR QUALITY MANAGEMENT SYSTEMS

Quality management has become a way of life in the manufacturing sector, in fact, in every sector, whether service, logistic, road building, or any other sector, and its importance is widely understood. This calls for a quality management system for unified standards for evaluating the processes. Without a system in place to establish procedure, monitor progress, and evaluate performance, it is nearly impossible to consistently deliver a quality product to your customer. By utilizing quality management systems, problems are identified and corrected as they arise, allowing you to be proactive and minimize the likelihood of reoccurrence, which benefits all parties involved.

34.2   INTERNATIONAL ORGANIZATION FOR STANDARDIZATION

The International Organization for Standardization (ISO), its central office located in Geneva, is the world's largest developer and publisher of international standards. It forms a platform for several countries to establish quality systems. It has a network of National Standards Institutes in 163 participating countries, including the Bureau of Indian Standards (BIS) of India. ISO standards are constantly advancing to meet the needs of growing sectors within the industry.

The vast majority of the ISO international standards are highly specific to product, material, and process, ensuring regulation from start to finish.

34.3   ISO 9000 SERIES OF QUALITY STANDARDS

The ISO 9000 family addresses various aspects of quality management and contains some of ISO's best known standards. The standards provide guidance and tools for companies and organizations who want to ensure that their products and services meet customers' requirements, and that quality is consistently improved.

34.4   EVOLUTION OF ISO 9000 FAMILY OF STANDARDS

1926—Formation of International Federation of the National Standardizing Associations (ISA).

1944—Formation of United Nations Standards Coordinating Committee (UNSCC).

1947—Birth of ISO on the basis of the above two Associations.

1959—United States Department of Defense drafts MIL-Q-9858 standard.

1969—Revising the above into NATO AQAP series of standards.

1974—British Standards Institute adapts these into BS 5179 series of guidance standards.

1979—BS 5750 Part I, II, III series of requirement standards for system control in all areas of management which directly or indirectly affect quality.
1987—ISO 9000 family of Standards by ISO.
1988—Bureau of Indian Standards IS 4000 series.
1994—First Revision of ISO 9000 series.

In 1987, ISO released ISO 9000, which deals with the fundamentals of quality management systems and also gives guidelines for selecting the standards.

Subsequent standards in the 9000 series were released as follows:

- ISO 8402:1994—Quality management and quality assurance—vocabulary
- ISO 9001:1987—*Model for quality assurance in design, development, production, installation, and servicing.* This basically is for companies and organizations whose activities included the creation of new products.
- ISO 9002:1987—*Model for quality assurance in production, installation, and servicing* has basically the same material as ISO 9001, but without covering the creation of new products. This is applicable to the organizations who manufacture, supply, and service products or services as per specifications given by the customer.
- ISO 9003:1987—*Model for quality assurance in final inspection and test.* This covers only the final inspection of finished product, with no concern for how the product was produced.
- ISO 9004:2009—Introduced in 2009, this focuses on how to make a quality management system more efficient and effective. This has seven sections, each giving guidelines on services, processed materials, quality improvement, project management, quality plans, etc.
- ISO 19011:2011—Introduced in 2011, this sets out guidance on internal and external audits of quality management systems.
- ISO 10006—Quality management—guidelines for quality management in projects
- ISO 10007—Quality management—guidelines for configuration management
- ISO 10012—Quality assurance requirements for measuring equipment
- ISO 10013—Quality management—guidelines for developing quality manuals
- ISO 10014—Quality management—guidelines for economic effects of quality
- ISO 10015—Quality management—guidelines for continuing education and training

These standards have been revised periodically, in 1994, 2000, and 2008. The next revision was due in Dec. 2015.

It may be of interest that during mid-2015, American Society of Quality (ASQ) had initiated a blog on the LinkedIn website, where several interested professionals, including this author, could contribute their opinions on this revision. ISO did revise this standard in November 2015, which is further detailed in Section 34.17.

34.5   ISO/TS16949

The ISO/TS16949 is a process-oriented approach to the technical specification aiming at the development of a quality management system that provides for continual improvement, emphasizing prevention of defects and reduction of variation and waste in the supply chain. It is based on the ISO9001, but developed by International Automotive Task Force (IATF) with the key requirement of fulfillment of customer-specific requirements, set up by the automotive manufacturer in addition to the quality management system of their suppliers.

34.6   QS-9000 SERIES

QS-9000titled *Quality System Requirements* is similar to ISO-9000, International Quality System Standard, interpreted by US automotive giants Chrysler and Ford, applicable particularly to the US automotive industry, but now adapted world over. QS-9000 is made up of all the three sections: an ISO-9000 based requirement, a sector-specific requirement, and a customer-specific requirement. These requirements ensure that each supplier procures a good quality product. Furthermore, by developing QS-9000, we will be able to improve our product, customer satisfaction, and supplier relations as well.

Similarly:

(a) AS 9000, TL 9000, and PS 9000, too, were adapted as industry specific interpretation of the guidelines applicable for the aerospace, telecommunication, and pharmaceutical industries, respectively.
(b) ISO 13485:2003 is the medical industry's equivalent of ISO 9001:2000.
(c) ISO/TS 29001 are quality management system requirements for the design, development, production, installation, and service of products for the petroleum, petrochemical, and natural gas industries.

The following table indicates the parameters covered by each of these two series:

Quality System Requirements	ISO 9001	QS 9000
Management responsibility	X	X
Quality system	X	X
Contract review	X	X
Design control	X	X
Document and data control	X	X
Purchasing	X	X

Control of customer-supplied product	X	X
Product identification and tractability	X	X
Process control	X	X
Inspection and testing	X	X
Control of inspection, measuring, and test equipment	X	X
Inspection and test status	X	X
Control of nonconforming product	X	X
Corrective and preventive action	X	X
Handling, storage, packaging, preservation, and Delivery	X	X
Control of quality audits	X	X
Training	X	X
Servicing	X	X
Statistical techniques	X	X
Production parts approval process	X	
Continuous improvement	X	
Manufacturing capability	X	
Customer-specific requirement	X	

34.7   REQUIREMENTS AS SPECIFIED BY ISO 9000

A. General
1. Scope and purpose of the standard
2. Nomenclature reference such as ISO 9000:2008
3. Terms and definitions, for all items referred in the QMS
4. Requirement specifications as indicated in the table above
5. Documentation:
    (a) Statement of quality policy, vision, mission, and quality objectives
    (b) Quality manuals
    (c) Procedure manuals
    (d) Control documents
    (e) Control records
B. Management responsibility
    (a) Commitment
    (b) Customer focus
    (c) Quality policy
    (d) Organization

    (e) Responsibility and authority
    (f) Communication—internal and external
    (g) Delegation of management responsibility, specifying the duties, responsibilities, and authority
    (h) Management review
C. Resource management
    (a) Human resources
- General
- Competitive
- Awareness
- Training

    (b) Infrastructure
    (c) Work environment
D. Product realization
    (a) Planning the product
    (b) Planning the process
    (c) Procedure followed for QFD
- Determination of requirement of procedure
- Review of requirement
- Customer communications

    (d) Design and development
- Inputs outputs
- Review
- Verification
- Validation
- Control of design and development changes

    (e) Purchasing
- Purchasing process
- Purchasing information
- Verification of purchased product

    (f) Production and service provision
- Validation of the processes
- Identification and traceability
- Customer priority

E. Control of marking and measuring devises (measurement, analysis, and improvement)
    (a) Planning and measuring customer satisfaction
    (b) Internal audit
    (c) Marking and measuring the process parameters
    (d) Marking and measuring the products and services
    (e) Control of nonconforming products
    (f) Analysis of the data
F. Contractual improvement.

Fig. 34.1 illustrates relationship among the quality management systems (QMS) requirements as specified in ISO 9000.

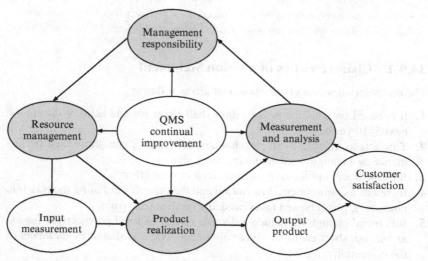

FIG. 34.1   Inter relationship among ISO requirement parameters.

34.8   BUREAU OF INDIAN STANDARDS

BIS was originally established in 1947 as Indian Standards Institution to fulfill the role of standardization in gearing industry to competitive efficiency and quality production. On Apr. 1, 1987, it became BIS as a statutory body. BIS started the QMS certification as per ISO 90001.

34.9   VISION AND MISSION STATEMENT

In today's competitive environment, it is not enough for an enterprise just to be a step ahead of others. They should advance by leaps and bounds. The first step in this advancement is the vision statement.

Vision is the basic requirement for an enterprise. It is reflected by your dream and creates a focus for the organization, inspires all stakeholders, and transforms the purpose into actions and draws people into the process of performance.

A vision statement helps the organization to focus on what is real as we go about doing our daily work and dealing with the day-to-day hassles that plague all organizations. Vision is set by the management with a foresight into the future, and states not only what the company wants to be in the near future, but also the avenues by which it aims to achieve this vision, generally in one sentence. A typical vision statement of an industrial organization can be cited as:

> *By the end of the decade, we shall be the industry leaders in the country, by doubling our turnover every fifth year through new products, market penetration, launching of new and contemporary products, and world-class customer service.*

34.9.1   Characteristics of a Vision Statement

The characteristics of a vision statement are as follows:

1. It is based on what the organization shall strive for and achieve during the next 5–10 years.
2. This acts as lodestar to all stakeholders, so that they can orient their performance as a team toward achieving the vision.
3. It inspires and uplifts everyone involved in your effort.
4. It creates a commitment. Motivation and the drive for initiating the mission, objectives, projects, and tasks needed to realize the vision.
5. It is broad enough to include a diverse variety of local perspectives for example, but short enough to be easily understood and shared by members of the community.

Before finalizing the vision statement, sufficient time shall be allowed for brainstorming among participants and then categorizing the random thoughts to ensure a comprehensive and realistic vision statement (Fig. 34.2).

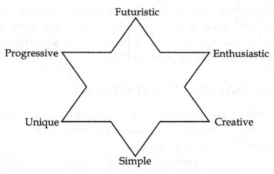

FIG. 34.2   Characteristics of a vision statement.

34.9.2   General Guidelines in the Formulation of a Vision Statement

1. The vision statement should be short and realistic to the present situation.
2. It should project the growth of the organization.

3. The intention in achieving the goal should be highlighted in the statement.
4. It should also reflect the value system that would be adapted for growth.
5. Total employee involvement shall be highlighted in the statement.

34.9.3   Extracts From Some of the Vision Statements of Famous Companies Are Illustrated in the Following Section

1. *A computer on every desk*—Microsoft.
2. *To bring inspiration and innovation to every athlete in the world*—Nike.
3. *A computer in the hands of everyday people*—Apple.
4. *Helping investors help themselves*—Charles Schwab.
5. *Make people happy*—Disney.
6. *To provide access to the world's information in one click*—Google.

34.9.4   Seven Tips on How to Write the Business Vision Statement

http://www.graystoneadvisors.com/7-ways-create-inspiring-vision-statement suggests the following seven tips on how to write the business vision statement

1. Define your future state
2. Make it powerfully memorable
3. Keep in synch
4. Gain consensus
5. Make it achievable
6. Make it visible
7. Align with your goals

34.10   MISSION STATEMENT

A mission statement is a written declaration of an organization's core purpose and focus that normally remains unchanged over time (Business Dictionary).

34.10.1   Features of Mission Statement

A mission statement:

- Serves as filters to separate what is important from what is not,
- Clearly states which markets will be served and how,
- Communicates a sense of intended direction to the entire organization.
- Converts the broad dreams of your vision into more specific, action-oriented terms.

- Explains your goals to interested parties in a clear and concise manner.
- Enhances your organization's image as being competent and professional, thus reassuring funding sources that their investment was (or would be!) a smart choice.

A mission statement defines the company's goals, ethics, culture, and norms for decision making. They highlight the company's goals in three dimensions:

- what it does for its customers,
- what it does for its employees, and
- what it does for its owners.

A nonprofit mission is never static. Especially in the first years, an organization's mission shifts and changes as the organization develops.

34.10.2   A Clear Mission Statement Would

The process of writing your mission statement is much like that for developing your vision statements. The same brainstorming process can help you develop possibilities for your mission statement. Remember, though, that unlike with vision statements, you will want to develop a single mission statement for your work. After having brainstormed for possible statements, you will want to ask of each one:

- Does it describe *what* your organization will do and *why* it will do it?
- Is it concise (one sentence)?
- Is it outcome-oriented?
- Is it inclusive of the goals and people who may become involved in the organization?

34.10.3   Five Steps for Drafting a Mission Statement

The website of bplans.com indicates the following five steps in drafting a mission statement:

1. Start with a market-defining story
2. Define how your customer's life is better because your business exists
3. Consider what your business does for employees
4. Add what the business does for its owners
5. Discuss, digest, cut, polish, review, and revise

34.11   OBJECTIVES, GOALS, AND ACTION PLANS

Vision and mission statements are followed by the statement of objectives, goals, targets, and action plans as illustrated in the Fig. 34.3 given in the following section.

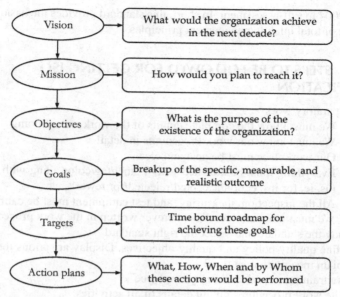

FIG. 34.3   Vision based activities per ISO 9001.

34.12   SOP—STANDARD OPERATING PROCEDURES

Standard operating procedures (SOP) are the instructions that cover operational parts. Initially, an SOP is based on army-wide publications and then modified to use local operating conditions and command policies as a guideline. The scope of SOP is extensive and varies. It provides the major instructions for all division elements of operational features.

In general, there are two formats for an SOP to follow:

- A format that publishes all comprising documents which details the function and the responsibilities of subordinate units.
- A format that is published as a basic document which includes general instructions to all units. This kind of format has specific instructions for each individual unit. It is more detailed and easier to use.

34.13   SPECIFIC FEATURES OF ISO 9004

ISO 9004:2009 is a guidance document for the sustained success of an organization with a quality management approach. It emphasizes the effectiveness and efficiency of a quality management system, and consequently the potential for improvement of the performance of an organization. The objectives of customer satisfaction and product quality are extended to include the satisfaction of interested parties and the performance of the organization. It spells out the benefits of ISO 9004 to self-assessment. In short, this standard provides the basic understanding of total quality management principles.

34.14   STEPS TO BE FOLLOWED FOR GETTING ISO CERTIFICATION

1. Preparatory work
   (a) We must know the complete details of the work we are doing
   (b) We must know our work Procedures in detail
   (c) Our workplace must be neat and clean
   (d) Everything must be attached with a label/tag/color along with its status, ie, for inspection/accepted/rejected/for rework, etc.
   (e) All the inspection, measuring, and test equipment must be calibrated
   (f) We must maintain records, wherever written in the work procedures
2. Study the standard and select the right standard
3. Define quality policy and quality objectives. Display at various locations, explain meaning/intent to all
4. Give training to each and every employee
5. Write work procedures giving details of all activities
6. Make changes in work "practices" wherever necessary
7. Conduct "Internal Audits" to find out whether work is carried out as per written procedures
8. Implement suggestions/observations of internal audit
9. Conduct an audit by external auditors
10. Implement suggestions received from them
11. For certification audit, call world-famous, well-known auditors for audit
12. Implement their suggestions
13. Achieve certification

34.15   BENEFITS OF ISO 9001-2000 AND TS 16949 QUALITY SYSTEMS

1. Improved communication at all levels.
2. Decreasing trend in rejections, reworks, and customer complaints.
3. Decreasing trend in inventories
   - Raw materials
   - W.I.P.
   - Finished goods
   - Lead time reduction
4. Customer relation improvement. Trust/confidence enhancement.
5. Subcontractor relations improvement. Long-term association.
6. Improved housekeeping.
7. Improved contacts with overseas buyers.
8. People development.

9. Improved health of people.
10. Reduced rejection/rework.
11. Improved housekeeping.
12. Increased morale of the company.
13. Ensured quality and after sales service to customers.
14. Improved team work.
15. It saves cost by:
    - avoiding repetition of work
    - avoiding unnecessary records
    - monitoring processes
16. Opportunities for export market.
17. Due to increased confidence of customers in you, you get more and more orders.
18. Improved discipline in the organization.
19. Increased credit in the market.

34.16   ISO 9000:2005

ISO 9000:2005 defines eight quality management principles in the quality management approach for the sustained success of an organization

Principle 1—customer focus
Principle 2—leadership
Principle 3—involvement of people
Principle 4—process approach
Principle 5—system approach to management
Principle 6—continual improvement
Principle 7—factual approach to decision making
Principle 8—mutually beneficial supplier relationships

34.17   2015 REVISION OF ISO 9000 SERIES

Subsequent to the publication of ISO 9001:2008 and ISO 9004: 2009, ISO's Technical Committee TC 176/TC has been developing new standards and periodically releasing the supporting documents, giving guidelines to the users. All ISO standards are reviewed every five years to establish if a revision is required to keep it current and relevant for the marketplace. ISO 9001:2015 is designed to respond to the latest trends and be compatible with other management systems, such as ISO 14001.

Accordingly, ISO had now revised the ISO 9000 *Quality Management Systems*, the world's leading quality management standard, which was last revised in 2008.

It may be noted that the ASQ initiated discussions on ISO 2015 on its Twitter and LinkedIn, ASQ during the mid-2015, when several professionals, including this author participated and discussed the benefits and limitations of the 20125 revision, a reference to which would yield several interesting criticisms.

Wikipedia records that the number of standards released in this series increased from 409,421 in 2001 to 1,138,155 in 2014. While Chands is first with 297,037 certificates issued in 2010, India stands eighth with 33,250, and the United States stands ninth with 25,101 certificates.

34.18 THE SIX STAGES OF THE RELEASE OF THE 2015 REVISION

In 2012, its silver jubilee year, ISO 9001 decided to create a new QMS model for the next 25 years, starting with the new QM principles. The revised standard ISO 9001:2015 was published by ISO on Sep. 23, 2015 in the stages as indicated in the following section:

1. The proposal stage commenced in May-Jun. 2012
2. The preparation stage commenced in Jun.-Oct. 2012
3. The Committee discussion commenced in Jun.-Sep. 2013
4. The enquiry and discussions stages went on until May-Oct. 2014
5. It was approved in Jul. 2015
6. Final standard text was released in Sep. 2015

34.19 REVISION OF ISO 9000 IN 2015

The new version follows a new, higher-level structure to make it easier to use in conjunction with other management system standards, with increased importance given to risk. More information about the changes can be found in the news area. It is specifically intended to ensure that ISO continues to adapt to the changing environments in which the organizations operate.

More importantly, the structure and core terms were modified to allow the standard to integrate more easily with other international management systems standards.

The 2015 revision focuses on performance by combining the process approach with Plan-Do-Check-Act and risk-based thinking such as failure mode and effect analysis (FMEA), which are described more earlier chapters and other safety standards, such as ISO 13485 on Food Safety, or Aerospace.

We can summarize the key changes as:

1. Greater emphasis on building a management system suited to each organization's particular needs.
2. A major focus is given on achieving value for the organization and its customers.
3. Revisions allow ISO 9001 to be more applicable by "service-based" organizations.
4. A requirement that those at the top of an organization be involved and accountable, aligning quality with wider business strategy.
5. Risk-based thinking throughout the standard, emphasizing the use of FMEA, and safety standards such as ISO 13485, makes the whole

management system a preventive tool and encourages continuous improvement. Identification of risk and risk control is now a requirement.

6. Addresses supply chain management more effectively.
7. Less prescriptive requirements for documentation: The organization can now decide what documented information it needs and what format it should be in.
8. Alignment with other key management system standards through the use of a common structure and core text.
9. Is more user-friendly for service and knowledge-based organizations.
10. Use of simplified language and a common structure of terminology by standardizing core text, structure, and definitions, to help organizations using multiple management systems, such as those for the environment, health and safety, or business continuity.

34.20   CONCLUSION

ISO 9000 deals with the fundamentals of quality management systems, including the eight management principles. Over one million global organizations are independently certified for ISO standards, making ISO 9001 one of the most widely-used management tools in the world today. However, ISO itself does not certify organizations. The third-party certification bodies provide independent confirmation that organizations meet the requirements of ISO 9001. The specific point of interest with ISO is the periodic revisions made in the standards, the latest being of ISO 9000 in 2015, as discussed earlier.

On the Lighter Side

He is bragging, he is lying, and he is giving three measures.

During the 1962 Tamilnadu elections, DMK party's election campaign was to give three measures (each measure equals 1.4 kg) of rice per rupee. The counter campaign of the Congress party was *'Arukkaraan, puzhukkaran, moondru padi kudukkuraan'* which in Tamil means *'He is bragging, he is lying, and he is giving three measures'*.

FURTHER READING

[1] Hoyle D. ISO 9000 quality systems handbook. Butterworth and Heinemann; 2001.
[2] Indian Standard on Quality Management Systems—Guidelines for Performance improvements (IS/ISO 9004: 2000—Bureau of Indian Standards—2000).
[3] www.bsigroup.com/en-IN/ISO-9001.
[4] www.lrqa.in/ISO9001.
[5] http://www.asrworldwide.com.
[6] www.iso.org/iso/catalogue.

[7] http://www.iso.org/iso/iso_9001_-_moving_from_2008_to_2015.pdf.
[8] http://asq.org/learn-about-quality/iso-9000/iso-9001-2015.
[9] ASQ's twitter discussions in Linked-in.
[10] http://en.wikipedia.org/wiki/ISO_9000.
[11] http://www.iso.org/iso/survey2008.pdf.
[12] http://the9000store.com/what-is-iso-9004.aspx.

Chapter 35

ISO 14000 Quality Systems

Chapter Outline

35.1 INTRODUCTION

Total quality management had evolved almost at the same time as the interest in environmental issues began to emerge and as such, it has built in similar issues as those relevant to the environment. A new concept of TQEM (total quality environmental management) has been initiated which clearly reflects the interdependency of quality and environmental problems.

The industrial revolution and the consequent development of new chemical processes and other manufacturing processes involving chemicals have resulted in a lot of pollution of the environment. This pollution was considered as a necessary evil in a country's economic development and was not taken seriously until the middle of the 20th century.

Multiplication of this problem resulting in issues such as the depletion of the ozone layer has created a high level of awareness, as well as a movement among all concerned. This movement sought to control the introduction of toxic and unwanted substances into the atmosphere, and to ensure a healthy environment for people. This pollution has not only affected the quality of life, it has also affected the quality of production of industrial goods, specifically in process industries.

In view of this, ISO formed the *Strategic Advisory Group on Environment (SAGE)*. The purpose of formation of this group was a worldwide emphasis on management of environmental issues as a part of quality management systems.

35.2   EVOLUTION OF THE ISO STANDARDS ON ENVIRONMENTAL ISSUES

Let us first give an overview of the evolution of the *ISO* standards on environmental issues by starting with some of the definitions of the terms related to environment management.

Environment is the surroundings in which an organization operates, including the air, water, land, natural resources, flora, fauna, humans, and their interrelationships. In this context, surroundings extend from within the organization to the global systems.

Environmental management system is the overall management system that includes organizational structure, planning activities, responsibilities, practices, procedures, processes and resources for developing, implementing, achieving, reviewing, and maintaining the environmental policy.

Environmental management system audit is a systematic and documented verification process of objectively obtaining and evaluating evidence to determine whether an organization's environment management systems (EMS) conforms to the EMS audit criteria set by the organization or to the international standards and for communicating the result of this process to the management.

Environmental policy is a statement from the organization of its intentions and principles in relation to its overall environmental performance, which provides a framework for action and for setting its environmental objectives and targets.

Sustainable development: The medieval concept of industrial development was that if development had to take place, a little of environment had to be sacrificed. If the use of chemicals was necessary for a better process, discharging the effluents of these chemicals into the open was inevitable, so the concept, rather the misconception, said. During the mid-20th century, global consciousness of the need to sustain the environment, despite the development, was created. In simple terms, *sustainable development* is the development process without any destructive activity, thus integrating development with environment. It is the process of meeting the basic needs of the present generation without affecting the life and scope of the future generations.

Environmental engineering: Environmental engineering is the application of engineering principles during all stages of setting up manufacturing units with a view of sustaining, or not adversely affecting the quality of the environment. All efforts for enhancing the public health and welfare must be taken. This includes green design, green manufacture, and development of environment-friendly processes and systems for disposal of solid, liquid, and gaseous wastes.

Green design: Green design involves systematic consideration of environmental objectives and factors during all stages of developing products, processes, and services. This also includes planning for recyclability, biodegradability, and eco-friendly disposal of the products after their life cycle. This is also known as design for environment (DOE).

Environmentally conscious manufacture: Environmentally conscious manufacture (ECM) is to reduce the harmful effects of manufacturing processes and methods, by reducing the hazardous wastes, emissions, and energy consumption. The principal issue is the elimination or reduction of the occupational hazards. One such step now adapted is the elimination of use of lead as an additive in petroleum refining or in paint manufacture. Another example is the mandatory phasing out of asbestos sheet manufacture. While the principles of green design as explained later involved the design stage, the ECM involves the manufacturing stage.

35.3 GLOBAL ENVIRONMENTAL ISSUES

- Global warming (or green house effect)
- Acid rain
- Ozone layer depletion
- Transboundary movement of hazardous wastes
- Ocean contamination
- Decrease in diversity of wild life
- Deforestation
- Desertification
- Pollution of rivers, lakes, and ponds
- Water-related diseases among humans as well as animals
- Increasing respiratory illnesses
- Unmanageable solid and hazardous waste generation

35.4 MAGNA CARTA ON ENVIRONMENT

At the United Nations Conference on Human Environment held in Stockholm on Jun. 5, 1972, the famous Declaration called the *Magna Carta on Environment* as given in the following section was made.

- Man has the fundamental right to freedom, equality, and adequate conditions of life in an environment of quality that permits a life of dignity and well-being.
- Man bears a solemn responsibility to protect and improve the environment for the present and future generations.

The U.N. General Assembly adapted the above resolution in Dec. 1972 and declared Jun. 5th as *World Environment Day*, which is celebrated all over the world year after year, including by the Institution of Engineers (India).

35.5 INTERNATIONAL INITIATIVES ON ENVIRONMENTAL ISSUES

The other global initiatives are:

- Charter on Economic Rights and Duties of States in 1974
- United Nations Habitat Conference on Human Settlement held in Vancouver in 1976
- United Nations Desertification Conference in Nairobi in 1977
- Inter-Governmental Environmental Education Conference in Georgia in 1977
- United Nations World Water Conference for "Clean drinking Water and Sanitation for All" in Nov. 1980
- World Conference on Environment Development held in 1984 where the expression "Sustainable Development of Environment" was conceived. An Earth Summit for 1992 in Rio de Janeiro was planned
- Montreal Protocol in 1987
- Clean Air Act amendments of USA in 1990
- Copenhagen Amendments in Jun. 1992
- United Nations Conference on Environment Development (UNCED) Earth Summit on Environment was held in Rio De Janeiro in 1992, the most strategic international effort for the preservation of the environment, where the Rio Declaration on Environment and Development was signed by all participating countries
- Kyoto protocol in 1997 required the industrial nations to reduce emission of carbon dioxide and other green house gases (GHG) by an average of 5.3% below the 1990 level by 2012, and this has come into force as late as Feb. 20, 2005. All the participating countries ratified this except the United States, which was blamed for its big brotherly attitude by all other nations, including European Union.
- In 1976, the Constitution of India was amended to reflect environmental priorities, motivated in part by the potential threat of natural resource depletion to economic growth.
- Earth summit 2002—Rio + 10 (+10 signifying the 10th year after the 1992 Rio Declaration) was held in Johannesburg, from Aug. 26 to Sep. 4, 2002.
- United Nations Conference on Sustainable Development, also known as "Rio 2012" was held in Rio de Janeiro from Jun. 13 to 22, 2012 as a 20-year follow-up of the 1992 Rio Declaration.
- Bonn Climate Change Conference was held in Bonn, Germany, Oct. 19–23, 2015 as a precurser to the Paris Climate Change Conference.
- 2015 United Nations Climate Change Conference, COP 21 or CMP 11 was held in Paris, France, from Nov. 30 to Dec. 12, 2015.

Some of the positive impacts of the international initiatives are:

- Categorizing the ozone depleting substances such as chlorofluorocarbons (CFCs) and halogens according to their ODP (ozone depletion potential) and gradual reduction of the production of these items.

- Reduction of the use of high ODP materials in automobile and refrigeration industries.
- Removal of air conditioners, foam insulation materials, and CFC's from cars before crushing them.
- Use of warning labels on the products containing High ODP materials.
- Banning of nonessential products using ozone depleting substances.

35.6   EVOLUTION OF ISO 14000 SERIES

ISO was established in 1947

- To facilitate international trade by developing international standards in resonance to individual country standards.
- Has 115 members representing 95% of the world's industrial production.
- Has 200 technical committees.
- Staffed by 3000 experts.
- Has so far developed over 8500 standards.

In 1992, BSI published the world's first environmental management systems standard, BS 7750. This provided the basic foundation for the development in 1996 of ISO 14000 series which refers to a series of standards related to environmental management to help organizations minimize the negative impact of their operations, processes, etc., on the environment. It stresses that the industries comply with applicable laws, regulations, and other environmentally-oriented requirements, and also to continually improve on the above. This series is similar to ISO 9000 quality management, which pertains to the process of how a product is produced, rather than to the product itself.

35.6.1   Formation of TC207

The TC207 was formed after the 1972 Earth Summit in Rio de Janeiro, which appointed seven subcommittees to work on different aspects of EMS as follows:

- SC1 stationed in the United Kingdom, concentrating on EMS (environment management systems)
- SC2 stationed in the Netherlands, concentrating on EMS Audit
- SC3 stationed in Australia, concentrating on environmental labeling
- SC4 stationed in the United States, concentrating on environmental performance and evaluation
- SC5 stationed in France, concentrating on life cycle analysis
- SC6 stationed in Norway, concentrating on terms and definitions
- SC7 stationed in Germany, concentrating on environmental aspects of product standards

The following standards were drafted

- ISO 14001—*environmental management system. Specification with guidance for use.* This serves as the basis for certification.

- ISO 14004—*environmental management system—general guidelines on principles, systems, and supporting techniques.*
- ISO 14006—*environmental management systems—guidelines for incorporating eco design*
- ISO 14010—*guidelines for environmental auditing, general principles.*
- ISO14011—*guidelines for environmental auditing, audit procedures, auditing of environmental systems*
- ISO14012—*guidelines for environmental auditing—qualification criteria for environmental auditors.*
- ISO 14014—*initial review.*
- ISO 14015—*site assessment guidelines.*
- ISO 14020, ISO 14021, ISO 14022, ISO 14023, and ISO 14024—basically *for environmental labeling.*
- ISO 14030—discusses post-production environmental assessment.
- ISO 14031—*environmental performance evaluation.*
- ISO 14040, ISO 14041, ISO 14042, ISO 14043—for *life cycle assessment.*
- ISO 14046—sets guidelines and requirements for water footprint assessments of products, processes, and organizations. Includes only air and soil emissions that impact water quality in the assessment.
- ISO 14050—*terms and definitions.*
- ISO 14060—*product standards.*
- ISO 14062—discusses making improvements to environmental impact goals.
- ISO 14063—environmental communication—Guidelines and examples.
- ISO 14064—measuring, quantifying, and reducing greenhouse gas emissions.
- ISO19011—single audit protocol for both 14000 and 9000 series standards together.

35.6.2   What is ISO 14001?

ISO 14001is a generic management system standard relevant to improving and managing resources more effectively and applicable to any organization from:

- single-site to large multi-national companies
- high-risk companies to low-risk service organizations
- manufacturing, process, and the service industries, including local governments
- all industry sectors, including public and private sectors
- original equipment manufacturers and their suppliers

All standards are periodically reviewed by ISO to ensure they still meet market requirements. It is interesting to note that ISO 14001 encourages a company to continually improve its environmental performance by Deming's P-D-C-A methodology described in more detail in the first chapter.

35.7   WATER FOOTPRINT

A water footprint is a concept introduced by UNESCO in 2002 to indicate the amount of fresh water needed by individuals, groups, or companies in order to make goods or provide services used by the community. This term was introduced in line with that of carbon footprint which indicates the total units of greenhouse gas emissions caused by an organization, event, product, or person. This in turn, originated from the ecological footprint concept of the 1990s that created worldwide awareness of the ecological impact of the industrialization. The Water Footprint Network is a Dutch-based International Learning Community which serves as a platform for connecting communities interested in sustainability, equitability, and efficiency in water usage.

35.8   THE BENEFITS OF ISO 14000

The benefits of ISO 14000 specifically of ISO 14001 include:

- Reduced cost of waste management
- Savings in consumption of energy and materials
- Lower distribution costs
- Improved corporate image among regulators, customers, and the public

The Kaizen Consulting Group on their website http://www.kcg.com.sg/benefits-iso14000.html, indicate the following benefits that can be achieved by complying with ISO 14000 standards.

1. Operational benefits
    - Efficiency, discipline, and operational integration with ISO 9000
    - Greater employee involvement in business operations with a more motivated workforce
    - Easier to obtain operational permits and authorizations
    - Assists in developing and transferring technology within the company
    - Helps reduce pollution
    - Fewer operating costs
    - Savings from safer workplace conditions
    - Reduction of costs associated with emissions, discharges, waste handling, transport, and disposal
    - Improvements in the product as a result of process changes
    - Safer products
2. Environmental benefits
    - Minimizes hazardous and nonhazardous waste
    - Conserves natural resources—electricity, gas, space and water with resultant cost savings
    - Prevents pollution and reduces wastage

3. Marketing benefits
   - Demonstrates to customers that the firm has met environmental expectations
   - Meets potential national and international government purchasing requirements
   - Delivers profits from marketing "green" products
   - Provides a competitive marketing tool
   - Improves international competitiveness
4. Financial benefits
   - Improves the organization's relationship with insurance companies
   - Elimination of costs associated with conformance to conflicting national standards
   - Process cost savings by reduction of material and energy input
   - Satisfying investor/shareholder criteria
   - Helps reduce liability and risk
   - Improved access to capital

35.9   ENGINEER'S ROLE IN ENVIRONMENT PROTECTION

Having discussed the roles played by the international communities, let us now view the role that should be played by the engineer in protecting the environment.

1. As an experimenter involved in environmental issues, the engineer must be aware of his role in environmental protection.
2. He should have full knowledge and confidence in their projects and should be meticulously careful to foresee the environmental effects of the project activities.
3. He should have sincere concerns about the environment during the project planning and execution stage and ask the following questions:
   - How does the industry effect the environment?
   - How far can such ill effects, if any, be controlled?
   - Is political or physical regularization needed?
   - Whether reasonable protective measures are available for immediate implementation?
   - Whether the engineer as an individual can ensure a safe and clean environment?
4. He should preplan all the activities and processes and the control systems without frequent re-planning or redesigns.
5. He should plan for safe exits.
6. He must budget the funds required for these control systems including, the safe exits, etc.

35.10   PRINCIPLES OF GREEN DESIGN

The following are some of the principles that should be considered while designing any engineering product.

- Consider the physical and chemical structure of the material before selection.
- Avoid use of toxic substances.
- Evaluate manufacturing processes with reference to environmental impact.
- Design for longer life.
- Design for ease of assembly and disassembly.
- Incorporate source reduction.
- Design for inter changeability and recyclability.
- Avoid use of throw-away-after-use materials such as ultra-thin plastic bags.

35.11 BASIC APPROACHES FOR RESOLVING ENVIRONMENTAL PROBLEMS

35.11.1 Cost Oblivious Approach

All efforts are to be made to make the environment as clean as possible, whatever may be the cost to do so. No level of environmental degradation is accepted. This approach is somewhat similar to that of *rights and duty ethics*.

Though ideal, this approach has two obvious problems.

- It is difficult to define exactly what is *as clean as possible* and
- In the highly competitive world of Indian industry, where every rupee counts, industries try to use the above indeterminate parameter as a loop-hole, and only try to do minimal expenditure to provide short run measures, enough to create an impression with the public that they are protecting the environment, which may not be true in the long run.

35.11.2 Cost-Benefit Approach

The problems are analyzed in terms of the benefits derived by reducing the pollution problems. The costs and the benefits are weighed to determine the optimum combination. Here the target is not to achieve a completely clean environment, but an economically viable environment protection. This can be compared to the *Utilitarian theory* of professional ethics which says that acts are morally right if they produce the most good to the most people, giving equal consideration to everyone that is affected.

35.11.3 Difficulties of Cost-Benefit Approach

This approach too suffers from four major difficulties.

- It is difficult to assess the true cost of human life or loss of a species, or environmental protection.
- It is difficult to assess accurately the costs and benefits, and much guess work or safety factors have to go into the calculations.

- This approach does not necessarily specify who should bear the cost and who should get the benefit.
- The cost-benefit analysis does not take morality and ethics into account. The decision is simply based on mathematical simulations and calculations, and there is no room for a discussion whether what is done is right or wrong.

A combination of the above with a sincere application of professional and personal ethics:

- Unlike professional decisions like bridge designs, projects involving environment affects the engineer personally, even as a member of the public. Hence the need to apply his decision-making with reference to his personal ethics.
- Whatever the approach may be, it is essential that the engineer applies both his professional and personal ethics, at the same time meticulously the following laws and regulations of the state.
- From the perspective of human health, the engineer's responsibility to protect is clear, which must be balanced between the consideration of the well-being of his employer, the public, and the community.

35.12   GUIDELINES FOR SOCIAL RESPONSIBILITY

On November 1, 2010, ISO launched an International Standard providing guidelines for social responsibility (SR) named ISO 26000, or simply ISO SR. As per the ISO website, its goal is to contribute to global sustainable development, by encouraging business and other organizations to practice SR to improve their impacts on their workers, their natural environments, and their communities. It is a voluntary guidance standard. There is no certification process for this, but organizations are allowed to state that they have used ISO 26000 as a guide to integrate SR into our values and practices.

35.13   5 RS OF WASTAGE UTILIZATION

Environmentalists all over the world profess 4 Rs in ensuring the reduction of solid waste pollution. These are *Reduce, Re-use, Recycle,* and *Replace (or Remanufacture)*. This author would like to add another R, viz, *Recover*. These 5 Rs are illustrated in Fig. 35.1 and supported by the subsequent case study.

An interesting incident of the 1970s can be cited to illustrate the 5th R (Recover):

"A medium-scale industry reported the loss of a small batch of work-in-progress components which were last seen at the inspector's table. After a couple of days, they were located in the scrap yard, perhaps being swept off by the sweepers. Since then the company started the practice of a weekly scrap yard visit by a supervisor for identifying any such recoverable, non-scrap item. We can call this as a pre-cursor to the Gemba walk practiced today."

	R	Symbol	Explanation
1	Reduce	R	Design your product or process to reduce waste generation at all stages. Value Analysis and DFSS are best illustrations.
2	Reuse	R	Reuse equipment parts and fixtures. Also reuse packaging materials (boxes/bags), where possible. Install tramp oil removers that enable reuse of coolant fluids.
3	Recycle	R	Sort scrap metal, wood and plastic from industrial waste according to what can be recycled and what that cannot be.
4	Replace	R	Choose environmentally friendly and biodegradable alternatives where possible. Use durable items instead of disposable items like use of cloth/jute bags in place of thin plastic bags.
5	Recover	R	During the above sorting identify if any good component or other item is found to be wrongly scrapped and recover it after due testing.

FIG. 35.1   5 Rs of wastage utilization.

35.14   CONCLUSION

Having understood the scenario of the environment issues and the evolution of the environment management systems, and the efforts of ISO and other organizations in instilling the SR of the engineers for environmental protection, it is hoped that the engineer perceives any process, or the industrial project from the environment perspective and applies his mind in achieving his objective of the project with minimal environmental degradation.

On the Lighter Side

A person who likes sweets very much but is very calorie conscious would apply sugar-taste analysis to match the taste he enjoys to the sugar intake. This is akin to the cost-benefit analysis we discussed in para 35.11.

FURTHER READING

[1] http://en.wikipedia.org/wiki/Carbon_footprint.

[2] http://www.iso.org/iso/home/standards/management-standards/iso14000.html.

[3] http://www.qualitydigest.com/oct/iso14000.html.

[4] Kiran DR. Professional ethics and human values. 2nd ed. New Delhi: McGraw Hill, Higher Education; 2013.

[5] Kiran DR. Maintenance engineering and management—precepts and practices. BS Publications; 2014.

[6] Kiran DR. Environmental engineering and principles of green design environmental. In: Seminar at Velammal Engineering College; 2004.

[7] Kiran DR. Evolution of ISO 14000. In: Proceedings of 11th NIQR National Convention; 2005.

[8] Kiran DR. Energy audit. In: Chief Guest address at ENFUSE Seminar at Sairam Engineering College; 2005.

Chapter 36

Terminology Used in Japanese Management Practices

Chapter Outline

36.1 INTRODUCTION

During the post-industrial revolution era, management thought and practices were developed mostly in the Western countries, especially in the United States. That is the reason why during the 1950s and 1960s, we studied only the terminologies used by writers such as Taylor and Peter Drucker.

However, during the post-World War II era, Japan emerged as a strong industrial nation creating awe among the Western world in view of its highly successful management practices. This resulted in most of the management consultants and authors wanting to understand and use the Japanese management terms.

It may be noted that most of these practices were in use in the Western world also, but the emphasis in Japan was the importance given to the core worker, which was absent in the Western world. It may hence be said that the use of Japanese terms in place of English terms created interest, and indirectly helped young managers to better understand and appreciate these practices.

This paper highlights the meaning and origin of some of the Japanese management terms used, so as to provide a lucid insight into the Japanese concept of World-Class Management Practices.

36.2 SOME OF THE TERMINOLOGIES CITED IN THIS CHAPTER

- Quality circles
- *Kaizen*
- CREW
- 4 wives and 1 husband
- 5 management objectives of factory management
- 5 Zus
- *GenchiGenbutsu*
- Heizunka
- *Nemawashi*
- 3 Mu (*Muda, Muri, Mura*)
- *Poka Yoke*
- *Hanedashi*
- *Andon*
- Jidhoka
- *ChakuChaku*
- 5 S
- *Gemba walk*
- *WarusaKagen*
- Single minute exchange of die
- Just-in-time
- *WarusaKagen*
- *Kanban*
- *HoshinKanri*
- *NichijoKanri*
- *Kata*
- 6 Sigma
- Total productive maintenance
- *Pecha-Kucha*
- *DakaraNani*
- *Kanso, Shizen*, and *Shibumi*
- *OkyaKusoma*

36.3   HISTORY OF DEVELOPMENT OF JAPANESE MANAGEMENT PRACTICES

- Before World War II, Japan was not a highly industrialized nation. Most of the electrical and electronic goods were imported from the United States and Europe. So it was playing second fiddle to the United States in commerce and trade.
- Japan's decision to side with Hitler alienated them against the United States and Japan's raid on Pearl Harbor infuriated the United States, in dropping of atom bombs on Hiroshima and Nagasaki, one of the moist inhumane acts ever committed in the history of mankind.
- Consequently Japanese wanted to pay back the Americans in their own coin.
- They knew it couldn't be in a war and hence, decided to beat the Americans in world trade, by producing more quality goods and capturing the international market currently reigned by Americans. Japanese being highly patriotic by nature, this desire has percolated into the minds of every national, specifically into all categories of personnel in Japanese industry.
- Higher productivity became the initial buzz word. Yet they realized higher productivity without quality products could take the nation nowhere, and the subsequent buzz word was high quality production. Quality in all aspects of manufacture was given high priority. That's how the Japanese industry got the momentum for quality-oriented higher productivity in order to capture, not only the domestic, but the international market.
- Around 1950, JUSE (Japanese Union of Scientists and Engineers) team visited the United States to study the US's industrial practices. During their visit, they invited Dr. Deming, and subsequently, Dr. Juran, to visit Japan and train their engineers.

36.4   QUALITY CIRCLES

The principle behind quality circles is dealt more in detail in Chapter 15.

36.5   KAIZEN

- In Japanese,

Kai	means	Change	and
Zen	means	Good	Thus
Kaizen	means	Change for the good	

Kaizen is dealt more in detail in Chapter 22.

36.6   GENCHIGENBUTSUGENJITSU

Genchi means actual place, *Genbutsu* means actual product, and *Genjitsu* means actual solution. Thus *GenchiGenbutsuGenjitsu* literally means "Go to the source to find the facts to make correct decisions." In other words, it is the old industrial engineering principle to conduct a study and collect information on the job directly, instead of resorting only to available statistical data.

36.7   MONOZUKURI AND HITOZUKURI

Zukuri (*tsukuri*) means the process of making something. *Monozukuri* is the spirit of producing excellent products and the ability to constantly improve a production system and process, as defined by The Japanese Institute for Trade and Organization (JETRO). *Hitozukuri is* the need to educate and train a person to become expert in *Monozukuri*. Together, *monozukuri* and *hitozukurican* provide the basis for a balanced approach to using technology and enhancing human capacities.

36.8   NEMAWASHI

Nemawashi means to make decisions slowly by consensus, thoroughly considering all options. This is the basic management principle cited in several chapters of this book, especially the chapter on the TQM tools.

36.9   HEIJUNKA

Heijunka implies to level out the work, or to redistribute the work in such a way that every operator gets more or less the same workload. This is the very principle behind line balancing of assembly lines of small products such as ceiling fans that are assembled by operators sitting in either side of the belt conveyor that moves the work-in-progress. This line balancing, illustrated in the case study in Section 22.20a in the chapter on *kaizen*, involves

- Splitting each operation into transferable elements
- Timing of each element by work measurement (the term time study, which created a negative impression during the latter half of 20th century, is avoided, though in principle both mean the same)
- Considering the elements that can either be eliminated or redistributed to other operatives in the line, so that each gets more or less the same workload in terms of the mean operational time.

36.10   3 MU CHECKLISTS

Kaizen practitioners have developed a system of checklists to help workers and management to be constantly mindful of areas of improvement, similar to several checklists developed and used by industrial engineers all over the world since the 1950s. In Japan, these are called the 3 Mu Checklists.

The 3 Mus are

- *Muda* (Signifying waste)
- *Muri* (Signifying strain)
- *Mura* (Signifying discrepancy).

The fields where these 3 Mus can be applied are:

- Method of operation
- Process involved
- Facilities
- Jigs and tools
- Materials
- Production volume
- Inventory
- Place
- Manpower
- Technique
- Ways of thinking
- etc. …etc.

36.11   FOUR WIVES AND ONE HUSBAND

This originates from a popular Japanese saying and as explained in Chapter 22, it highlights the principle of a questioning technique which is similar to critical analysis technique or Cost Reduction through Elimination of Waste (CREW) adapted by industrial engineers the world over, even prior to the 1950s. Though they are not Japanese terms, they are cited here for comparison with similar Japanese techniques.

- The 4 Ws (Wives) are What, Where, Why, and When
- The 1 H (Husband) is How.

"Why" is the most significant and is the very basis for the success of industrial engineers. In fact, just to highlight the significance of *why*, this author prefers to choose the second letter H of Why to call it the husband and the third letter W of How to group it under Wives.

If you do not wish to deviate from the significant use of the first letter, let us illustrate the importance of the third word from the Indian mythology of Mahabharat by considering the simile that Satyabhama, the third wife of Krishna is the most powerful among his wives, or to Ramayana where Kaikeyi, the third wife of Dasaratha, is the most powerful among his wives.

Remember the famous poem by Rudyard Kipling?

I had six stalwart serving men,
They taught me all I know,
Their names were What and When,
And Where and Why and How and Who.

36.12   CREW

In contrast, *CREW* (*Cost Reduction through Elimination of Waste*) is popular in the Western world. Though this is not a Japanese tern, it is included here in view of its similarities with *kaizen*, etc. Canon has identified 9 waste categories as:

Waste category	Nature of waste	Type of economization
Work-in-progress	Stocking items not immediately needed	Inventory management
Rejection	Producing defective products	TQM
Facilities	Having idle machineries	Increase capacity utilization ratio
Process time	High production costs	Method improvement studies
Production delays	Non-smooth flow of work-in-progress	Production planning and control
Downtime of machinery	Excessive break time or tool set-up time	TPM, SMED, etc.
Expense	Over-investing or over-expenditure	Technical audit
Indirect labor	Excessive personnel	Effective job classification
Design	Products with more functions than needed	Value analysis
Operator talent	Highly skilled workers employed for routine operations	Effective job assignment

36.13   5 MANAGEMENT OBJECTIVES OF FACTORY MANAGEMENT

The five key points set forth by the Mitsubishi Corporation are:

- Achieve maximum quality with maximum efficiency.
- Maintain minimum inventory.
- Eliminate hard work.
- Use tools and facilities to maximize quality and efficiency and minimize effort.
- Maintain a questioning technique and open-minded attitude for constant improvement based on teamwork and cooperation.

36.14   5 ZUS

In Japanese, Zu, as a suffix means *don't*

like	*Math*	in	Hindi,
or	*Vaddu*	in	Telugu
or	*Vendam*	in	Tamil

The five don'ts or the things the operators should avoid doing with respect to defects are:

Uketorozu	meaning	don't accept defects
Tsurazu	meaning	don't make defects
Baratsukasazu	meaning	don't create variations
Kuriakalsazu	meaning	don't repeat mistakes
Nagasazu	meaning	don't supply defects

36.15   POKA YOKE

- *Poka Yoke* or *AUTO-NO-MATION* is based on the philosophy "to err is human." That is, instead finding out who erred, find out why it happened, and ensure it does not happen again.
- It ensures that the machine stops automatically whenever there is an error. That is, the machine *automatically says no to further operation*, from which the author coined the term AUTO-NO-MATION. This is also called "mistake proofing." Most CNC machines provide striking examples of this.
- It may be noted that even in 1950s, the textile looms in Bombay were provided with a mechanical interlocking system that, whenever any thread snaps off during weaving, a thin reed supported by the thread slips down into the mechanism and the machine stops automatically.

36.16   ANDON AND HANEDASHI

Andon is an indication to stop work manually in case of any problem. While *Poka Yoke* involves automatic stoppage of the machine, *Andon* involves the manual stoppage by the vigilant worker.

Hanedashi is the use of auto-eject devices to unload the parts automatically after the operation is over. This is similar to *poka yoke*, but is applied after the operation.

36.17   JIDHOKA

Jidhoka is based in the philosophy that all individuals are responsible for the services they provide.

36.18   CHAKUCHAKU

ChakuChaku means "Load, Load," referring to the positioning of all the machines as per the operation sequence and very close to each other. In other words, it implies Product Layout.

36.19   5 S

5 S is a method for organizing the workplace like a shop floor or an office space. It advocates what to keep, where to keep it, and how to keep it (maintaining, cleaning, etc.). It also instills a sense of ownership among the workers to be more accountable for their work place. In Chapter 23, 5S is dealt with more in detail, but the basics are repeated here.

SEIRI	SORTING	Distinguish between necessary and unnecessary
SEITON	SYSTEMIZING	A place for everything and everything in its place
SEIKO	SHINING	Keep the workplace clean
SEIKETSU	STANDARDIZING	Maintain a good environment
SHITSUKE	DISCIPLINE	Follow the rules of the company

Further reference may be made to the chapter on 5S.

36.20   SIX SIGMA

Six sigma is a business management strategy originally developed by Motorola, US in 1981. However, because Japanese adapted this principle initially, and was more successful than the rest of the world, and the Japanese industries have been the benchmarks for the six sigma concept, this is cited here to highlight its basic principle. Chapter 24 on six sigma is devoted fully to this concept.

Since the 1920s, the word "sigma" has been used by mathematicians and engineers as a symbol for a unit of measurement in product quality variation. However, the engineers of Motorola in the United States used "Six Sigma" as an informal name for an in-house initiative for reducing defects in production processes.

The concept with which we use the term σ in statistics is different from that which is used in TQM, as can be clearly understood by the explanation given in Chapter 24.

Hence, it is important to call it six sigma (at the most 6-sigma), and not by the symbol 6σ. Also while referring to the extent of its application, to use the term "level" or "performance level," and not "value." That is to say, "I achieved six sigma level quality," and not "I achieved a quality of six sigma value."

36.21 GEMBA WALK

Gemba walk is the practice of senior managers to tour several places in the factory, along with the concerned operatives, with the basic purpose of identifying the areas for improvements. Each and every *Muda* noticed, or suggestions offered by anyone for improvements, would be recorded and analyzed and posted on the notice board to motivate the operatives. This is generally done during the afternoon hours or Sundays to create a friendly and holiday atmosphere and may be followed by snacks or lunch. Preferably the Chief Executive Officer accompanies them to instill interest among the workers.

36.22 WARUSAKAGEN

WarusaKagen implies that things are not problematic now, but may soon develop into a problem unless controlled now. It is a caution noticed by a vigilant worker in the system and hence, a starting point for several improvement activities.

This can be understood and remembered better by the Tamil metaphor—WarusaikkuAghum—(or Varusakuagunu in Telugu), both implying *"will be next in the line."*

36.23 SINGLE MINUTE EXCHANGE OF DIE

Single Minute Exchange of Die (also known as SMED), is the Lean Manufacturing tool used to create very fast changeovers and setups to reduce machine downtime and increase throughput. SMED was developed by Shigeo Shingo and successfully introduced at Toyota Motors, reducing machine changeover times from hours to less than 10 min. The success of this program contributed directly to just-in-time manufacturing, which is part of the Toyota Production System. SMED makes load balancing much more achievable by reducing economic lot size and thus stock levels. This is dealt with more in detail in Chapter 21 on modern seven management tools.

36.24 JUST IN TIME

Just in time (JIT) is a production and inventory control technique to ensure that the inventory level, either as stocks in the store or as work-in-process on the shop floor is reduced to a minimum, almost to a zero level.

JIT purchasing is to ensure that the supplies are received in small quantities just in time for production, by establishing an agreement with vendors

JIT on the shop floor is to ensure that each machine produces just what is required for the next machine in quantities, but not more than what is required. This is also called a pull-system of production. *Kanban* as explained below is a part of this concept.

Prior to the 1980s, during the days of manual inventory control by Kardex system, the critical items with low stocks were identified by a thin red strip inserted in the card, so that procurement action can be initiated once a month to avoid stock outs. But today, with computerization, such control can be done on a day-to-day basis even for most A and B items, enabling application of JIT concept to reduce the average inventory to 1 or 2 day's level.

36.25   KANBAN

- This is JIT application in production planning and control and has become synonymous with JIT system.
- *Kanban*, in Japanese language, means visible signboards, cards, or chits.
- The *Kanban* can be a card, a container, or an electronic signal.
- Every machine operator tends to produce only those quantities required for the next operation and keeps a *Kanban* in the container of the components as an indication to the next operator that the required semi-processed material is ready. He slows down his pace for the next lot if the container is still not drawn by the next operator.
- When the subsequent operator finishes his operation, he draws the material returning this *Kanban*, forming a signal to produce a further lot of the required quantity.
- Hence, this is also called the *pull-system* of production.
- The underlying principle is that the needed parts should be received just in time for further processing.

36.26   HOSHINKANRI

HoshinKanri is a Japanese term for strategic planning.

HoshinKanri can be broken down into four parts,

Ho	direction
shin	shining needle, used in a compass

So *Hoshin* means progress towards a goal

Again

Kan	control or channeling the progress
ri	reason or logic

So *HoshinKanri* means achieving the organization's direction, focus or goal, by logically controlling the progress.

In other words, *HoshinKanri* represents the management planning and control towards the achievement of the goal. It is a method devised to capture and cement strategic goals, as well as to provide insight about the future and develop the means to bring these into reality. This is called strategic planning in the western world as explained further in Chapter 7 on strategic planning.

As Dr. Yoji Akao puts it,

With HoshniKanri, the daily crush of events and quarterly bottom line pressures do not take precedence over strategic plans, rather, these short term activities are determined and managed by the plans themselves.

36.27 NICHIJOKANRI

Nichijo means daily routine and *kanri* means management and control, similar to *kanri* of *hoshinkanri*. Thus *nichijokanri* covers all the day-to-day aspects of operations planning and is complementary to *hoshinkanri*, which refers to the long-range or strategic planning.

36.28 KATA

Kata is a descriptive term for the organizational routines. It can be defined as *behavior patterns, routines, or habits of thinking and doing that are practiced over and over every day.*

Kata as a term became popular at Toyota and provides a level of clear insight into the key behaviors underlying Toyota Culture in a way that can be easily understood and applied. Mike Rother in his book *Kata in Toyota*, refers to "improvement *kata*," "coaching *kata*," etc.

We may hence equate the term *Kata* (the story of the routines) to the Indian term *Katha*, but it is something more than a *Katha*. This is the very philosophy, which all the employees of Toyota breathe, day in and day out.

36.29 TOTAL PRODUCTIVE MAINTENANCE

Like any other Japanese practice, total productive maintenance (TPM) emphasizes that the base worker shall be entrusted with the task of performing the routine maintenance activities for the machine he operates. The maintenance problems, too, are discussed by the quality circles to ensure that the machine upkeep is more effective.

It may be noted that while the principles and procedures for preventive maintenance originated in the West, the Japanese emphasis on making the operator the central focus for the routine maintenance, combining with the 5 S practices, made all the difference in the effective overall maintenance of the equipment. And this resulted in TPM, which is explained further in Chapter 13.

Today world over, it has been realized that this Japanese practice of worker-oriented routine machine maintenance is much more effective than the maintenance conducted by maintenance department workers.

36.30 PECHA-KUCHA

Literally meaning chatter or chit-chat, it emphasizes the need to plan contents and time management of the power point or other presentations, thereby avoiding *Pecha-kucha* during the presentation. If you have 20 slides and only less than seven minutes to complete the presentation before discussions, the slides advance automatically before 20 sec, forcing you to limit presentation of each slide for 20 sec only. This concept is hence called $20 \times 20/6.40$ (min). Started in Tokyo in Feb. 2003 as an event for young designers to present their work, it has turned into a massive celebration and *Pecha-kucha Nights* are held in hundreds of cities around the world.

36.31   DAKARANANI

Literally meaning "so what?," it implies that while planning any slide, be prepared for the audience to question "so what?" for every step and prepare yourself for a convincing response or edit the material accordingly.

36.32   KANSO, SHIZEN, AND SHIBUMI

These are the three vital elements of presenting your report in a meeting. *Kanso* means "simplicity" that is achieving maximum with minimum means, using ideal concepts, visual elegance, and perfect communication. Use of overelaborate deigns and excessive refinement is avoided. *Shizen* means Naturalness in presentation of ideas that suit the particular audience. This point is also emphasized by Taichii Ohno in his book, *Toyota Production Systems*. *Shibumi* means elegant simplicity in visual communication and graphic design.

36.33   OKYAKUSOMA

This may not be a management practice as such, but means *honorable guest*, a term often used for the customer in Japan. Thus, this may be called a marketing strategy.

36.34   CONCLUSION

As explained in the synopsis, the author makes an attempt to enlighten the reader on some of the Japanese terminology and practices, the application of which has resulted in a major revolution in industrial management all over the world during the latter half of the 20th century, next in importance only to the Industrial Revolution of the 17th century.

Of course, during the early part of 21st century, China has overtaken Japan in industrial success and maybe we have to retune ourselves to the Chinese Management terminology and practices. We should hope for the day when India overtakes the rest of the world in successful management practices, when every other country starts using Sanskrit words on management as their regular vocabulary on management terminology.

ON THE LIGHTER SIDE

Japanese PM asks Japanese women to die.

In Japanese Language Shinu means the verb die and shin-e (pronounced as sheeney) is an expression of the command to die. The Japanese Prime Minister Shinzo Abe, in one of his speeches to encourage women to shine (meaning to perform well) in their work places, used the English word shine but pronounced it phonetically as sheeney without realizing its implication.

...This news item appeared in the Times of India of 28th June 2014

FURTHER READING

[1] D.R. Kiran, Japanese management practices, A PP presentation 2010.
[2] http://en.wikipedia.org/wiki/Japanese_management_culture.
[3] Heinz Weihrich, http://www.usfca.edu/fac-staff/weihrichh/docs/management_practices.pdf.
[4] Martin JR et al., maaw.info/ArticleSummaries/ArtSumMartin92.htm.

Annexure I

University Syllabi

1 ANNA UNIVERSITY – BE (MECH/PROD) - GE 406 - TOTAL QUALITY MANAGEMENT

UNIT 1: Introduction

Definition of Quality, Dimensions of Quality, Quality Planning, Quality costs – Analysis Techniques for Quality Costs, Basic concepts of Total Quality Management, Historical Review, Principles of TQM, Leadership – Concepts, Role of Senior Management, Quality Council, Quality Statements, Strategic Planning, Deming Philosophy, Barriers to TQM Implementation.

UNIT 2: TQM Principles

Customer satisfaction – Customer Perception of Quality, Customer Complaints, Service Quality, Customer Retention, Employee Involvement – Motivation, Empowerment, Teams, Recognition and Reward, Performance Appraisal, Benefits, Continuous Process Improvement – Juran Trilogy, PDSA Cycle, 5S, Kaizen, Supplier Partnership – Partnering, sourcing, Supplier Selection, Supplier Rating, Relationship Development, Performance Measures – Basic Concepts, Strategy, Performance Measure.

UNIT 3: Statistical Process Control (SPC)

The seven tools of quality, Statistical Fundamentals – Measures of central Tendency and Dispersion, Population and Sample, Normal Curve, Control Charts for variables and attributes, Process capability, Concept of six sigma, New seven Management tools.

UNIT 4: TQM Tools

Benchmarking – Reasons to Benchmark, Benchmarking Process, Quality Function Deployment (QFD) – House of Quality, QFD Process, Benefits, Taguchi Quality Loss Function, Total Productive Maintenance (TPM) – Concept, Improvement Needs, FMEA – Stages of FMEA.

UNIT 5 - Quality Systems

Need for ISO 9000 and Other Quality Systems, ISO 9000:2000 Quality System – Elements, Implementation of Quality System, Documentation, Quality Auditing, QS 9000, ISO 14000 – Concept, Requirements and Benefits.

2 ANNA UNIVERSITY FOR MBA - GE2022 - TOTAL QUALITY MANAGEMENT

UNIT I: Introduction

Introduction – Need for quality – Evolution of quality – Definition of quality – Dimensions of manufacturing and service quality – Basic concepts of TQM – Definition of TQM – TQM Framework – Contributions of Deming, Juran and Crosby – Barriers to TQM.

UNIT II: TQM Principles

Leadership – Strategic quality planning, Quality statements – Customer focus – Customer orientation, Customer satisfaction, Customer complaints, Customer retention – Employee involvement – Motivation, Empowerment, Team and Teamwork, Recognition and Reward, Performance appraisal – Continuous process improvement – PDSA cycle, 5s, Kaizen – Supplier partnership – Partnering, Supplier selection, Supplier Rating.

UNIT III: TQM Tools & Techniques I

The seven traditional tools of quality – New management tools – Six-sigma: Concepts, Methodology, applications to manufacturing, service sector including IT – Bench marking – Reason to bench mark, Bench marking process – FMEA – Stages, Types.

UNIT IV: TQM Tools & Techniques II

Quality circles – Quality Function Deployment (QFD) – Taguchi quality loss function – TPM – Concepts, improvement needs – Cost of Quality – Performance measures.

UNIT V: Quality Systems

Need for ISO 9000 – ISO 9000-2000 Quality System – Elements, Documentation, Quality auditing – QS 9000 – ISO 14000 – Concepts, Requirements and Benefits – Case studies of TQM implementation in manufacturing and service sectors including IT.

3 JAWAHARLAL NEHRU TECHNOLOGICAL UNIVERSITY – HYDERABAD

Mechanical Engg. Common to all courses

ME 05554: Total Quality Management – Elective for ME

UNIT I

TQM – overview – history – stages of evolution – elements – definitions – continuous improvement – objectives – internal and external customers.

UNIT II

Quality Standards – need for standardization – institutions – bodies of standardization – ISO 9000 series – ISO 14000 series – other contemporary standards.

UNIT III

Quality Measurement Systems – Developing and implementing QMS – non-conformance database.

UNIT IV
Problem solving – problem solving process – corrective action – order of preference – system failure – analysis approach – flow chart – fault tree analysis – failure mode assessment – and assignment matrix – organizing failure mode analysis – pedigree analysis.

UNIT V
Quality circles – organization – focus team approach p0- statistical process control – process chart – Ishikawa diagram – preparing a designing control charts.

UNIT VI
Quality function deployment (QFD) – elements of QFD – benchmarking – Taguchi Analysis – loss function – Taguchi design of experiments.

UNIT VII
Value improvement elements – value improvement assaults – supplier teaming.

UNIT VIII
Six sigma approach – application of six sigma approach to various industrial situations.

4   VISVESVARAYA TECHNOLOGICAL UNIVERSITY, BELGAUM - 06IM72 TOTAL QUALITY MANAGEMENT

PART A
UNIT I
OVERVIEW OF TOTAL QUALITY MANAGEMENT: History of TQM. Axioms of TQM, contributions of Quality Gurus – Deming's approach, Juran's quality trilogy, Crosby and quality treatment, Imai's Kaizen, Ishikawa's company wide quality control, and Fegenbaum's theory of TQC.

UNIT II
EVOLUTION OF QUALITY CONCEPTS AND METHODS: Quality concepts. Development of four fatnesses, evolution of methodology, evolution of company integration, quality of conformance versus quality of design from deviations to weaknesses to opportunities. Future fitness's, four revolutions in management thinking, and four levels of practice.

UNIT III
FOUR REVOLUTIONS IN MANAGEMENT THINKING: Customer focus, Continuous Improvement, Total participation, and Societal Networking. FOCUS ON CUSTOMERS; Change in work concept marketing, and customers.

UNIT IV
CONTINUOUS IMPROVEMENT: Improvement as problem solving process; Management by process, WV model of continuous improvement, process control, process control and process improvement, process versus creativity. Reactive Improvement; Identifying the problem, standard steps and tools, seven steps case study, seven QC tools.

PART B
UNIT V
PROACTIVE IMPROVEMENT: Management diagnosis of seven steps of reactive improvement. General guidelines for management diagnosis of a QI story, Discussion on case study for diagnosis of the seven steps. Proactive Improvement; Introduction to proactive improvement, standard steps for proactive improvement, semantics, example-customer visitation, Applying proactive improvement to develop new products – three stages and nine steps.

UNIT VI
TOTAL PARTICIPATION: Teamwork skill. Dual function of work, teams and teamwork, principles for activating teamwork, creativity in team processes, Initiation strategies, CEO involvement Example strategies for TQM introduction. Infrastructure for mobilization. Goal setting (Vision/Mission), organization setting, training and E education, promotional activities, diffusion of success stories, awards and incentives monitoring and diagnosis, phase-in, orientation phase, alignment phase, evolution of the parallel organization.

UNIT VII
HOSHIN MANAGEMENT: Definition, phases in hosing management-strategic planning (proactive), hoshin deployment, controlling with metiers (control), check and act (reactive). Hoshin management versus management by objective, hoshin management and conventional business planning, an alternative hoshin deployment system, hoshin management as "systems Engineering" for alignment.

UNIT VIII
SOCIETAL NETWORKING: Networking and societal diffusion – Regional and nationwide networking, infrastructure for networking, openness with real cases, change agents, Center for quality Management case study, dynamics of a societal learning system. TQM as learning system, keeping pace with the need for skill, a TQM model for skill development, summary of skill development.

5 PUNE UNIVERSITY - 406D - QUALITY MANAGEMENT

UNIT I
Quality, Strategic Planning, and Competitive Advantage: Brief History – Modern Developments in Quality – A Race Without a Finish Line. Definitions of Quality. Quality in Manufacturing and Service Systems. Economic Issues – Quality and Price – Quality and Market Share – Quality and Cost – The Taguchi Loss Function. Quality & Competitive Advantage. Perspectives on Leadership for Quality – The Balridge View of Leadership.

UNIT II
Principles of Total Quality Management: Introduction – Elements of Total Quality Management – Strategic Planning and Leadership – A Customer Focus – Fact-Based Management – Continuous Improvement – Teamwork and Participation. Malcolm Baldrige National Quality Award Criteria. Benefits of Total Quality

Management. The Deming Management Philosophy – Profound Knowledge – The Impact of Profound Knowledge – Deming's 14 Points for Management. The Juran Philosophy – The Juran Quality Trilogy. The Crosby Philosophy.

UNIT III

Customer Focus: The Customer-Driven Quality Cycle – Identifying Customer Needs – Achieving Customer Requirements in Production – Implications of the Customer-Driven Quality Cycle. Quality Function Deployment – The Quality Function Deployment Process – Building the House of Quality – Implementing Quality Function Deployment. Designing Quality into Services – Service Needs Identification – Service System Design. Customer Satisfaction Measurement Techniques – Customer Relationship Management Techniques.

UNIT IV

Quality Control and Quality Assurance: Concept of Quality Control – Concept of Process Variation – Acceptance Sampling – Sampling Inspection vs. 100% Inspection – Attributes and variable sampling plans – OC Curves – Producer and Consumer Risk – AQL, RQL, TQL, AOQL and AOL.

UNIT V

Statistical Process Control: Control Charts – X-R, P, np and C Charts – Benefits of Control Charts and Applications.

UNIT VI

Quality Management Assistance Tools: Ishikawa Fish Done diagram – Nominal Group Technique – Quality Circles – Flow Charts – Pareto Analysis – Pokka Yoke (Mistake Proofing).

UNIT VII

Reliability: Concept and Components – Concepts of failure – Reliability of system – Success and Failure models in series and parallel – Methods of achieving higher reliability – Concept of maintainability and availability – Comparison with reliability.

UNIT VIII

Managing and organization for quality: Quality Policy – Quality Objectives Leadership for Quality – Quality and organization culture – Change Management. Team Building. Partnerships – Cross-Functional Teams – Supplier/Customer Partnerships.

UNIT IX

Quality Management Standards: (Introductory aspects only)

(a) The ISO 9001:2000 Quality Management System Standard
(b) The ISO 14001:2004 Environmental Management System Standard
(c) ISO 27001:2005 Information Security Management System 110
(d) ISO/TS16949:2002 for Automobile Industry
(e) CMMI Fundamentals & Concepts
(f) Auditing Techniques – Planning for an audit – Developing a Check-list – Conducting an Audit – Writing an Audit Report – Auditor Ethics – Value – addition process during Internal Audit – Mock Audits – Quiz. (8)

6   SIVAJI UNIV. KOLHAPUR, BE MECH, TOTAL QUALITY MANAGEMENT

Section I

1. **Quality Basic Concepts:** Various definitions of quality and their implication, ISO definition of Quality, Quality cost-estimation and reduction.
2. **Quality Assurance System:** Basic concepts-QA input-process-output, Significance of feedback for QA, Internal customer approach. Statistical Quality Control- Basic philosophy, Significance of N-D curve, Control charts for attributes and variables, Process capability analysis, Concept of six sigma.
3. **Acceptance Sampling:** Inspection standards, OC curve, Sampling plans (single, double, multiple only)
4. **Product And System Reliability:** Basic concepts, prediction and evaluation of component and system reliability.
5. **Taguchi's Quality Engineering:** Loss function, Orthogonal arrays, Signal to noise ratio, Parameter design and tolerance design.

Section II

6. **Overview of TQM:** Concept and definition, Fundamentals, TQM Verses Management relationship, Elements of TQM, approaches to TQM, TQM models, Zero defect concept.
7. **Contributions of Quality Gurus:** Deming's approach, Jurans quality trilogy, Crossby and quality improvement, Ishikawas company wide quality control, Fegenbaum theory of TQC.
8. **Revolution in Management Thinking:** Customer focus, problem solving QC tools, Continuous improvement (Kaizen), Customer satisfaction, Kanos model, Customer retention.
9. **Quality Circles:** Total Employee Involvement (TEI), Employee empowerment, Employee suggestion scheme.
10. **Creating Quality Culture:** Requisite changes to implement Quality culture, developing TQM culture.
11. **Total Quality in Service Sector.**
12. ISO 9001-2000 series of standards, overview of ISO 9000-1993 series standards, structure of ISO 9000-2000 series standards, clauses, contents, and interpretations, implementation. 03
13. **Quality System-policy and objectives.**

7   UTTAR PRADESH TECHNICAL UNIVERSITY - EME-041: TOTAL QUALITY MANAGEMENT

UNIT I

Quality Concepts: Evolution of Quality control, concept change, TQM Modern concept, Quality concept in design, Review off design, Evolution of proto type.

Control on Purchased Product: Procurement of various products, evaluation of supplies, capacity verification, Development of sources, procurement procedure.

Manufacturing Quality: Methods and Techniques for manufacture, Inspection and control of product, Quality in sales and services, Guarantee, analysis of claims.

UNIT II

Quality Management: Organization structure and design, Quality function, decentralization, Designing and fitting organization for different types products and company, Economics of quality value and contribution, Quality cost, optimizing quality cost, seduction programme.

Human Factor in Quality: Attitude of top management, co-operation, of groups, operators attitude, responsibility, causes of operators error and corrective methods.

UNIT III

Control Charts: Theory of control charts, measurement range, construction and analysis of R charts, process capability study, use of control charts.

Attributes of Control Charts: Defects, construction and analysis off-chart, improvement by control chart, variable sample size, construction and analysis of C-chart.

UNIT IV

Defects Diagnosis and Prevention: Defect study, identification and analysis of defects, corrective measure, factors affecting reliability, MTTF, calculation of reliability, Building reliability in the product, evaluation of reliability, interpretation of test results, reliability control, maintainability, zero defects, quality circle.

UNIT V

ISO 9000 and its concept of Quality Management: ISO 9000 series, Taguchi method, JIT in some details.

8 M.J.P. ROHILKHAND UNIVERSITY, BAREILLY: MBA(GEN.) CN-405 TOTAL QUALITY MANAGEMENT

UNIT I: Introduction

Quality, Total quality, Rationale for total quality, key elements of total quality, quality circles, quality gurus.

UNIT II: Quality Control and Improvement Tools

Check Sheet, Histogram, Pareto Chart, Cause and Effect diagram, Scatter diagram, Control chart, Graph, Affinity diagram, Tree diagram, Matrix diagram, Process decision program chart, Arrow diagram, Acceptance Sampling, Process capability studies, Zero defect program (POKA-YOKE).

UNIT III: Benchmarking and Kaizen

Benchmarking, Rationale of benchmarking, Approach and process, Prerequisites of benchmarking, Benefits of benchmarking, Obstacles to successful benchmarking, perpetual benchmarking.

Concept of Kaizen, Kaizen vs Innovation, Kaizen and management, Kaizen practice.

UNIT IV: TQM Models

Demings Award criteria, Malcolm Baldridge national quality award, European quality award, Australian quality award, Confederation of Indian Industries award.

UNIT V: Quality Management System & Quality Audit

Quality Systems, Quality management principles, ISO 9001: 2000, ISO 14000, Future of quality system audit, Audit objectives, types of quality audit, Quality Auditor, Audit performance.

9   VTU - TOTAL QUALITY MANAGEMENT

PART A

UNIT I

Introduction to TQM. Quality movement in Japan, US & India. Definition of quality. Small q & Big Q, Quality characteristics – weaves, Dimensions, determinants. Quality & profitability.

UNIT II

QUALITY & MANAGEMENT PHILOSOPHIES – Deming Philosophy – Chain reaction, 14 points for management, triangle theory of variance, deadly diseases & sins, Demings wheel. Juran Philosophy – 10 steps for quality improvement, quality trilogy, universal breakthrough sequence. Crosby Philosophy – Crosby's 6 C's, Absolutes of quality, Crosby's 14 points for quality, Crosby triangle. Comparison of 3 major quality philosophies.

UNIT III

MANAGING QUALITY – traditional vs Modern quality management, the quality planning, road map, the quality cycle. Cost of quality – Methods to reduce cost of quality, Sampling plans, O.C. curve.

UNIT IV

QUALITY CONTROL – Objectives of quality control, Strategy & policy. Company wise quality control. Quality Assurance – Definition, concepts & objectives. Economic models for quality assurance. Statistical methodology in quality assurance. Process capability ratio, 6 sigma in quality assurance.

PART B

UNIT V

QUALITY IMPROVEMENT – Principles of Total Quality, Evolution of Total Quality Control & Principles. TQM – Basic concepts & overview. Necessity of TQM. Elements of TQM, benefits of TQM, TQM in services, ISO 9000 & ISO 14000 in quality management system.

UNIT VI: FOCUSSING ON CUSTOMER

Importance of customer satisfaction, Kano's model of customers satisfaction, customers driven quality cycle, understanding customers needs & wants, customers retention.

UNIT VII: LEADERSHIP

Introduction, characteristics of quality leaders, role of TQM in leadership. Tools & Techniques of TQM, Just in time system-Concepts, objectives, overview, characteristics, benefits. Benchmarking – Introduction, process of bench marking, benefits, advantages & limitations. Hours

UNIT VIII: SUPPLY CHAIN MANAGEMENT

Objectives, process tools, supply chain management for manufacturing organization & service organization world class manufacturing – becoming world class, relevance of TQM in world class manufacturing. World class supplier, world class customer, present global business conditions, world class companies in 21st century. Future of TQM.

10   MAHATMA GANDHI UNIVERSITY, MEGHALAYA

NRAI SCHOOL OF MASS COMMUNICATION & MANAGEMENT
MGT 301: TOTAL QUALITY MANAGEMENT

1. An Overview and Role of TQM – Classical Definitions of Quality – Bhagawan Baba's Definition of Quality – Product Satisfaction & Product Dis-satisfaction
2. Trends in Change Management and role of TQM.
3. Philosophical Approaches to TQM – Eastern & Western Approaches – Bhagawan Baba's Teachings
4. Methodological Approaches to TQM – Deming, Juran, Crosby, Others
5. Tools of TQM – Diagnostics – 7 Tools of Analysis – old and new, Cybernetic Analysis
6. Overview of Other Developments in TQM – QFD, ISO, CMM, Benchmarking, Six Sigma
7. Leadership requirements for TQM
8. Integration & Implementation of TQM in Organizations.
9. Application of TQM.
10. Video tapes on TQM
11. Case Studies & Exercises.

11   WEST BENGAL UNIVERSITY - ME 821: TOTAL QUALITY MANAGEMENT

Basic concepts, definitions and history of quality control. Quality function and concept of quality cycle. Quality policy and objectives.

Economics of quality and measurement of the cost of quality. Quality considerations in design.

Process control: Machine and process capability analysis. Use of control charts and process engineering techniques for implementing the quality plan.

Acceptance Sampling: single, double and multiple sampling, lot quality protection, features and types of acceptance sampling tables, acceptance sampling of variables and statistical tolerance analysis. Quality education, principles of participation and participative approaches to quality commitment.

Emerging concepts of quality management: Taguchi's concept of off-line quality control and Ishikawa's cause and effect diagram.

12 MADRAS UNIVERSITY FOR MASTER OF BUSINESS ADMINISTRATION

TOTAL QUALITY MANAGEMENT

Introduction to quality control – quality and cost consideration – statistics and its implication in quality control – sampling inspection in engineering manufacturing – statistical quality control by the use of control charts – methods of inspection and quality appraisal – reliability engineering – value engineering and value analysis. Theory of sampling inspection – standard tolerancing – ABC analysis – defect diagnosis and prevention. Recent techniques of quality improvements – zero defect – quality motivation techniques – quality management system and total quality control.

Section of ISO model and implementation of ISO 9000. Human resource development and quality circles – Environmental management system and total quality control.

13 TAMIL NADU OPEN UNIVERSITY MBA - MSP 61 - TOTAL QUALITY MANAGEMENT PAPER

BLOCK I Syllabus: Introduction to Total Quality Management – Leadership – Information and Analysis – Strategic Quality Planning.

BLOCK II Syllabus: Human Resource Development and Management – Management of process quality – Customer focus and satisfaction – Bench marking.

BLOCK III Syllabus: Organising for Total Quality Management – Productivity and Quality – The Cost of Quality.

BLOCK IV Syllabus: Processes and Quality tools – The Concept of a process – Total Quality Management and data – Quality improvement tools – Understanding process variation.

BLOCK V Syllabus: Criteria for Quality programs – ISO 9000: Universal Standards of Quality – Reengineering.

14 INDIAN INSTITUTE OF PLANT ENGINEERS - DIPLOMA IN PLANT ENGINEERING & MANAGEMENT

TOTAL QUALITY MANAGEMENT

1. Total quality management (TQM) – Principles, Characteristics of TQM – Definition of quality, Quality Control, Quality assurance – Methods of inspection – Quality planning and Quality policy – What are 5s ?
2. TQM Principles and practices-leadership of TQM-Strategic Quality planning, Customer focus, Customer satisfaction, Employee involvement, Bench marking, Continuous process improvement

3. TQM tools & techniques – The seven traditional tools of Quality, New seven tools, Major approaches to TQM – Quality Circles, PDCA, JIT, SPC, Cross functional approach, Quality function Deployment, TPM, Six sigma principle,
4. Total productive maintenance (TPM) – Principles and strategy – Multi-skill development – TPM concepts and steps for implementing TPM – Gains and benefits of ZTPM, Autonomous maintenance
5. Quality systems-ISO 9000-ISO 14000 concepts, Environmental management system concepts, purpose and steps for implementation, Barriers for implementation and how to overcome-requirement and benefits

15   MIDDLE EAST TECHNICAL UNIVERSITY

Total Quality Management
Topic 1
- Introduction
- Evolution of Quality Concepts and Quality Paradigms

Topic 2
- Quality and Team Organization

Topic 3
- Teams' Thinking and Communication

Topic 4
- Problem Solving and Decision Making Process

Topic 5
- Leadership and Empowerment
- Benchmarkiing

Topic 6
- TQM Implementation

Topic 7
- Achieving quality by planning: QFD, Hoshin planning e

Topic 8
- Quality through improvement and control: SPC

Topic 9
- Quality through design: Robust design

Topic 10
- Quality through IT: Customer Relationship Management

Topic 11
- Achieving Quality By Innovation: TRIZ

16   PRINCE SULTAN UNIVERSITY

IS 470 TOTAL QUALITY MANAGEMENT (TQM)
COURSE TOPICS:

1. Introduction
2. Quality basics and history – Quality advocates – Quality improvement
3. Total Quality Management

4. Deming – Juran – Crosby – Quality Management
5. Quality Improvement Techniques
6. Pareto Diagrams – Cause-Effect Diagrams – Scatter Diagrams – Run Charts – Cause and Effect Diagrams
7. Statistical Concepts
8. Definitions – Measures of Central Tendency – Measure of Dispersion – Concepts of Population and Samples – Normal Curves
9. Control Charts for Variables
10. Definitions – Variation: Common vs. Special Causes – Control Chart Techniques – X-bar and R chart Correlation – X-bar and S charts
11. Control Chart Interpretation and Analysis
12. Using Charts to Pinpoint Problems – Process Capability
13. Other Variable Control Charts
14. Individuals and Moving Range Charts – Moving Average and Moving Range Charts – Charts for Individuals – Median and Range Charts
15. Fundamentals of Probability
16. Basic Concepts and Definitions – Discrete Probability Distributions – Continuous Probability Distributions
17. Control Charts for Attributes
18. Definitions – Control Charts for Non-conforming Units – Control Charts for Counts of Non-conforming Units
19. Reliability
20. Product Life Cycle – Measures of Reliability
21. 11. Quality Costs
22. Quality Cost Measurement – Utilizing Quality Costs for Decision-Making
23. Advanced Topics
24. Quality Function Deployment – Design of Experiments
25. Quality Systems: ISO 9000, Six Sigma
26. Certification Requirements – Evolving Standards
27. Benchmarking and Auditing
28. Reaching World Class Standards

17   ST. MARTIN UNIVERSITY, WASHINGTON STATE

TOTAL QUALITY MANAGEMENT
   COURSE TOPICS:
1. Introduction
2. Quality basics and history – Definitions of quality – Major contributors to quality – Deming – Juran – Crosby – Ishikawa – Taguchi – Feigenbaum – Shewhart
3. Strategic Quality Management
4. STQM – Dimensions, measures, and metrics – Garvin's approach to operationalizing quality dimensions
5. Designing Quality Into Products and Services

6. Seven management tools – Quality function deployment (QFD) – Design for six sigma (DFSS) – Robustness – Reliability – Risk assessment (FMEA and FTA)
7. Creativity in Quality
8. Breakthrough improvement – Designing the innovative organization
9. Quality Systems and QS Auditing
10. Quality management systems – ISO 9000 – Baldridge
11. Product, Process, and Materials Control
12. Identification of materials and status – Traceability – Supplier management
13. Quality Improvement Tools
14. Seven quality tools – PDSA – DMAIC – Benchmarking
15. Metrology, Inspection, Testing
16. Gauging – Precision and accuracy – Non-destructive testing
17. Statistical Process Control
18. Probability and statistics – Variation
19. Variable Control Charts
20. x-bar and R-bar charts – Moving Average and Moving Range Charts – Charts for Individuals – Median and Range Charts
21. Control Charts for Attributes
22. Non-conforming – Non-conformities (defects)
23. Quality Costs
24. Quality Cost Measurement – Utilizing Quality Costs for Decision-Making
25. Human Factors in Quality
26. Barriers to quality improvement efforts – Employee involvement

18   UNIVERSITY OF KOKYBĖS VADYBOS (LITHUANIAN UNIVERSITY)

TOTAL QUALITY MANAGEMENT
Course unit content The course studies cover these topics:

- Quality role managing organizations.
- Quality management concepts.
- Prerequisites and stages of Total quality management evolution.
- Total quality management gurus.
- Business excellence models.
- Organization's commitment to satisfy customer's needs.

19   UNIVERSITY OF HRADEC KRALOVE & UNIVERSITY OF PARDUBICE (CZECHOSLOVAKIA)

CURRENT TOPICS IN TOTAL QUALITY MANAGEMENT

- Academic readings (based in Eastern European and American topics, affiliations, applications, experience)

- Case studies (based in Eastern European and American topics, affiliations, applications, experience)
- Quality management gurus
- Quality management
- Quality planning
- Frameworks to improve Organizational performance (ISO 9000, Malcolm Baldrige Quality Award, Six Sigma)
- SPC and control charts
- Customer satisfaction

20   CORK INSTITUTE OF TECHNOLOGY

SUBJECT/MODULE TITLE: QUALITY MANAGEMENT Subject Code: 4CIS.5B
Syllabus Content Time (%)

1. **Introduction to Quality Systems 20%:** Overview of quality, history of quality, competitive advantage, industrial perspective, total quality system, Taguchi "Loss Function" concept.
2. **Statistical Process Control 25%:** Process Control Chart calculations, Extraction of information, Capability Index, Individual and Moving Range Charts, Implementation of Statistical Process Control, Control Charts for Attributes.
3. **Problem solving techniques for Quality Management 15%:** Pareto Analysis, Ishikawa (Cause/Effect) Diagrams, Failure Modes and Effects Analysis, Brainstorming, Program for Quality Improving, Deming, Juran.
4. **Strategic Quality Management 10%:** Total Quality Management (TQM), total company involvement, technical and managerial, Implementation of TQM, Philosophies of TQM.
5. **Reliability 20%:** Defining Reliability, Product Life Characteristic Curve, Reliability Function, Reliability Engineering.
6. **Health and Safety 5%:** Classification of Hazards, ISO 9000, Codes of Practice, Company Safety Statement.

21   A. AU & BPGTQM AS A COURSE WITH 3 QUALITY RELATED PAPERS

MA 9105, 9112, 9122
MA 9105 PROBABILITY AND STATISTICAL METHODS
UNIT I: ONE DIMENSIONAL RANDOM VARIABLES
Random variables – Probability function – moments – moment generating functions and their properties – Binomial, Poisson, Geometric, Uniform, Exponential, Gamma and normal distributions – Functions of a Random variable.
UNIT II: TWO DIMENSIONAL RANDOM VARIABLES
Joint distributions – Marginal and conditional distributions – Functions of two dimensional random variables – Regression Curve – Correlation.

UNIT III: ESTIMATION THEORY

Unbiased Estimators – Method of moments – Maximum likelihood Estimation – Curve fitting by principle of least squares – Regression lines.

UNIT IV: TESTING OF HYPOTHESIS

Sampling distributions – Type I and Type II errors – Tests based on normal, t, χ^2 and F distributions for testing of mean, variance and proportions – Tests for Independence of attributes and Goodness of fit.

UNIT V: MULTIVARIATE ANALYSIS

Covariance matrix – Correlation Matrix – Multivariate Normal density function Principal components – Sample variation by principal components – Principal components by graphing.

22   B QE 9112 TOTAL QUALITY MANAGEMENT

UNIT I: INTRODUCTION

Need for TQM, evolution of quality, Definition of quality, TQM philosophy – CONTRIBUTIONS OF Deming Juran, Crosby and Ishikawa, TQM models.

UNIT II: PLANNING

Vision, Mission, Quality policy and objective Planning and Organization for quality, Quality policy Deployment, Quality function deployment, introduction to BPR and analysis of Quality Costs.

UNIT III: TQM PRINCIPLES

Customer focus, Leadership and Top management commitment, Employee involvement – Empowerment and Team work, Supplier Quality Management, Continuous process improvement, Training, performance Measurement and customer satisfaction.

UNIT IV: TQM TOOLS AND TECHNIQUES

PDSA, The Seven Tools of Quality, New Seven management tools, Concept of six sigma, FMEA, Bench Marking, JIT, POKA YOKE, 5S, KAIZEN, Quality circles.

UNIT V: QUALITY SYSTEMS

Need for ISO 9000 Systems, clauses Documentation, Implementation, Introduction to ISO 14000 and OSHAS18000, Implementation of TQM, Case Studies.

23   C QE 9122 QUALITY BY DESIGN

UNIT I: INTRODUCTION

Perception of quality, Taguchi's definition of quality – quality loss function, tolerance using loss function, quality and process capability, Planning of experiments, design principles, terminology.

UNIT II: FACTORIAL EXPERIMENTS

Design and analysis of single factor and multi-factor experiments, tests on means, EMS rules.

UNIT III: SPECIAL DESIGNS
K Factorial designs, Fractional factorial designs, Nested designs, Blocking and Confounding.

UNIT IV: ORTHOGONAL EXPERIMENTS
Selection of orthogonal arrays (OA's) OA designs, conduct of OA experiments, data collection and analysis of simple experiments, Modification of orthogonal arrays.

UNIT V: ROBUST DESIGN
Variability due to noise factors, Product and process design, Principles of robust design, objective functions in robust design – S/N ratios, Inner and outer OA experiments, optimization using S/N ratios, fraction defective analysis, case studies.

Bibliography

While this bibliography lists the books that can be referred in general in relation to all the chapters of the book, chapter specific references and websites are also indicated at the end of the chapters.

[1] Evans JR, Lindsay WM. The management and control of quality. Mason, OH: Thomson South Western Publication; 2005.

[2] Besterfield DH, et al. Total quality management. N.Y.: Pearson Education Publ; 2003.

[3] Feigenbaum. Total quality control. 3rd ed. McGraw Hill; 1995.

[4] Ho SK. TQM: an integrated approach. London: Kogan Page Publ; 1995.

[5] Juran JM, Gryna FM. Quality planning and analysis. New York: McGraw Hill Publ; 1980.

[6] Deming WR. Out of crisis. Chambers University Press; 1993.

[7] Murthy MN, editor. Excellence through quality & reliability. Chennai: Applied Statistical Centre; 1989.

[8] Kume Hitishi. Management by quality. Chennai: Productivity Press; 1995.

[9] Chang RY, Neidzwiecki ME. Continuous improvement tools. Wheeler Publ; 1998.

[10] Drummand H. Quality Systems Handbook. New York: Nicolas Publ.; 1994.

[11] Omachony VK, Ross JE. Principles of total quality. Cambride, MA: Kogan & Page; 1995.

[12] Hoyle D. ISO 9000 quality systems handbook. Butterworth & Heinemann Publ; 2001.

[13] Ishikawa K. Introduction to quality control. Tokyo: 3A Corpn. Publ; 1989.

[14] Buckford J. Quality. London: Routledge; 1998.

[15] Pyzdek T. The six sigma project planner. New York: McGraw Hill Publ; 2003.

[16] In: Clarke G, editor. Managing service quality. Bedford: IFS Publications; 1990.

[17] Band W. Creating value for customers. New York: John Wiley Publications; 1991.

[18] Akiyama K. Function Analysis. Tokyo: Japan Standards Association; 1989.

[19] Drucker P. Management challenges for the 21st century. Addison Wesley, Reading (MA): Harper Business Publ; 1980.

[20] Juron JM. Juron on leadership for quality, an executive handbook. New York: The Free Press; 1989.

[21] Garvin D. Managing quality. New York: Free Press; 1988.

[22] Ireson W, Grant E. Handbook of industrial engineering & management. Englewood Cliffs, NJ: Prentice Hall; 1971.

[23] Maynard HB, editor. Industrial engineering handbook. 3rd ed. New Delhi: McGraw Hill.

[24] Certo S. Modern management. 9th ed. Prentice Hall; 1971.

[25] Scholtes PR, et al. The team handbook—how to use teams to improve quality. Madison, WI: Joiner Associates; 1998.

[26] Hackman JR, Oldham GR. Motivation through design of work. Reading, MA: Wesley; 1989.

[27] Joseph P, Furr D. Total quality in managing human resources. Boca Raton, FL: St. Lucie Press; 1995.

[28] Kemp RL. Handbook of strategic planning. New York: Cummings & Hathaway; 1995.

[29] Burkhart PL, Reuss S. Successful strategic planning. Newbury Park, CA: Sage Publications; 1993.

[30] Bradford RW, Duncan JP. Simplified strategic planning. Worcester, MA: Chandler House; 2000.

[31] Snape, Wilkinson, Marchington, Redman. Managing human resources for TQM.

[32] Deming WE. Out of the crisis. Cambridge, MA: MIT Press; 2000.

[33] Grant EL, Leavenworth RS. Statistical quality control. 6th ed. New York: McGraw Hill; 1988.

[34] Roy RK. A primer in Taguchi methods. Dearborn, MI: Society of Manufacturing Engineers; 2010.

[35] Schnaars SP. Megamistakes. New York: Free Press; 1989.

[36] Ariely D. Predictably irrartional. New York: Herper Perenniel; 2008.

[37] Kiran DR. Maintenance engineering and management: precepts and practices. BS Publishers; 2014.

[38] Kiran DR. Professional ethics and human values. 2nd ed. New Delhi: McGraw Hill; 2013.

[39] Gulfreda JJ, Maynard LA, Lytie L. Employee involvement in the quality process. Hyderabad: BS Publications; 2014.

[40] McGregor D. The human side of enterprise. New York: McGraw Hill; 1960.

[41] Montgomery D. Introduction to statistical quality control. New York: John Wiley & Sons; 2004.

[42] Stamatis DH. Failure mode and effect analysis. Chennai, India: ASQ Publ; 1997.

[43] Kiran DR. Resistance to change. Excell Superv 1985; [of NPC].

[44] Kiran DR. Participative management. J Manuf Technol Manag 2000.

[45] Kiran DR. How to be more creative. Ralli Group J 1975.

[46] Kiran DR. Value engineering—a case study. Ind Eng J 1975.

[47] Kiran DR. Evolution of ISO 14000. In: Proceedings of 11th NIQR Convention; 2005.

[48] Kiran DR. Environmental engineering & principles of green design. In: Environmental Seminar; 2004.

[49] Kiran DR. Energy audit. Chief Guest address at ENFUSE, 2005.

[50] Simmons, Shadim, Arthur. Integrated TQM and HRM. No. 17/3. New Delhi: 1995.

[51] Rubinenstein. QC circles & US participation movements. In: Proceedings of ASQC Technical Conference; 1992.

[52] Blackburn R, Rosen B. Total quality & human resources management. Acad Manag Exec 1993;7(3):49–66.

[53] Orsburn, Jack D, Moran, Linda, Musselwhite, ED, Zenger, John H. Self directed work teams. Homewood, IL: Business One Irwin, 1990.

[54] Kondo Y. Human motivation a key factor for management. Tokyo: 3A Corpn; 1989.

[55] Cochran C. Customer satisfaction, the elusive Quarry. Qual Prog 2001.

[56] Rosenberg J. The five myths about customer satisfaction. Qual Prog 1996; 29:57–60.

[57] Vavra T. Is your satisfactory service creating dissatisfied customers? Qual Prog 1997.

[58] Finch B. A new way to listen to the customer. Qual Prog 1997; 30:73–76.

[59] Gardner R. What do customers value? Qual Prog 2001.

[60] Brecka J. The American customer satisfaction index (ACSI). Qual Prog 1994; 27(10):41–44.

[61] Horowitz J. Putting service quality into gear. Qual Prog 1991.

[62] Jeffrey J. Preparing the front line. Qual Prog 1995.

[63] Duray R and Milligan, GW. Improving customer satisfaction through mass customization. Qual Prog 1999; 32(8):60–66.

[64] Labowitz G. Keeping your internal customers satisfied. Wall Street J 1987.

[65] Virginal Baldwin Hic. Technology is redefining the meaning of customer service. St Louis Post Dispenser 1999.

[66] Aman S. The essence of TQM—customer satisfaction. J Ind Technol 1994; 10(3):2–4.

[67] Scot Madison Patton. Unhappy employees and unhappy customers. Quality Digest 1999; 4.

[68] Harrington J. Looking down at the customer. Quality Digest 2001; 24.

[69] Godfrey B. Beyond satisfaction. Quality Digest 1996.

[70] Kiran DR. Customer satisfaction. In: Proceedings of the NIQR/IIPE seminar; 2006.

[71] Grewal, Chopra. Development of quality costing system in small scale industry. J Ind Eng 2006.

[72] Indian Standard on Quality Management Systems. Guidelines for performance improvements (IS/ISO 9004:2000). Bureau of Indian Standards; 2000.

Index

Note: Page numbers followed by *f* indicates figures and *t* indicates tables.

Printed in the United States
By Bookmasters